Connections in Environmental Science

A Case Study Approach

POLITICAL

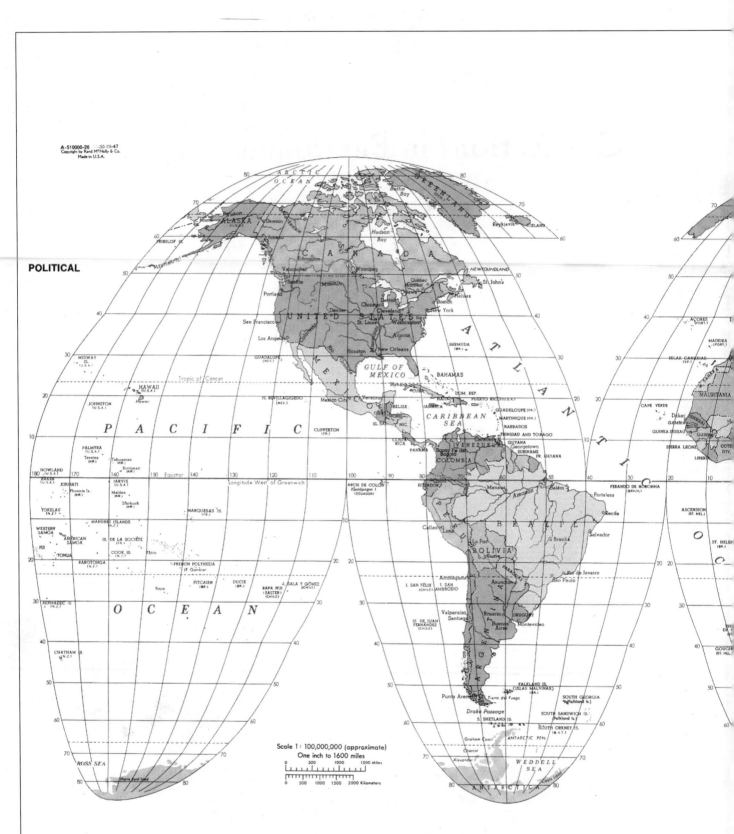

Scale 1 : 100,000,000 (approximate)
One inch to 1600 miles

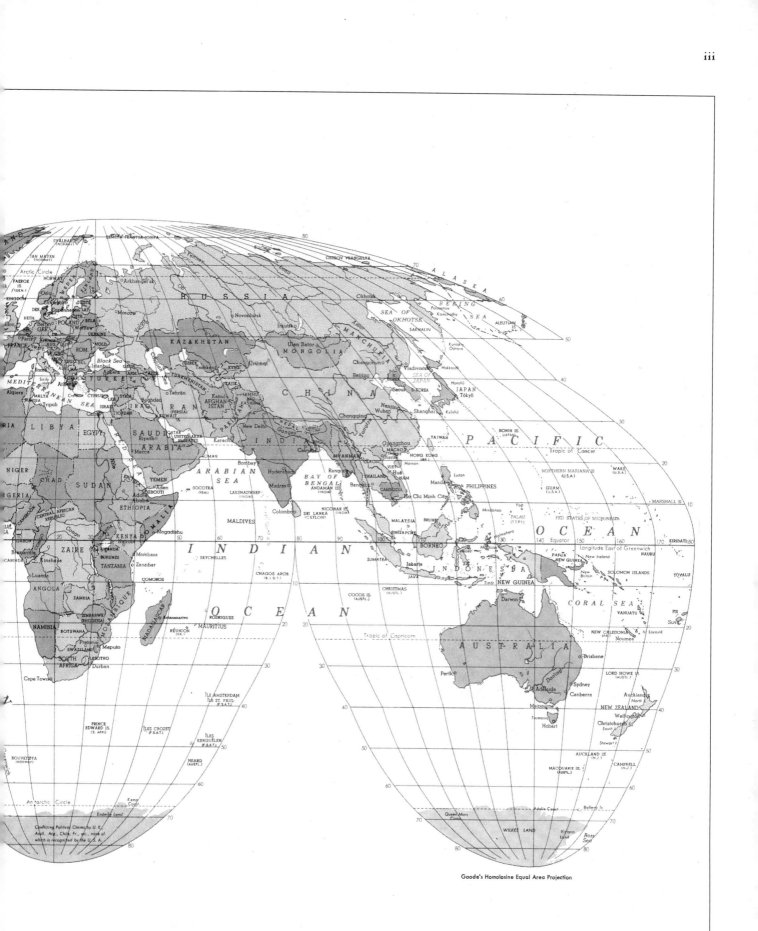

Goode's Homolosine Equal Area Projection

Connections in Environmental Science

A Case Study Approach

J. Richard Mayer
Western Washington University

Boston   Burr Ridge, IL   Dubuque, IA   Madison, WI   New York   San Francisco St. Louis
Bangkok   Bogotá Caracas   Kuala Lumpur   Lisbon   London   Madrid   Mexico City   Milan
Montreal   New Delhi   Santiago   Seoul   Singapore Sydney   Taipei Toronto

McGraw-Hill Higher Education

A Division of The McGraw-Hill Companies

CONNECTIONS IN ENVIRONMENTAL SCIENCE:
A CASE STUDY APPROACH
International Edition 2001

10 09 08 07 06 05 04 03 02 01
20 09 08 07 06 05 04 03 02 01
BJE   BJE

The credits section for this book begins on page 316 and is considered an extension of the copyright page.

Library of Congress Cataloging-in-Publication Data

Mayer, J. Richard.
   Connections in environmental science : a case study / J. Richard Mayer. – 1st ed.
      p.   cm.
   Includes index.
   ISBN 0-07-229726-3
     1.   Environmental sciences. I. Title.
   GE105.M38   2001
   363.7—dc21   00-024393
                                     CIP

www.mhhe.com

When ordering this title, use ISBN 0-07-118941-6

Printed in Singapore

Writing this book has been a joy and has given me a

deep sense of satisfaction. It would not have been

possible to devote so much time and energy to this

work without the patience and understanding of my

family. The book is dedicated to them—my wife,

children, and grandchildren:

June

Alisa, Mili, Jim, Jon

Derek, Jamie, Alex, Zack, Luke, Connor, Jimmy,

Kamarin, and ?

J. Richard Mayer

J. Richard Mayer was born and raised in New York State attending public schools both down-state and up-state including a one-room school house in Cross River, New York. He received a B.S. in Chemistry from Union College in Schenectady, New York and an M.A. in Chemistry one year later from Columbia University. He then served as a Research Fellow at Yale University in New Haven, Connecticut earning a Ph.D. in Organic Chemistry. After working as a research chemist for Sterling Drug Research Institute in Albany, New York, he spent eight years as a program director with the National Science Foundation in Washington, D.C. assisting in the process of awarding research and research facilities grants to colleges and universities across the United States. He began his teaching and research career as Director of the Environmental Research Center at State University of New York in Fredonia, New York where he taught environmental science and led faculty and student field studies on Lake Erie and Chautauqua Lake. In 1978, together with his wife June and their four children, he moved to Bellingham, Washington in the Pacific Northwest serving as Dean of Huxley College at Western Washington University for seven years. Today, as Professor of Environmental Studies he teaches courses in environmental science, water quality, and environmental chemistry. He has published numerous scientific papers on agricultural pesticide runoff and groundwater quality. He received a Superior Performance Award from the National Science Foundation and an Excellence in Teaching Award from Western Washington University.

Brief Contents

Contents

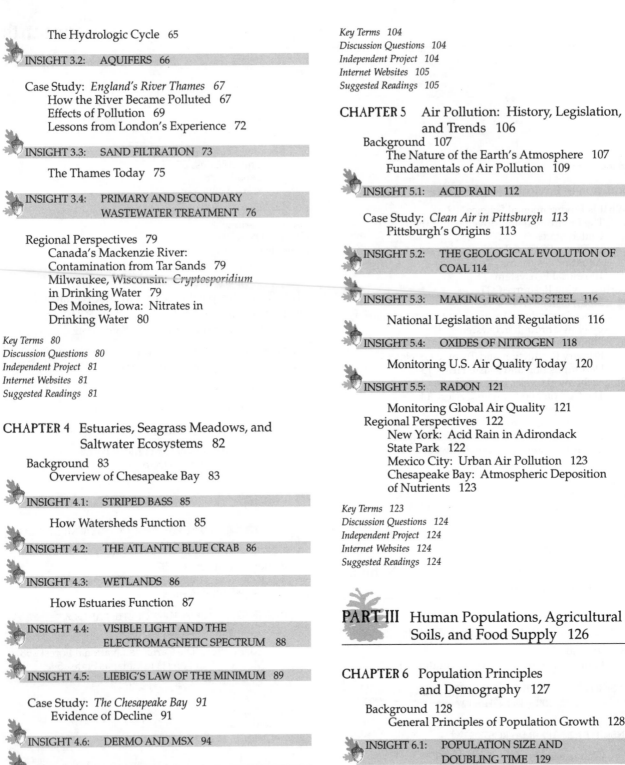

PART IV   Energy Production and
Consumption 186

WHAT IS ENVIRONMENTAL SCIENCE?

Environmental science is more than just an academic subject. It is a way of thinking about the world we live in and our connections to it. Never before have so many shared such a keen interest and deep concern for the well being of our planet — its majestic ecosystems and biomes, its treasure of natural resources, and its rich diversity of living resources. As a result, worldwide attention is focused on the earth's biosphere because we now know that many of the planet's life-support systems are being threatened.

The environmental problems we face today are many. They include human population growth, deforestation, water pollution, rising demands for non-renewable resources, toxic effects of hazardous chemicals, global climate change, and stratospheric ozone depletion. The aim of environmental science is to understand these problems and learn how to manage, if not solve, them. It is an immensely important and exciting venture because life on earth may well "hang in the balance."

WHY I DECIDED TO WRITE THIS BOOK

I have been teaching college-level environmental science courses for thirty years. It is an adventure in learning for myself, and my students, because environmental problems and issues are "front-page" news almost every day. Engaging them in reasonable and rational ways requires reliable insights to environmental science principles. Gaining these insights can be strongly reinforced through major case studies — carefully researched, real-world narratives describing environmental problems and issues that have been or are being successfully addressed.

I am sure that most educators use case studies in one form or another. Often, they are brief narratives focused on a particular point being made. What I have found is that up-to-date, in-depth case studies and regional perspectives capture student attention, underscore scientific principles, clarify technical concepts, and illustrate alternative approaches to problem solving. I believe that this approach helps students to comprehend critically important connections in environmental science subject matter.

Connections in Environmental Science: A Case Study Approach is different from other textbooks. Reviewers state that it is well written, comprehensive, technically penetrating in its development, and unbiased in its treatment of scientific questions and issues. Educators who choose to adopt the book will find that their students actually read it!

ACKNOWLEDGMENTS

I am deeply appreciative and most grateful to McGraw-Hill Publishers for their decision to help develop and publish this book. I am particularly indebted to:

Marge Kemp, Senior Sponsoring Editor, who from the very start demonstrated faith in the concepts that were to shape the book and lead the way to McGraw-Hill's publication decision.

Kathy Loewenberg, Senior Developmental Editor, who orchestrated the evolution of the book and its principal features, and whose expertise and wisdom are reflected in the book itself.

Kennie Harris, Copyeditor, whose brilliant ideas, literary excellence, and perseverance guided the book's progress, unfolding, and final configuration.

Jayne Klein, Senior Project Manager, who skillfully directed the editing, final assembly, and production of the book and all of its elements.

I am grateful as well to the many others at McGraw-Hill who together have made the publication of *Connections in Environmental Science: A Case Study Approach* a successful venture. They include:

Dianne Berning, Editorial Assistant
LouAnn Wilson, Permissions Researcher
Mary Reeg, Photo Research & Permissions Editor
Michelle Watnick and Heather Wagner, Marketing Directors
Tara McDermott, Advertising Coordinator

The following individuals have made valuable technical contributions to the development of this book by providing information, perspective, data, and special insights to different case studies and regional perspectives. I am deeply grateful to them for their generous assistance.

Kay W. Whittenburg
Tennessee Valley Authority
Chattanooga, Tennessee

Dr. Roger Rowe
Deputy Director General of Research
CIMMYT, Mexico City, Mexico

Dr. Alfred E. Slinkard
Senior Research Scientist
Crop Development Centre
University of Saskatchewan
Saskatoon, SK. Canada

Dr. Joel A. Tarr
Professor of Urban Environmental
History and Policy
Carnegie Mellon University
Pittsburgh, Pennsylvania

Clarence and Ray Bauml
Organic Farmers
Marysburg, Saskatchewan

Mary E. Garren
Remedial Project Manager
Massachusetts Superfund Section
Boston, Massachusetts

John C. Miles
Professor and Director
Center for Geography and Environmental Social Sciences
Western Washington University
Bellingham, Washington

Lee Jackson
Extension Agronomist
Agronomy & Range Science
University of California, Davis

Lauren Wenzel, Planner
Coastal Zone Management Program
Maryland Department of Natural Resources
Annapolis, Maryland

Richard A. Oatley
Senior Pollution Officer
Environment Agency
Thames Region, UK

Elmer Laird
Organic Farmer
Davidson, Saskatchewan

John F. McLaughlin
Professor of Environmental Science
Western Washington University
Bellingham, Washington

Mark Rodekohr, Ph.D.
Director Division of Energy Markets
and Contingency Information
Energy Information Administration
U.S. Department of Energy

Ed Best
Corporate Library
Tennessee Valley Authority
Knoxville, Tennessee

I am also indebted to the many reviewers who provided detailed analysis of the textbook during its development. In the midst of their busy schedules, they took the time to read the manuscript and offer advice that greatly improved the final effort.

Mark C. Belk
Brigham Young University

Del Blackburn
Clarke College

Guy E. Farish
Adams State College

David G. Fisher
Maharishi University of Management

Kevin M. Fitzsimmons
University of Arizona

Stephen Fleckenstein
Sullivan County Community College

Lesley C. Garner
University of West Alabama

GUIDED TOUR

Organization and Principal Features

The book is organized in five parts:

- Understanding and Protecting Natural Environments
- Water and Air: Fundamental Resources
- Human Populations, Agricultural Soils, and Food Supply
- Energy Production and Consumption
- Global and Regional Environmental Problems.

Introductory Chapter

An Introductory Chapter presents the book's major themes and identifies key environmental science principles.

INTRODUCTION TO ENVIRONMENTAL SCIENCE

When you dip your hand into nature, you find that everything
is connected to everything else.

John Muir, American naturalist

PhotoDisc, Inc./Vol. 16

---

188   PART IV   Energy Production and Consumption

The *Exxon Valdez* oil spill injured not only fish and wildlife populations and their

habitats, but also human use of the affected areas. Some people, such as fishers and

recreation guides, could no longer work at their regular occupations.... Some peoples'

whole life style changed, especially those who relied on subsistence in the spill area.

Finally, many people who have never been to the spill area, even people who have never

been to Alaska, felt a loss because a pristine area was degraded.

Alaska's Wildlife, January/February 1993

The fossil fuels are natural resources that have shaped the modern world. Petroleum, natural gas, coal, and products derived from them fuel the lifestyle to which most of us have become accustomed, enabling us to drive our cars, heat our homes, and cook our food. Indeed, the use of fossil resources has led to the manufacture of hundreds of contemporary "necessities" as diverse as rubber, plastic, perfume, pharmaceuticals, and ink. However, like all natural resources, supplies are limited; they are also nonrenewable. All of this is especially true for petroleum, whose underground reserves need to be managed judiciously—not only to conserve them but also to keep drilling and refining processes from harming the natural environment.

Ever since the use of petroleum became common, instances of pollution, particularly oil spills, have occurred. The 1989 *Exxon Valdez* oil spill in Prince William Sound off the coast of southern Alaska is especially notable because it was the largest in U.S. history and had devastating effects on local communities and area wildlife. However, there were also positive results because the incident triggered the formulation and enforcement of the first effective state and federal regulations for transporting petroleum products. This chapter traces the discovery and use of petroleum and describes the impact of the historic *Exxon Valdez* oil spill.

BACKGROUND

Origin and Composition of Petroleum

Although petroleum has many modern-day uses, its origins are ancient. Petroleum was formed slowly through geologic eras spanning 60 to 600 million years. The process began when organic matter derived from marine plants and animals settled to the ocean floor and accumulated in sediments. The buildup of successive sediment layers exerted increasing pressure on this organic matter, creating temperatures as high as 100 °C (212 °F). A series of biological, chemical, and physical reactions took place, and ultimately petroleum and natural gas collected in porous sandstone and limestone formations within the earth's crust (Figure 9.1).

The earth's crust consists of a solid layer of continents (**continental crust**) and ocean floor (**oceanic crust**). Continental crust underlies the planet's continents and is about 35 kilometers (≈ 22 miles) thick. Oceanic crust lies beneath the ocean floor and is about 5–7 kilometers (≈ 3–4 miles) thick. Oil and natural gas frequently migrate to porous, permeable rock formations within the earth's crust, displacing water originally trapped there.

Petroleum is a thick, dark oil composed of a complex mixture of organic chemicals, most of which are

Chapter Openers

Each chapter begins with a Table of Contents, an Opening Quote, and an Introduction.
A detailed Background discussion provides the necessary foundation for students to comprehend and appreciate the upcoming case study.

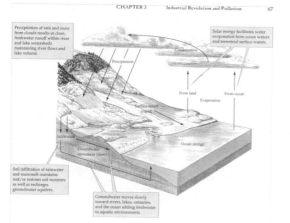

Figure 3.5   The hydrologic cycle.

Aquatic System	Percent of Total Water
TABLE 3.1 Distribution of Water on Earth	
Seawater and estuaries	97
Glaciers, ice fields, and ice caps	2
Freshwaters	1*

SOURCE: Data from *The Water Encyclopedia, Second Edition,* 1990.
*The 1 percent of freshwater is distributed as follows: groundwater aquifers, 97 percent; surface waters, 3 percent.

about 71 percent of the earth's surface and account for 97 percent of the planet's total water supply. Glaciers and other frozen water resources account for another 2 percent. Freshwaters add up to only 1 percent of the total. Approximately 97 percent of the planet's freshwater

is stored in groundwater aquifers. This means that only 0.03 percent of the earth's water supply is available as surface freshwater.

CASE STUDY: ENGLAND'S RIVER THAMES

How the River Became Polluted

Industrial expansion and population increase led to the pollution of the Thames estuary. Historians connect the rise of England's Industrial Revolution with the development of the steam engine by Thomas Savery in 1698, Thomas Newcomen in 1705, and James Watt in 1770. One of the earliest uses of the steam engine, which generated power by burning coal to boil water, was to pump out water that had seeped into coal mines, making them unsafe and difficult to work in. Later, steam engines transformed factory work so that

Line Drawings

Unique tutorial Line Drawings have helpful explanatory boxes directly in the art to provide students with key information where they need it most. This makes the art an active part of the teaching process.

Figure 9.7   The Trans Alaska Pipeline System is engineered to minimize pipeline disruption and oil spills due to earthquakes. *(Reprinted courtesy of ARCO Alaska, Anchorage, Alaska.)*

when a 22-year-old ban prohibiting the export of North Slope oil to other countries was lifted. In 1999, approximately 20 percent of domestically produced petroleum originated from oil fields on Alaska's North Slope and was transported through the Trans Alaska Pipeline System.

Benefits to Alaska   North Slope oil development has proven to be of great economic value to Alaska and the rest of the United States. Construction of the pipeline and its infrastructure created a large number of well-paid jobs. Oil-drilling leases are yielding billions of dollars to the State of Alaska, funds that in turn can be used to build schools, roads, highways, clinics, and hospitals. In the 1990s, 85 percent or more of Alaska's total state revenues were derived from oil and natural gas royalties and production taxes. The fact that there is no state income tax in Alaska is largely due to the state's oil industry. In addition, the Alaska Permanent Fund, established by the state legislature in 1980, distributes annual dividends to Alaska citizens based on income earned from oil industry lease and tax arrangements. In

1998, every eligible Alaska man, woman, and child received a dividend check of $1,540.88.

However, despite these benefits and the many precautions that were taken to protect the environment, oil-industry-related accidents have occurred; one of the most significant is discussed in the following case study.

CASE STUDY: THE *EXXON VALDEZ* OIL SPILL

It was a typical night departure from the Port of Valdez as the supertanker *Exxon Valdez,* loaded with North Slope crude oil, left port. Three football fields long (= 300 meters or 1,000 feet), the tanker was headed for Long Beach, California. As the tanker entered Prince William Sound, icebergs were reported in the shipping lane. Captain Joseph Hazelwood made a change in course, turned on the automatic gyro pilot, and left his command post, placing a Third Officer in charge. Later, an agreed-upon change in direction was made, but not in time. Twenty-five miles from port, the *Exxon Valdez* hit submerged rocks just off Bligh Reef. It was 12:03 A.M. March 24, 1989.

Impacts of the Spill

In the collision, eight of the ship's 11 cargo tanks were ruptured, and about 260,000 barrels (= 11 million U.S. gallons) of crude oil spilled into Prince William Sound's pristine waters. Crude oil saturated and spoiled *intertidal* and *subtidal* habitats and damaged submerged eelgrass meadows, which are critical habitat for many species of finfish and shellfish.

Intertidal habitats occupy coastal marine beach areas exposed between high tides and low tides.

Subtidal habitats lie below low-tide levels.

The three days following the spill were calm. Massive amounts of oil spread slowly but remained near the tanker. Little was done to recover the spilled oil, largely because there was almost no oil-spill response equipment on hand. Then, on March 27, a major windstorm drove oil slicks southwest toward Smith, Naked, and Knight Islands. Soon the oiled areas included open waters, pristine beaches, and mouths of rivers in the Chugach National Forest, Kenai Fjords National Park, and a number of other protected parks and wildlife refuges. Shorelines as far as 600 miles southwest of Bligh Reef were damaged. In all, close to

Case Study Approach

No other environmental science textbook has this Case Study Approach! Students learn the important concepts of environmental science through reading about actual events. The Case Study narratives do not dwell on disaster stories. They identify environmental problems and issues, discuss them in the light of alternative solutions, and examine the progress being made in discovering and implementing acceptable answers.

INSIGHT 2.2
Ecological Succession

Succession is a series of distinct changes in plant and animal communities that take place over time in a particular ecosystem. Such changes generally lead to more complex communities.

Primary succession depends first on rebuilding soils lost through destructive events such as volcanic explosions and lava flows. Very slow processes, including weathering of rocks and minerals and decomposition of organic matter, must take place to rebuild soils before vegetation can recur. **Secondary succession** refers to recolonization of a particular landscape by shrubs, trees, and animal life following a disturbance such as deforestation or a forest fire. It is a natural process of restoration that rebuilds on soils that remain intact and are capable of supporting plant life.

Both primary and secondary succession occur in stages that are marked by changing flora and fauna. Opportunistic **pioneer organisms**, including grasses, ferns, mosses, and lichens, are well represented during early successional stages. More complex plant life appears later, followed over time by a mature, self-regulating, steady-state community of living things called the **climax ecosystem.** In many cases, a climax system resembles the original ecosystem before any disturbances occurred.

While distinct stages of plant growth appear during succession, the emergence of each stage is thought to be independent of prior communities. Vegetative and tree emergence is a function of differing rates of seed propagation, germination, and growth of each species present.

Rica. This number is *twenty times* greater than earlier estimates of the total number of insect species on the entire planet! A single tropical tree in Costa Rica may well have more than 50 different epiphytic plant species growing on it, including algae, lichens, ferns, vines, mosses, bromeliads, and exotic flowering plants like orchids. Similar temperate-region trees support fewer than 100 epiphyte species. More than 100 species of bats have been identified in Costa Rica. In comparison, only 40 bat species are known in all of North America. Conservative estimates of species biodiversity in Costa Rica are given in Table 2.5.

Characteristics of Costa Rican Forests

In the early 1500s, when Costa Rica became a Spanish colony, tropical forests covered virtually the entire territory. But, as in many tropical countries, the country's hardwood trees have been cut and exported for over a hundred years. They are one of Costa Rica's chief economic assets. In the 1800s, the establishment of coffee plantations added to deforestation in the Central Plateau where fertile volcanic soils and an equable climate prevail. In the 1900s, additional forest lands gave way to banana plantations in the country's Atlantic and Pacific lowlands as well as cattle-grazing pastures in what were northwestern dry forests. Today, most of Costa Rica's original forest cover has been lost. What is left are smaller, scattered forests, remnants of earlier luxuriant tropical ecosystems.

TABLE 2.5 Estimates of Numbers of Plant and Animal Species in Costa Rica	
Insects, spiders	365,000
Other invertebrates	85,000
Bacteria, viruses	35,000
Vascular plants	10,000
Fungi	2,500
Vertebrates	1,500
Fish	1,013
Birds	845
Reptiles	218
Mammals	205
Amphibians	160
TOTAL	≈ 500,000

SOURCE: Data from Instituto Nacional de Biodiversidad de Costa Rica (INBio), 1993.

In Costa Rica's remaining rain forests, high canopies tower 50 to 60 meters (= 160 to 200 feet) above the forest floor (see Figure 2.1). The broad crowns and interwoven branches of tall, dominant trees create an almost unbroken carpet of vegetative cover high above the forest floor. A smaller number of towering trees, such as the ceiba or silk-tassel tree, stand as high as 75 meters (= 250 feet), punctuating the high canopy.

Introspective Boxed Readings

"Insights" Boxed Readings, Charts, and Tables all support the main text by adding significant, applicable information.

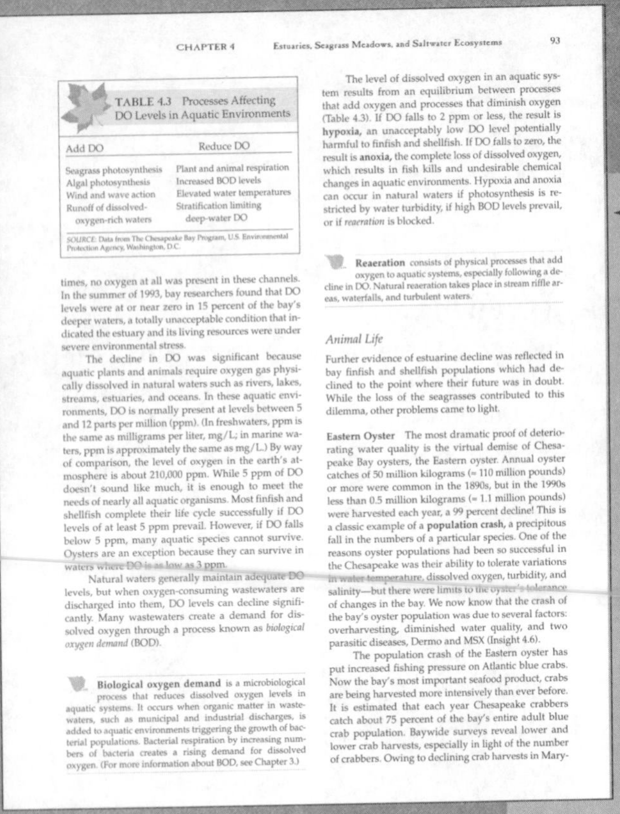

CHAPTER 4 Estuaries, Seagrass Meadows, and Saltwater Ecosystems 93

TABLE 4.3 Processes Affecting DO Levels in Aquatic Environments

Add DO	Reduce DO
Seagrass photosynthesis	Plant and animal respiration
Algal photosynthesis	Increased BOD levels
Wind and wave action	Elevated water temperatures
Runoff of dissolved-oxygen-rich waters	Stratification limiting deep-water DO

SOURCE: Data from The Chesapeake Bay Program, U.S. Environmental Protection Agency, Washington, D.C.

times, no oxygen at all was present in these channels. In the summer of 1993, bay researchers found that DO levels were at or near zero in 15 percent of the bay's deeper waters, a totally unacceptable condition that indicated the estuary and its living resources were under severe environmental stress.

The decline in DO was significant because aquatic plants and animals require oxygen gas physically dissolved in natural waters such as rivers, lakes, streams, estuaries, and oceans. In these aquatic environments, DO is normally present at levels between 5 and 12 parts per million (ppm). (In freshwaters, ppm is the same as milligrams per liter, mg/L; in marine waters, ppm is approximately the same as mg/L.) By way of comparison, the level of oxygen in the earth's atmosphere is about 210,000 ppm. While 5 ppm of DO doesn't sound like much, it is enough to meet the needs of nearly all aquatic organisms. Most finfish and shellfish complete their life cycle successfully if DO levels of at least 5 ppm prevail. However, if DO falls below 5 ppm, many aquatic species cannot survive. Oysters are an exception because they can survive in waters where DO is as low as 3 ppm.

Natural waters generally maintain adequate DO levels, but when oxygen-consuming wastewaters are discharged into them, DO levels can decline significantly. Many wastewaters create a demand for dissolved oxygen through a process known as *biological oxygen demand* (BOD).

Biological oxygen demand is a microbiological process that reduces dissolved oxygen levels in aquatic systems. It occurs when organic matter in wastewaters, such as municipal and industrial discharges, is added to aquatic environments triggering the growth of bacterial populations. Bacterial respiration by increasing numbers of bacteria creates a rising demand for dissolved oxygen. (For more information about BOD, see Chapter 3.)

The level of dissolved oxygen in an aquatic system results from an equilibrium between processes that add oxygen and processes that diminish oxygen (Table 4.3). If DO falls to 2 ppm or less, the result is **hypoxia,** an unacceptably low DO level potentially harmful to finfish and shellfish. If DO falls to zero, the result is **anoxia,** the complete loss of dissolved oxygen, which results in fish kills and undesirable chemical changes in aquatic environments. Hypoxia and anoxia can occur in natural waters if photosynthesis is restricted by water turbidity, if high BOD levels prevail, or if *reaeration* is blocked.

Reaeration consists of physical processes that add oxygen to aquatic systems, especially following a decline in DO. Natural reaeration takes place in stream riffle areas, waterfalls, and turbulent waters.

Animal Life

Further evidence of estuarine decline was reflected in bay finfish and shellfish populations which had declined to the point where their future was in doubt. While the loss of the seagrasses contributed to this dilemma, other problems came to light.

Eastern Oyster The most dramatic proof of deteriorating water quality is the virtual demise of Chesapeake Bay oysters, the Eastern oyster. Annual oyster catches of 50 million kilograms (≈ 110 million pounds) or more were common in the 1890s, but in the 1990s less than 0.5 million kilograms (≈ 1.1 million pounds) were harvested each year, a 99 percent decline! This is a classic example of a **population crash,** a precipitous fall in the numbers of a particular species. One of the reasons oyster populations had been so successful in the Chesapeake was their ability to tolerate variations in water temperature, dissolved oxygen, turbidity, and salinity—but there were limits to the oyster's tolerance of changes in the bay. We now know that the crash of the bay's oyster population was due to several factors: overharvesting, diminished water quality, and two parasitic diseases, Dermo and MSX (Insight 4.6).

The population crash of the Eastern oyster has put increased fishing pressure on Atlantic blue crabs. Now the bay's most important seafood product, crabs are being harvested more intensively than ever before. It is estimated that each year Chesapeake crabbers catch about 75 percent of the bay's entire adult blue crab population. Baywide surveys reveal lower and lower crab harvests, especially in light of the number of crabbers. Owing to declining crab harvests in Mary-

Key Terms

Important Key Terms are highlighted to signify definitions and help students review the chapter material.

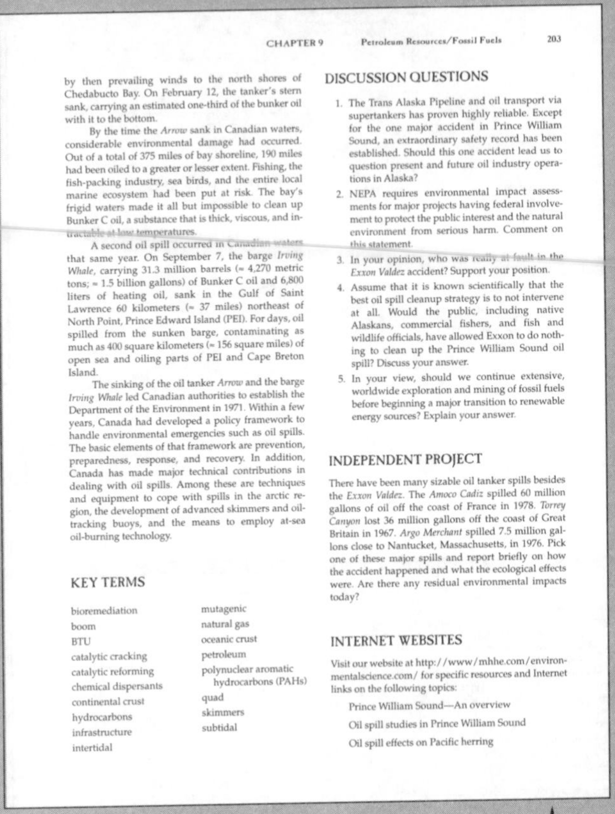

CHAPTER 9 Petroleum Resources/Fossil Fuels 203

by then prevailing winds to the north shores of Chedabucto Bay. On February 12, the tanker's stern sank, carrying an estimated one-third of the bunker oil with it to the bottom.

By the time the *Arrow* sank in Canadian waters, considerable environmental damage had occurred. Out of a total of 375 miles of bay shoreline, 190 miles had been oiled to a greater or lesser extent. Fishing, the fish-packing industry, sea birds, and the entire local marine ecosystem had been put at risk. The bay's frigid waters made it all but impossible to clean up Bunker C oil, a substance that is thick, viscous, and intractable at low temperatures.

A second oil spill occurred in Canadian waters that same year. On September 7, the barge *Irving Whale*, carrying 31.3 million barrels (≈ 4,270 metric tons; ≈ 1.5 billion gallons) of Bunker C oil and 6,800 liters of heating oil, sank in the Gulf of Saint Lawrence 60 kilometers (≈ 37 miles) northeast of North Point, Prince Edward Island (PEI). For days, oil spilled from the sunken barge, contaminating as much as 400 square kilometers (≈ 156 square miles) of open sea and oiling parts of PEI and Cape Breton Island.

The sinking of the oil tanker *Arrow* and the barge *Irving Whale* led Canadian authorities to establish the Department of the Environment in 1971. Within a few years, Canada had developed a policy framework to handle environmental emergencies such as oil spills. The basic elements of that framework are prevention, preparedness, response, and recovery. In addition, Canada has made major technical contributions in dealing with oil spills. Among these are techniques and equipment to cope with spills in the arctic region, the development of advanced skimmers and oil-tracking buoys, and the means to employ at-sea oil-burning technology.

KEY TERMS

bioremediation
boom
BTU
catalytic cracking
catalytic reforming
chemical dispersants
continental crust
hydrocarbons
infrastructure
intertidal
mutagenic
natural gas
oceanic crust
petroleum
polynuclear aromatic hydrocarbons (PAHs)
quad
skimmers
subtidal

DISCUSSION QUESTIONS

1. The Trans Alaska Pipeline and oil transport via supertankers has proven highly reliable. Except for the one major accident in Prince William Sound, an extraordinary safety record has been established. Should this one accident lead us to question present and future oil industry operations in Alaska?
2. NEPA requires environmental impact assessments for major projects having federal involvement to protect the public interest and the natural environment from serious harm. Comment on this statement.
3. In your opinion, who was really at fault in the *Exxon Valdez* accident? Support your position.
4. Assume that it is known scientifically that the best oil spill cleanup strategy is to not intervene at all. Would the public, including native Alaskans, commercial fishers, and fish and wildlife officials, have allowed Exxon to do nothing to clean up the Prince William Sound oil spill? Discuss your answer.
5. In your view, should we continue extensive, worldwide exploration and mining of fossil fuels before beginning a major transition to renewable energy sources? Explain your answer.

INDEPENDENT PROJECT

There have been many sizable oil tanker spills besides the *Exxon Valdez*. The *Amoco Cadiz* spilled 60 million gallons of oil off the coast of France in 1978. *Torrey Canyon* lost 36 million gallons off the coast of Great Britain in 1967. *Argo Merchant* spilled 7.5 million gallons close to Nantucket, Massachusetts, in 1976. Pick one of these major spills and report briefly on how the accident happened and what the ecological effects were. Are there any residual environmental impacts today?

INTERNET WEBSITES

Visit our website at http://www/mhhe.com/environmentalscience.com/ for specific resources and Internet links on the following topics:

Prince William Sound—An overview
Oil spill studies in Prince William Sound
Oil spill effects on Pacific herring

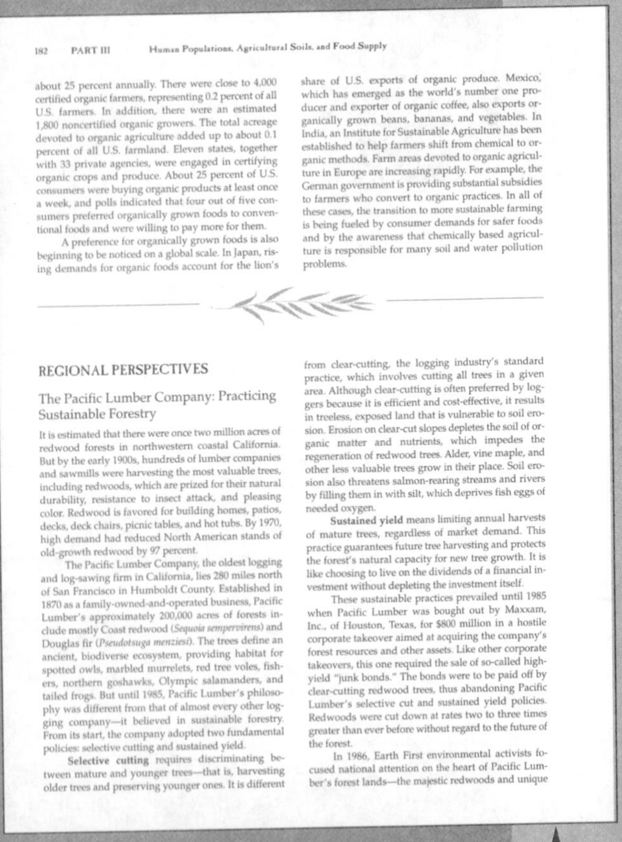

182 PART III Human Populations, Agricultural Soils, and Food Supply

about 25 percent annually. There were close to 4,000 certified organic farmers, representing 0.2 percent of all U.S. farmers. In addition, there were an estimated 1,800 noncertified organic growers. The total acreage devoted to organic agriculture added up to about 0.1 percent of all U.S. farmland. Eleven states, together with 33 private agencies, were engaged in certifying organic crops and produce. About 25 percent of U.S. consumers were buying organic products at least once a week, and polls indicated that four out of five consumers preferred organically grown foods to conventional foods and were willing to pay more for them.

A preference for organically grown foods is also beginning to be noticed on a global scale. In Japan, rising demands for organic foods account for the lion's share of U.S. exports of organic produce. Mexico, which has emerged as the world's number one producer and exporter of organic coffee, also exports organically grown beans, bananas, and vegetables. In India, an Institute for Sustainable Agriculture has been established to help farmers shift from chemical to organic methods. Farm areas devoted to organic agriculture in Europe are increasing rapidly. For example, the German government is providing substantial subsidies to farmers who convert to organic practices. In all of these cases, the transition to more sustainable farming is being fueled by consumer demands for safer foods and by the awareness that chemically based agriculture is responsible for many soil and water pollution problems.

REGIONAL PERSPECTIVES

The Pacific Lumber Company: Practicing Sustainable Forestry

It is estimated that there were once two million acres of redwood forests in northwestern coastal California. But by the early 1900s, hundreds of lumber companies and sawmills were harvesting the most valuable trees, including redwoods, which are prized for their natural durability, resistance to insect attack, and pleasing color. Redwood is favored for building homes, patios, decks, deck chairs, picnic tables, and hot tubs. By 1970, high demand had reduced North American stands of old-growth redwood by 97 percent.

The Pacific Lumber Company, the oldest logging and log-sawing firm in California, lies 280 miles north of San Francisco in Humboldt County. Established in 1870 as a family-owned-and-operated business, Pacific Lumber's approximately 200,000 acres of forests include mostly Coast redwood (*Sequoia sempervirens*) and Douglas fir (*Pseudotsuga menziesi*). The trees define an ancient, biodiverse ecosystem, providing habitat for spotted owls, marbled murrelets, red tree voles, fishers, northern goshawks, Olympic salamanders, and tailed frogs. But until 1985, Pacific Lumber's philosophy was different from that of almost every other logging company—it believed in sustainable forestry. From its start, the company adopted two fundamental policies: selective cutting and sustained yield.

Selective cutting requires discriminating between mature and younger trees—that is, harvesting older trees and preserving younger ones. It is different from clear-cutting, the logging industry's standard practice, which involves cutting all trees in a given area. Although clear-cutting is often preferred by loggers because it is efficient and cost-effective, it results in treeless, exposed land that is vulnerable to soil erosion. Erosion on clear-cut slopes depletes the soil of organic matter and nutrients, which impedes the regeneration of redwood trees. Alder, vine maple, and other less valuable trees grow in their place. Soil erosion also threatens salmon-rearing streams and rivers by filling them in with silt, which deprives fish eggs of needed oxygen.

Sustained yield means limiting annual harvests of mature trees, regardless of market demand. This practice guarantees future tree harvesting and protects the forest's natural capacity for new tree growth. It is like choosing to live on the dividends of a financial investment without depleting the investment itself.

These sustainable practices prevailed until 1985 when Pacific Lumber was bought out by Maxxam, Inc., of Houston, Texas, for $800 million in a hostile corporate takeover aimed at acquiring the company's forest resources and other assets. Like other corporate takeovers, this one required the sale of so-called high-yield "junk bonds." The bonds were to be paid off by clear-cutting redwood trees, thus abandoning Pacific Lumber's selective cut and sustained yield policies. Redwoods were cut down at rates two to three times greater than ever before without regard to the future of the forest.

In 1986, Earth First environmental activists focused national attention on the heart of Pacific Lumber's forest lands—the majestic redwoods and unique

End of Chapter

Key Terms, Discussion Questions, an Independent Project, and Suggested Readings at the end of each chapter provide an opportunity for students to test their understanding of the material just covered. Internet websites list key topics in the chapter that are repeated in the Online Learning Center as hyperlinks to websites that offer additional information and resources for continued research.

Regional Perspectives

Regional Perspectives take the concepts covered in the case study and discuss them within different geographical and cultural contexts. Seeing these applications helps students to better understand the core principles of environmental science.

SUPPLEMENTS

Instructor's Manual

• An Instructor's Manual accompanies this text and provides chapter outlines, key terms, test and discussion questions, suggestions for class activities, and resources for teaching aids.

Transparency Masters

• A set of 75 Transparency Masters is available to users of the text. The masters duplicate key figures from the text that are important to understanding major concepts in environmental science.

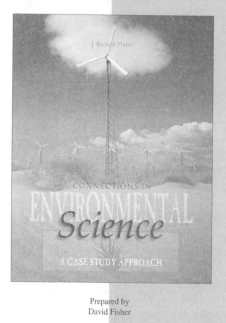

Website

• A comprehensive website at http://www.mhhe.com/environmentalscience (click on book cover) offers numerous resources for both students and instructors including additional case studies, practice quizzing, key term flashcards, current global environmental issues and events, and much more.

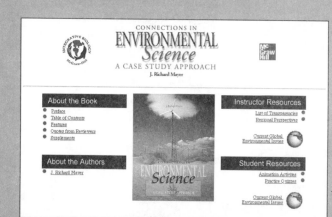

Visual Resource Library

• Environmental Science and Ecology Visual Resource Library CD-ROM. This classroom presentation CD includes images from two environmental science textbooks, in addition to an ecology text.

Essential Study Partner

• The Essential Study Partner CD-ROM is a valuable component to any student's learning package in the field of environmental science. This tutorial CD contains high-quality 3-D animations, interactive study activities, illustrated overviews of key topics in environmental science, and self-quizzes and exams for each important unit.

The topic menu contains an interactive list of the available topics. Clicking on any of the listings within this menu will open your selection and will show the specific concepts presented within this topic. Clicking any of the concepts will move you to your selection. You can use the UP and DOWN arrow keys to move through the topics.

The unit pop-up menu is accessible at anytime within the program. Clicking on the current unit will bring up a menu of other units available in the program.

To the right of the arrows is a row of icons that represent the number of screens in a concept. There are three different icons, each representing different functions that a screen in that section will serve. The screen that is currently displayed will highlight yellow and visited ones will be checked.

The film icon represents an animation screen.

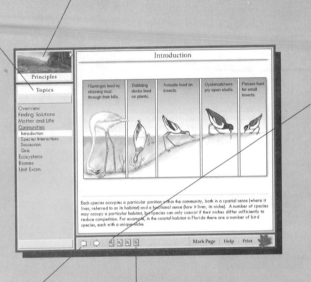

Along the bottom of the screen you will find various navigational aids. At the left are arrows that allow you to page forward and backward through text screens or interactive exercise screens. You can also use the LEFT and RIGHT arrows on your keyboard to perform the same function.

The activity icon represents an interactive learning activity.

The page icon represents a page of informational text.

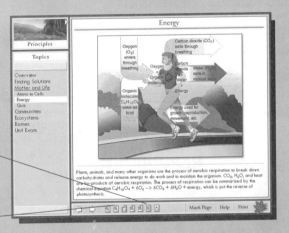

Contact your McGraw-Hill sales representative for more information or visit *www.mhhe.com*.

INTRODUCTION TO ENVIRONMENTAL SCIENCE

*When you dip your hand into nature, you find that everything
is connected to everything else.*

John Muir, American naturalist

PhotoDisc, Inc./Vol. 16

A drop of water on a leaf of goldenrod evaporates into the air, and then condenses, falling as rain. A grasshopper eats part of the leaf—and is in turn eaten by a pheasant. Later, the plant is cleared away, along with other grasses, shrubs, and trees in the field, to make way for a new factory. The grasshoppers, pheasants, and other wildlife lose their habitat. Eventually, a highway is built nearby, automotive traffic escalates, and air pollutants are released, causing smog that leads to health problems for people living in the area.

In the natural world, everything—living and nonliving—is connected. The earth is a place of complex relationships, and they become even more complex when the activities of humans are factored in. Not so long ago, people routinely dumped sewage wastes into rivers without realizing they were polluting their own drinking water; they allowed black smoke to pour into the very air they breathed; they cut down trees and hunted animals without regard for the uniqueness and intrinsic worth of individual species.

At first, such practices rarely caused noticeable effects on the environment because relatively small numbers of people were involved. But environmental problems became more acute as human populations grew, and people began to experience the consequences of not treating the natural world with respect. Gradually, over the course of the twentieth century, humans gained more appreciation for the environment and an awareness of their own impact upon it. Thus was launched an environmental movement to protect nature and natural ecosystems. Today, almost every time we pick up a newspaper or listen to a news broadcast, we are informed about yet another environmental problem, and we see more clearly that John Muir was right in saying, "When you dip your hand into nature, you find that everything is connected to everything else."

In the same way that the elements of the natural world are connected, so are the chapters of this book. Throughout, we will discuss the principles of environmental science, not as isolated facts, but in relation to other disciplines and in terms of environmentally significant events, both past and present. We will begin our study in this chapter by tracing the background of the environmental movement and introducing the concepts that form the basis for the environmental discussions in the rest of the book.

WHAT IS ENVIRONMENTAL SCIENCE?

The interrelationships among living things and their environment are the focus of **ecology.** Ecologists study individual organisms, *populations* of species, *communities* of species, and entire *ecosystems*, seeking to understand diversity, relationships, and interactions in nature.

A **population** is composed of all the individuals *of a single species* living in a defined area.

A biological **community** consists of interacting populations *of different species*—plants, animals, bacteria, fungi, and viruses—living in a defined area.

An **ecosystem** is a community of living organisms interacting in a defined physical environment where living and nonliving components are linked by chemical exchanges, biological interactions, and energy flow.

Environmental science, a relatively new discipline, builds upon ecology and strives to comprehend the nature and extent of human influences on natural systems. Environmental scientists identify problems that may affect living systems, including fish, wildlife, and humans, and try to solve those problems by applying knowledge from many disciplines, such as chemistry, biology, geology, geography, economics, political science, sociology, and anthropology.

Environmentalists and ecologists engage in a variety of activities. They teach, conduct research, assess alternative approaches to problems, work to establish more enlightened environmental policies, and promote sustainable conservation, agricultural, and energy practices. Like all scientists, they apply the **scientific method:** making observations, gathering data, and formulating, testing, and altering hypotheses. Methodical data-gathering and research can lead to important environmental breakthroughs. A physician in nineteenth-century London discovered the source of the polluted water causing cholera by gathering data, plotting it on a map, and then convincing the authorities to shut down a public well (see Chapter 3). Researchers exploring the cause of ecological decline in Chesapeake Bay discovered that overgrowths of microscopic algae on submerged vegetation blocked out sunlight and triggered massive declines in seagrass meadows (see Chapter 4). A biologist carefully traced the course of pesticides such as DDT, and her published findings detailing toxic effects on wildlife raised public awareness and gave impetus to the environmental movement (see Chapter 7). Scientists in Sweden monitoring the atmosphere alerted the world to a dangerous release of radioactive material from the nuclear accident at Chernobyl in the former Soviet Union (see Chapter 10). And systematic, long-term observation and record-keeping by Canadian scientists in Antarctica led to the discovery of the ozone hole at the South Pole (see Chapter 13).

The Environmental Movement in the United States

Throughout history, humans have taken advantage of the world's **natural resources**—its water, soil, and minerals as well as its wide variety of plant and animal life. And during most of this time they have usually been too preoccupied with ensuring their own survival to be concerned about the effects of their activities. For example, as the pioneers in the early United States cut down trees to build their cabins and to clear fields for growing their crops, they did not consider that they were also destroying animals' habitats and contributing to potential soil erosion. Many people took what they needed from the earth—and often did not think about giving anything back.

However, a few individuals in nineteenth-century North America were beginning to consider the relationship between man and nature. They realized that the earth's natural resources were limited and that mining, harvesting, and exploiting them could well alter entire ecosystems. American Indians were among the first to understand that the earth is vulnerable. This is reflected in a letter said to have been written by Chief Seattle, patriarch of the Duwamish and Squamish Indians of Puget Sound, to U.S. President Franklin Pierce in 1885:

> This we know: the earth does not belong to man, man belongs to the earth. . . . All things are connected like the blood that unites one family. . . . Whatever befalls the earth, befalls the sons of the earth. Man did not weave the web of life, he is merely a strand in it. Whatever he does to the web, he does to himself.

The American writer Henry David Thoreau advocated living in harmony with nature, and he lived out his belief by dwelling alone on the shore of Walden Pond in Massachusetts from 1845 to 1847. Isolated living was an experiment Thoreau had discussed with his friend Ralph Waldo Emerson—something he hoped would serve as an example to young people, inspiring them to think about the world and their place in it. In *Walden*, the book recording his experiences, Thoreau wrote:

> I went to the woods because I wished to live deliberately, to front only the essential facts of life, and see if I could not learn what it had to teach, and not, when I came to die, discover that I had not lived.

Conservation and Preservation

Thoreau's ideas about wilderness are echoed in the writings of others, including Ralph Waldo Emerson's essay "Nature" published in 1836 and George Perkins Marsh's *Man and Nature* written in 1864. They recognized the intrinsic value of wild things and the importance of *conserving* nature and natural resources.

Conservation means managing and protecting natural resources while enjoying and using them in ways that prevent their exploitation, depletion, and pollution so as to ensure their future use.

In actual practice, conservation means different things to different people. To the American naturalist John Muir (1838–1914), it meant **preservation**—protecting unique natural environments from human development. Muir viewed wild places as "cathedrals." His preservation philosophy led to the creation of Sequoia and Yosemite National Parks and the 1892 founding of the Sierra Club, initially an American organization dedicated to protecting the environment that in 1969 became active in Canada as well.

The conservation movement gained momentum during the presidency of Theodore Roosevelt (1901 to 1909). Roosevelt was an avid hunter of big game in western Montana, northwestern Wyoming, and the valley of the Little Missouri River. He loved wild things and wild places. After reading Muir's *Our National Parks*, published in 1901, Roosevelt visited the author in Yosemite, and together they planned a conservation program that became one of the hallmarks of Teddy Roosevelt's presidency. His administration's environmental accomplishments include:

- Creation of the U.S. Forest Service and the U.S. National Forest System. Also, the nation's public forests were transferred from the Department of the Interior to the Department of Agriculture.
- Expansion of U.S. national forests from 17 million hectares (≈ 42 million acres) to 70 million hectares (≈ 172 million acres).
- Establishment of the first federally protected wildlife refuge, Pelican Island on Indian River, Florida, as well as an additional 52 refuges.
- Designation of 18 national monuments through the American Antiquities Act, whose purpose is to preserve historic places and objects of scientific interest. A number of these monuments have since become national parks, including Grand Canyon National Park, Petrified Forest National Park, and Olympic National Park.

Following Roosevelt's presidency, Senator Robert LaFollette of Wisconsin said of him:

> His greatest work was actually beginning a world movement for staying terrestrial waste

Figure I.1 One of many magnificent rivers and waterfalls in Canada's Banff National Park.

and saving for the human race the things upon which alone a great and peaceful and progressive and happy race can be founded.

When John Muir wrote *Our National Parks*, there were only four U.S. national parks: Yellowstone, the world's first national park lying mainly within the northwest corner of Wyoming, and Yosemite, Sequoia, and General Grant (now a part of Kings Canyon National Park) in California. The idea of national parks had originated from Muir's urging that unique, unspoiled, spectacular areas be forever conserved. Conservation efforts were further advanced through the establishment of the National Park System in 1912, which was needed to manage the national parks and encourage education, preservation, interpretation, and recreation within them. Similar efforts in Canada had led to the 1885 founding of Banff National Park in Alberta, that country's first national park (Figure I.1). Today, there are more than 1,000 national parks in 120 countries.

Although conservation had become an accepted philosophy in the early years of the twentieth century, many individuals had distinctly different views. For example, Gifford Pinchot (1865–1946), the first head of the U.S. Forest Service under Roosevelt, viewed conservation as utilitarian resource management—in his words, "the use of the earth for the good of man." He believed:

The object of our forest policy is not to preserve the forests because they are beautiful . . . or because they are refuges for the wild creatures of the wilderness . . . but . . . [for] the making of

prosperous homes; . . . every other consideration becomes secondary.

However, Aldo Leopold, an authority on game management who served with the U.S. Forest Service in the 1920s, saw conservation in a different light. In his classic book, *A Sand County Almanac*, he wrote:

Conservation is getting us nowhere. . . . We abuse land because we regard it as a commodity belonging to us. When we see land as a community to which we belong, we may begin to use it with love and respect. There is no other way for land to survive the impact of mechanized man, nor for us to reap from it the aesthetic harvest it is capable, under science, of contributing to culture.

In the 1930s, when Dust Bowl calamities and economic failures hit North America, U.S. President Franklin Roosevelt authorized new federal agencies to put people back to work and at the same time move conservation efforts forward. One of these agencies, the Civilian Conservation Corps (CCC), engaged in reforestation, fire-lookout and fire fighting, and construction of recreational facilities and fish hatcheries. The Tennessee Valley Authority (TVA) built dams to control Tennessee River flooding, generate cheap hydroelectric power, and "electrify" America's Southeast (see Chapter 10). The Soil Conservation Service (SCS) helped farmers control soil erosion and adopt better conservation practices.

Perhaps the most important and far-reaching event signaling the reality of the conservation and preservation movement in the United States was the 1964 enactment of the Wilderness Act (see Chapter 1). This legislation established the National Wilderness Preservation System (NWPS), which initially included 9.1 million acres of Forest Service wilderness areas and has since been greatly expanded. The 1964 act defines wilderness as:

. . . an area where the earth and its community of life are untrammeled by man, where man himself is a visitor who does not remain . . . land retaining its primeval character and influence, without permanent improvements or human habitation, which is protected and managed so as to preserve its natural conditions and which (1) generally appears to have been affected primarily by the forces of nature, with the impact of man's works substantially unnoticeable; (2) has outstanding opportunities for solitude or a primitive and unconfined type of recreation; (3) has at least 5,000 acres of land or is of sufficient size as to make practicable its preservation and use in an unimpaired condition; and (4) may also contain ecological, geological, or other features of scientific, educational, scenic, or historical value.

Environmental Protection

In 1962, Rachel Carson, a U.S. Fish and Wildlife Service biologist, published the influential book *Silent Spring*. In it she described how chemical pesticides such as DDT were dispersed in the environment far beyond their point of application, bioaccumulated in fish, and then bioconcentrated to higher and higher levels in food webs, adversely affecting reproduction among bald eagles, peregrine falcons, and other birds. The numbers of raptors and songbirds were declining, foreshadowing a day, she wrote, when springtime might well become silent — the sounds of many birds gone forever. *Silent Spring*, perhaps more than any other publication, awakened North Americans to the reality of environmental *pollution*. Carson's influence and that of other environmental advocates ultimately led the United States, Canada, and other countries to adopt laws and regulations aimed at environmental protection.

Pollution is a detrimental change in the physical, chemical, or biological characteristics of air, soil, water, or any other environmental resource that adversely affects fish, wildlife, humans, or other living organisms. Sources of pollution are usually broken down into two types: **Point sources** are identifiable, local sites where pollution is discharged to the environment. Examples include wastewater treatment plant effluents, coal-burning power-plant emissions, and industrial discharges into rivers, bays, and estuaries. **Non-point sources** are areas (as opposed to single points) from which pollution enters the environment. They are often more difficult to identify than point sources. Examples include fertilizer runoff from farms, pesticide runoff from lawns and gardens, stormwater runoff in urban areas (including street debris, automotive oils, and eroded soils), and atmospheric fallout, such as wind-borne soils, fertilizers, auto emissions, and chemical pollutants

On April 22, 1970, the United States promoted environmental awareness and launched the contemporary environmental movement by observing the first Earth Day. An estimated 20 million Americans participated in Earth Day events, which included public speeches, parades, marches, rallies on college campuses, and "teach-ins." U.S. senators and congressmen returned to their home districts that day to take part in local events. Citizens were encouraged to walk to work or change their daily routines in ways that would benefit the environment. Earth Day helped bring home the message that every individual has a direct and lasting

impact on the environment. The effects were almost immediate: Environmental organizations grew in number and size; new Clean Air Act and Clean Water Act amendments were passed; the Resources Recovery Act was adopted; and the U.S. Environmental Protection Agency was created.

Subsequent Earth Days have witnessed increasing numbers of people becoming active environmentalists. Today, recycling paper, glass bottles, and aluminum cans is common practice, and many people purchase organically grown food, biodegradable products, and items made from recycled materials. Instead of living as though the earth exists solely for the benefit of humans and as though its resources are inexhaustible, many people now think in terms of *sustainability* and *stewardship*.

Sustainable practices allow us to meet present-day needs without compromising the ability of future generations to meet their needs. For example, many scientists and engineers believe that sources of sustainable energy, such as solar and wind power, will be able to satisfy the world's future energy requirements without polluting air and water resources—a dilemma we face today as coal and other fossil fuels are burned.

Stewardship is an ethical view that humans have a responsibility to care for and wisely manage the earth's natural resources, including all its living things. For example, current initiatives in sustainable agriculture are aimed in large part at protecting agricultural soils, including their microbiological organisms.

By 1972, the U.S. Environmental Protection Agency (EPA) had banned DDT as well as several other persistent, harmful pesticides in the United States. That same year, representatives of 113 nations assembled in Stockholm for the first United Nations Conference on the Human Environment. The Stockholm Conference put environmental issues on the international agenda and laid the foundation of the international environmental movement as we now know it. One result of the Stockholm meeting was the creation of the United Nations Environment Program (UNEP). UNEP has since become a major environmental agency, pursuing worldwide studies in population, agriculture, global climate change, and stratospheric ozone depletion.

Some of the ideas generated at Stockholm related to environmental sustainability were later crystallized in the Brundtland Commission report, *Our Common Future*, presented to the U.N. General Assembly in 1987. Prior to this report, most people considered environmental protection and economic development to be mutually exclusive—that is, they believed the pursuit of either one surely meant sacrificing the other.

However, in *Our Common Future* and in later agreements at the Rio De Janeiro Earth Summit in 1992, a totally different future was envisioned—one in which environmental quality was linked to economic development, thereby fostering both a healthy economy and environmental stewardship.

Environmental Policy

Gradually, ideas having to do with stewardship and *environmental ethics* led to *environmental policies* and practices, first in the United States and then in other countries as well.

Environmental ethics is a way of thinking and making value judgments about the earth's environment and the kind of world we want to live in and pass on to our children and grandchildren.

Environmental policies refer primarily to actions based on environmental laws and regulations. The adoption and enforcement of drinking water standards set by the U.S. Safe Drinking Water Act of 1974 and its amendments comprise an example of environmental policy.

The National Environmental Policy Act (NEPA), signed into law by President Richard Nixon in January 1970, became the cornerstone of environmental policy in the United States by setting forth principles and practices aimed at ensuring environmental protection. NEPA focuses national attention on proposed industrial, agricultural, and government-sponsored developments that might cause such adverse impacts as:

- Degradation of fish and wildlife habitats
- Deterioration of water quality in rivers, lakes, and estuaries
- Decline in air quality
- Pollution of natural resources
- Health problems in humans

Prior to NEPA, no means had existed in the United States at local, state, or federal levels to compel consideration of environmental issues connected with any project or development. NEPA requires that an environmental impact statement (EIS) be written for projects that may significantly affect the human environment—specifically, federal government projects and those requiring a federal permit or license. The EIS must explain why the project is being proposed, how it might impact air, water, soil, or living systems, how such impacts can be mitigated, and possible alternatives to the proposed project. One of NEPA's goals is to foster scientific study and risk assessment to shed light on a proposed project's benefits

and environmental impacts before it is carried out. EIS documents must be made available to the public for review and comment.

Following the passage of NEPA, Congress adopted a number of major legislative acts that define U.S. environmental policy today, including the following:

- Resources Recovery Act of 1970
- Clean Air Act of 1970
- Marine Mammal Protection Act of 1972
- Marine Protection, Research, and Sanctuaries Act of 1972
- Federal Insecticide, Fungicide, and Rodenticide Control Act of 1972
- National Coastal Zone Management Act of 1972
- Federal Water Pollution Control Act of 1972
- Ocean Dumping Act of 1972
- Endangered Species Act of 1973
- Safe Drinking Water Act of 1974
- Resources Conservation and Recovery Act of 1976
- Toxic Substances Control Act of 1976
- Comprehensive Environmental Response, Compensation, and Liability Act of 1980
- Alaskan National Interests Lands Conservation Act of 1980
- Nuclear Waste Policy Act of 1982
- Clean Water Act of 1987
- Oil Spill Prevention and Liability Act of 1990
- Clean Air Act of 1990
- Waste Reduction Act of 1990

In some cases, it is necessary to reauthorize a particular act in a subsequent year; in other cases, amendments are added to strengthen or broaden an act's provisions.

EARTH'S ECOSYSTEMS

When discussing the intricate natural world, environmental scientists and ecologists often find it helpful to classify its parts and focus on a manageable functional unit, commonly an ecosystem. Ecosystems may be defined in terms of natural landscapes, human-designated boundaries on maps, or even legislation. They include watersheds, forests, mountains, grasslands, wetlands, deserts, rivers, lakes, oceans, seas, and estuaries. An ecosystem can be as small as a rock with moss and microbes living on it or as large as an ocean abounding with marine life. But although ecosystems are sometimes discussed as isolated units of the environment, the more we study them, the more we see that they are interconnected and interdependent. The total of all of the earth's ecosystems can be thought of as the planet's *biosphere*.

The **biosphere** consists of the relatively thin layer of soil, water, and atmosphere where life is found on the earth. It is composed of **biotic factors** (living organisms), including individual plants and animals, and **abiotic factors** (materials and forces), such as temperature, humidity, elevation, precipitation, wind, light, shading, soil chemistry, and atmospheric pressure. The biosphere extends from Pacific Ocean Mariana Trench depths of about 10,000 meters ($\approx$ 32,800 feet) to Himalayan Mountain elevations of almost the same extent. Compared to the size of the earth itself, the biosphere is analogous to the skin of an apple—a very limited life-support system encircling the planet.

For purposes of discussion, the biosphere can be divided into aquatic ecosystems and terrestrial ecosystems. We will define and consider each of these briefly.

Aquatic Ecosystems

More than 70 percent of the earth is covered by water, and close to 97 percent of that water is saltwater in the planet's oceans, seas, and estuaries. The remaining 3 percent is *freshwater,* most of which is frozen in ice caps and glaciers (76 percent) or stored underground (23 percent).

Freshwater contains little in the way of dissolved salts, minerals, and other chemical substances. It is not too salty (saline) to drink or to use for crop irrigation. Humans, as well as many fish and wildlife populations, depend on freshwater for survival.

Only 1 percent of the earth's freshwater is free-flowing in rivers and lakes. Even smaller percentages are found in the soil, atmosphere, and living organisms. At any given time, only 0.03 percent of the earth's total water supply is available to plants, fish, wildlife, and humans as free-flowing freshwater. The rest is either saltwater, ice caps, glaciers, or groundwater. Over time, water resources are continuously recycled through the *hydrologic cycle*.

The **hydrologic cycle** is the natural recycling of water within and among earth's ecosystems through evaporation, atmospheric transport, precipitation, runoff, soil infiltration, and re-evaporation. Solar energy drives the hydrologic cycle. (For further information, see Chapter 3.)

The vast accumulation of oceanic marine waters along with surface freshwaters in streams, rivers, and lakes is constantly evaporating, leaving dissolved minerals and salts behind and adding large quantities of water vapor to the earth's atmosphere. Atmospheric water vapor is a "greenhouse gas," thereby contributing to the **greenhouse effect**, which maintains the warmth necessary for living things on the earth (see Chapter 12). Atmospheric moisture also accounts for recurring precipitation, which recharges freshwaters in streams, rivers, lakes, and groundwater.

Freshwater Systems

The most familiar freshwaters are streams, creeks, rivers, and lakes. Not only are they the most visible, but they play a vital role in supporting fish and wildlife as well as supplying drinking water for humans and livestock. For these reasons, early civilizations arose close to surface freshwater resources—for example, at the juncture of the Tigris and Euphrates Rivers in the Middle East and along the River Thames in England (see Chapter 3). In North America, the rise of cities on or near the Great Lakes is another example of settlement close to freshwater resources—cities such as Buffalo, New York; Erie, Pennsylvania; Cleveland, Sandusky and Toledo, Ohio; Detroit, Michigan; Chicago, Illinois; and Windsor, Ontario.

Most of the earth's freshwater is stored in groundwater **aquifers** and obtained by drilling wells, a practice that dates back to antiquity. Today, groundwaters serve as indispensable sources of drinking water for rural and farm families, suburban communities, and urban centers. They also provide irrigation water for farmland crops around the world. For example, groundwater pumped from the great Ogallala aquifer beneath the American Great Plains is used to water corn, wheat, soybeans, and other crops on thousands of farms (see Chapter 8).

Saltwater Systems

The fact that ocean ecosystems dominate our planet is most clearly evident in spacecraft digital imagery showing the earth as a blue, mostly water-covered planet suspended in space. Because of their immense mass and heat capacity, oceans and seas serve to buffer the temperature extremes that would occur without them. As a result, average temperatures on earth are continuously favorable to life at almost all latitudes, the only exceptions being the North and South Poles where only a few species thrive. The stored heat in marine waters prevents them from freezing except in shallow coastal embayments. Saltwater habitats are therefore beneficial year-round to marine plant and animal life. Even in relatively cold Antarctic Ocean waters, prolific populations of very small aquatic plants and animals play an important role in supporting fishes, squids, seals, and whales.

Earth's oceans also act as a *sink* for carbon dioxide, slowly dissolving quantities of the gas added to the planet's atmosphere by both natural processes and industrial developments. Although the earth's levels of atmospheric carbon dioxide are rising, due mainly to the combustion of fossil fuels, they would be rising even faster if oceans and seas were not performing this function (see Chapter 12).

Pelagic areas, high seas open-ocean environments, are often biological deserts because their distance from land deprives them of land-based nutrient runoff. Lack of nutrients slows photosynthesis in these waters, limiting the amount of food available to organisms at higher trophic levels. But **neritic** areas, coastal waters lying above relatively shallow continental shelves, are generally rich in nutrients and support an abundance of diverse marine life (Figure I.2). This productivity is often maximized in estuaries such as Chesapeake Bay, North America's largest estuary, which also serves as a prime example of their importance. For more than 400 years, watermen of Maryland and Virginia have worked the bay, first as subsistence fishers and later as highly successful commercial suppliers of oysters, clams, striped bass, and blue crab (see Chapter 4). Another important ocean ecosystem is Prince William Sound off the coast of Alaska. In addition to its natural grandeur, the sound is habitat for a profusion of Pacific herring, pink salmon, sea otters, bald eagles, harbor seals, harlequin ducks, orca whales, and other wildlife (see Chapter 9).

Terrestrial Ecosystems

Terrestrial ecosystems are land-based environments, either natural or artificial. Examples of natural ecosystems are Alaskan tundra, a Canadian coniferous forest, or an African desert. Examples of artificial ecosystems are a cornfield in Iowa, a tree plantation in England, or an urban center such as Toronto, Canada. Like aquatic ecosystems, terrestrial systems range from very small to very large. A small park or woodlot near your home—or all of Australia—can be considered a terrestrial ecosystem.

Ecosystems may be described as undisturbed or disturbed. Undisturbed ecosystems are those largely unaffected by human activities, including permanent settlement, home building, tree harvesting, mining, and agriculture. Disturbed ecosystems have been signifi-

Figure I.2 Coastal Pacific Ocean waters near Lincoln City, Oregon.

cantly changed either by a natural event such as a flood or volcano or by human actions such as cutting down a forest or building a highway or dam.

Examples of undisturbed ecosystems include some Pacific Northwest old-growth forests and parts of temperate-zone rain forests in Canada's Clayoquot Sound (see Chapter 2). Another important and yet-undisturbed terrestrial ecosystem is the Arctic National Wildlife Refuge, part of which is being considered for oil and gas development (see Chapter 1). Ecologically disturbed systems include England's River Thames (see Chapter 3), Chesapeake Bay (see Chapter 4), and Alaska's Prince William Sound (see Chapter 9).

Biomes

Similar land-based ecosystems exist on different continents and islands. For example, tundra ecosystems occur not only in the Far North of Alaska but also across northern latitudes in Canada's Yukon and Northwest Territories, northern Ontario and Quebec, Russia, Siberia, and Norway, Sweden, and Finland. The total of all of these circumpolar tundra ecosystems is called the tundra *biome.*

A **biome** is the totality of similar terrestrial ecosystems that may span many of the planet's continents and islands. Each biome has distinctive climatic, physical, and biological features. When represented on a map, the earth's biomes appear as a planetary patchwork quilt of like ecosystems.

Biomes are usually classified in terms of the most common kinds of plant life they support, which largely depends on rates of precipitation and prevailing climate. In this book we will consider eleven major biomes, though other classification systems may arrive

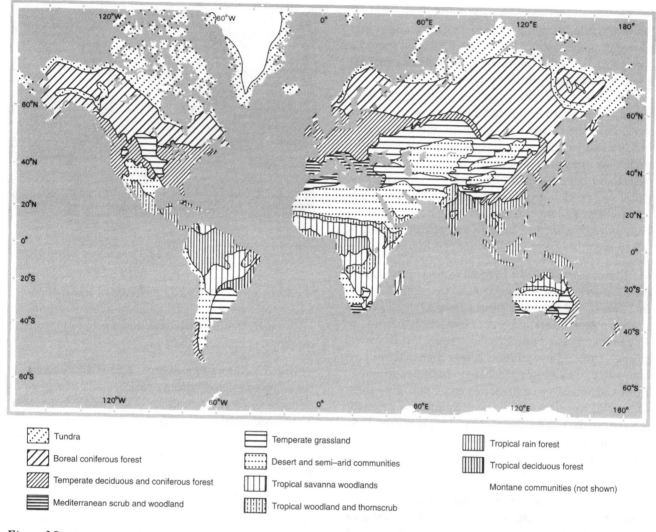

Figure I.3 The earth's major biomes.

(Adapted with permission from Environmental Systems, An Introductory Text, *by I.D. White, D.N. Mottershead, and S.J. Harrison, 2d ed., Chapman & Hall, London, 1992.)*

at slightly different numbers (Figure I.3). Keep in mind that the biome boundaries set forth on maps are inexact and exist only to make it easier to discuss and compare these natural areas.

Tundra The tundra biome exists mainly in latitudes north of the Arctic Circle. It is a mostly flat, almost treeless arctic coastal plain that receives less than 250 millimeters (≈ 10 inches) of annual precipitation. Tundra soils, which are frozen most of the year, thaw and support a rich diversity of grasses and shrubs during brief arctic summers. Permafrost underlies most tundra. Animal species native to the tundra biome include large mammals such as caribou, musk oxen, sheep, bear, and wolves. Smaller mammals are also found, such as lemmings, ground squirrels, and arctic foxes. Many bird species migrate to tundra grasslands each summer, including the snowy owl, arctic tern, peregrine falcon, and golden plover. Chapter 1 describes the Alaskan

tundra and includes a case study detailing efforts to preserve portions of this wilderness area. Chapter 9 discusses the Trans Alaska Pipeline System, part of which is constructed on the Alaskan tundra.

Boreal Coniferous Forest Boreal coniferous forests (also called taiga) circle the earth south of arctic tundra regions. They are characterized by warm summers and cold winters. Annual precipitation is about 500 millimeters (≈ 20 inches). Dominant trees include aspen and coniferous (cone-bearing) species such as black and white spruce and jack pine, from which the biome takes its name. Boreal coniferous forest serves as winter habitat for migrating caribou and year-round habitat for the moose, black bear, grizzly bear, lynx, wolverine, snowshoe hare, and red squirrel.

Temperate Deciduous and Coniferous Forest Temperate deciduous and coniferous forests are found in

Legend:
- Tundra
- Boreal coniferous forest
- Temperate deciduous and coniferous forest
- Mediterranean scrub and woodland
- Temperate grassland
- Desert and semi–arid communities
- Tropical savanna woodlands
- Tropical woodland and thornscrub
- Tropical rain forest
- Tropical deciduous forest
- Montane communities (not shown)

the eastern and western United States, British Columbia in Canada, Britain, eastern Europe, southern Russia, and southeastern China and Japan. They are characterized by hot summers, cold winters, and long growing seasons. Annual precipitation is about 1,000 millimeters (≈ 39 inches), and vegetation consists of broad-leaved hardwood deciduous trees (those that seasonally shed their leaves), including oak, maple, hickory, ash, and beech, and softwood evergreen trees, such as spruce, pine, hemlock, and cedar. Some of the oldest and tallest trees in the world—Douglas fir and Coast redwoods—grow in Pacific Northwest coniferous forests. Fauna include cougars, bears, deer, rodents, tree squirrels, birds, and stream-spawning salmon.

Mediterranean Scrub and Woodland   This temperate biome, also known as **chaparral,** occurs along Mediterranean seacoasts as well as coastal areas of California, southern Australia, and parts of South Africa. Both Los Angeles and San Francisco lie within the chaparral biome. Annual precipitation is less than 300 millimeters (≈ 12 inches). Typical trees include evergreen shrubs, drought-resistant scrub oaks, and pine trees. Chaparral vegetation produces flammable aromatic oils that make this biome prone to frequent fires. Resident animal species include wild sheep and goats near the Mediterranean, small antelope in southern Africa, and kangaroos in Australia.

Temperate Grassland   The temperate grassland biome, also called **prairie,** occurs in midcontinental areas of North America, South America, Europe, and Asia. Originally, these areas supported wide expanses of tall and short prairie grasses and prodigious herds of bison, pronghorns, and wild horses. However, few examples of the undisturbed biome are left. Most have been converted into agricultural grazing and crop lands; the American Midwest and the Canadian prairies are prime examples of what were once vast open prairie lands. A moderate climate prevails in these areas, with an annual precipitation of 250–750 millimeters (≈ 10–30 inches). Chapters 7 and 8 describe agriculture on today's North American prairie biome in Canada and the United States.

Desert and Semi-Arid Communities   Desert is an arid (dry) biome where annual precipitation is generally less than 250 millimeters (≈ 10 inches). Vegetation is sparse and includes creosote bushes, sagebrush, and cacti (Figure I.4). Desert and semi-arid communities typically exist close to 30 °N and 30 °S latitudes, where dry air originating from above the equatorial tropics descends. Deserts are found in southern Arizona and New Mexico, northern Mexico, coastal Chile and Peru,

Figure I.4   Mojave Desert near Barstow, California.

northern Africa, Saudi Arabia, parts of Iran, Mongolia, China, and central Australia. Examples of animal life adapted to the desert biome include buffalo, camels, donkeys, sheep, and goats.

Tropical Savanna Woodlands   Tropical savannas are open grasslands in eastern Africa, parts of Australia, Venezuela, Brazil, southern Asia, and Madagascar. Their climate is mostly hot, and they average 250–750 millimeters (≈ 10–30 inches) of annual precipitation. Savannas are known for their abundant tall grasses, often growing as high as 3 meters (≈ 10 feet), and their umbrella-shaped Acacia trees, a dominant species. Animal life on the African savanna includes zebras, wildebeests, lions, elephants, giraffes, and rhinoceroses.

Tropical Woodland and Thornscrub   Tropical woodlands consist of dry forests and scrublands occurring at lower elevations. They have summer and winter rainy seasons interspersed with long drought periods when little or no rain falls for six to eight months. Annual rainfall is generally 500 millimeters (≈ 20 inches) or less. Examples of tropical dry forests are found in western Costa Rica where some trees and shrubs are deciduous, losing all their leaves during prolonged dry periods to conserve moisture. Other trees are semi-deciduous, retaining their leaves, especially if the dry periods are not too long. Shrubs include columnar and organ pipe cacti. Their well-defined dry season limits plant growth and the activity of animals, which may include the white-faced monkey, American alligator, and Tasmanian devil. Tropical woodlands are found mainly in Central America, southern Asia (monsoon forests), and parts of Australia.

Tropical Rain Forest   The tropical rain forest biome is found near the equator, spanning continents and islands mainly between 23.5 °N and 23.5 °S latitude.

Annual rainfall ranges between 2,000 and 4,500 millimeters (≈ 80 to 180 inches), although dry seasons less than three months long also occur. The rain forest supports complex multi-level tree canopies and lush, diverse plant and animal communities. Although rain forests account for about 8 percent of the world's land area, ecologists believe 80 percent or more of all of the earth's species live in that biome. Rain forest trees, although not conifers, are called "evergreens," since leaf fall and regeneration occur year-round. Fauna include monkeys, bats, sloths, snakes, frogs, monkeys, and innumerable insect species. Major tropical rain forest formations exist in Brazil, Peru, Ecuador, Colombia, Panama, Costa Rica, Nicaragua, Honduras, Guatemala, Congo, Gabon, Guinea, Cameroon, Nigeria, Malaysia, the Philippines, Vietnam, Myanmar (formerly Burma), and Papua New Guinea. Chapter 2 examines environmental issues involving tropical rain forests.

Tropical Deciduous Forest Tropical deciduous forest, another equatorial biome, is similar to tropical woodland and thornscrub but occurs at moderate elevations where rainfall usually exceeds 500 millimeters (≈ 20 inches) a year. The trees, which are taller than thornscrub and cacti, are sensitive to temperatures approaching freezing. This biome occurs in southeastern Brazil and India. Animal life native to this biome includes monkeys, parrots, Asian tigers, and the American jaguar.

Montane Montane, the mountain biome, makes up one-fifth of the world's landscape. Its ecology is highly variable owing to significant changes in temperature, cloud, and rainfall patterns with increasing elevation (Figure I.5). Indeed, distinctly different vegetative zones are seen in mountainous environments, depending on their elevation above sea level. The sequence of changes in plant life in the progression to higher elevations mimics the pattern found from the equator to the poles. For example, temperate grasslands that normally occur north of the Tropic of Cancer and south of the Tropic of Capricorn occur also in alpine (high-elevation) ecosystems. Montane fauna differ greatly as well, depending on a mountain's latitude and elevation. Overall, the diversity of plant and animal life decreases with increasing elevation, paralleling what occurs at higher and higher latitudes in other biomes.

EARTH'S LIVING RESOURCES

Taxonomists, scientists who classify living things, have identified about 1.75 million species of plants and animals on the earth. However, in 1995 a United Nations team of scientists estimated the total number of

Figure I.5 Ancient bristle-cone pine trees in California's Inyo National Forest—an example of unique montane-biome high-elevation vegetation.

species to be a much higher number—as many as 14 million. If that estimate is correct, it means most existing species have not yet been found and classified. Furthermore, some scientists believe even the U.N. figure greatly underestimates the actual number of species on earth. E.O. Wilson of Harvard University assesses the total number of plant and animal species alive today at between 10 million and 100 million.

Biodiversity

The number of different plant and animal species living in a defined ecosystem or study area is referred to as the area's **biodiversity.** Biodiversity can be expressed mathematically as an index that includes both the number of different species in a particular ecosystem (*species richness*) and the relative abundance of each species present (*species equitability*). Abiotic factors such as soil chemistry and prevailing climate greatly influence biodiversity. For example, the high temperatures and copious rainfall in tropical rain forests support a far greater biodiversity than do the colder, drier conditions in tundra ecosystems.

Not only do many, if not most, plant and animal species alive today remain to be identified, but some are becoming extinct without our ever having been aware of their existence. It is believed that before humans populated the earth, only one species per one million species became extinct each year, a natural rate of extinction. But this annual loss was matched by an equal number of new species—one new species per one million species per year.

However, biologists tell us that during the 1990s, close to 1,000 species per one million species, mostly plants, became extinct each year. This large increase is referred to as the earth's Sixth Extinction. Five earlier extinctions of life on earth have been documented (the best-known being the demise of the dinosaurs about 65 million years ago), and as a result, 95 percent of all the species that ever lived on the planet are today extinct.

Environmentalists believe it is important to preserve biodiversity because it is nature's only blueprint of itself. Biodiversity is the ultimate gene bank to which we must repeatedly return to improve crop varieties, yields, and resistance to disease (see Chapter 7), as well as a priceless resource that we often tap for new medicinal agents, therapies, and cures. In addition, its very existence enlivens and enriches the human spirit, as Aldo Leopold expressed in the foreword to *A Sand County Almanac*:

> There are some who can live without wild things, and some who cannot. These essays are the delights and dilemmas of one who cannot.

Interrelationships Among Plants and Animals

As we discussed earlier, the word community refers to organisms living together in a particular ecosystem. The grasshopper described at the beginning of this chapter, along with the goldenrod that provides its food and shelter and all the other plants and organisms in the field, comprise a community. An organism's **habitat** is the place or places where it lives and reproduces, such as a field, pond, tree, or rock. Its role in the community is its *niche*.

Niche refers to the interactions and relationships an organism experiences—how it reproduces and where it obtains food, water, and shelter. An organism or population's niche is the more or less exclusive role of only one species in a given community. Therefore, according to ecological theory, two species cannot occupy the same niche at the same time. This is called the competitive exclusion principle.

Interactions among plants and animals are highly varied and complex, and populations of different species affect, and are affected by, other populations. Some of these interactions are cooperative, and some are competitive. However, most of them can be explained in terms of one or more of the following relationships:

- **Mutualism** is an intimate symbiotic association between two different species, wherein both species clearly benefit. An example is a lichen, in which algae live within fungi. The fungi provide an enclosure to retain moisture, while the algae produce needed food through photosynthesis.

- **Commensalism** is an intimate symbiotic association between two species wherein one species benefits while the other is unaffected. An example is a tropical orchid growing on a host tree but using the tree only for support, not as a source of nutrients.

- **Predation** is a relationship between two species, one the predator and the other the prey, wherein the predator benefits while the prey is either partially consumed and continues to live or is consumed to death. Types of predation include:

  Herbivory Consumption of green plants (prey) by herbivores (e.g., cows grazing on alfalfa).

  Carnivory Animals (prey) being consumed by other animals (e.g., arctic wolves killing and eating caribou).

  Parasitism A short- or long-term association between organisms wherein a parasite feeds off a host, normally without killing it (e.g., mosquitoes drawing blood from caribou that have migrated to Alaska's North Slope [see Chapter 1]).

- **Competition** involves interactions among affected organisms wherein limited resources, which may include space, water, or food, are divided among competitors whether they are of the same species (*intraspecific competition*) or of different species (*interspecific competition*). An interesting example of intraspecific competition was demonstrated by ecologists in 1931. A trench was dug and maintained in the midst of a mature white pine forest in New Hampshire, cutting hundreds of tree roots in the process. Competition among trees for soil moisture and nutrients was thereby limited to what each tree near the trench could obtain on its side of the trench. Another part of the forest was identified as a control study area. Eight years later, white-pine sapling growth was much greater in the trenched forest area compared to the untrenched control area, showing that mature pine trees exert intraspecific competition for resources, limiting the growth of new saplings.

- **Allelopathy** refers to an interaction between two plant species wherein one species produces and releases a specific chemical substance that inhibits the growth of the other species. An example of allelopathy is the release of natural pesticides by California Coast redwood trees, which inhibits attack by insect pests.

ENERGY FLOW THROUGH ECOSYSTEMS

Photosynthesis

All organisms require energy in order to live, grow, and reproduce. In the natural world, this energy is almost always obtained directly from sunlight captured by green plants. Plants convert solar energy into chemical energy, which is then stored as organic matter in a process known as **photosynthesis,** as expressed in the following equation:

$$6CO_2 + 6H_2O + Sunlight/Chlorophyll$$
$$\rightarrow C_6H_{12}O_6 + 6O_2$$

In other words, six molecules of carbon dioxide (CO_2) combine with six molecules of water (H_2O) in the presence of sunlight and chlorophyll to form one molecule of the simple sugar glucose ($C_6H_{12}O_6$) and six molecules of oxygen gas (O_2). Although photosynthesis itself produces simple sugars, plants are able to transform some of them into more complex food products such as carbohydrates, proteins, and lipids, which higher organisms need to maintain life. The overall process is known as **primary production.**

Primary Production and Trophic Levels

Green plants convert inorganic sources of carbon, such as carbon dioxide, into organic compounds—foods that they and other organisms need. Thus, plants are called **primary producers,** or **autotrophs** (self-feeders: *auto-* = "self"; *-troph* = "nutrition"). All other organisms depend on sources outside themselves for their food, and so are called **consumers,** or **heterotrophs** (*hetero-* = "other").

Primary production and consumption can be visualized in terms of **trophic levels** within a biotic "pyramid" (Figure I.6). As primary producers, green plants occupy the first trophic level of the pyramid. Animals that eat green plants are called **herbivores,** and

they occupy the second trophic level. The third and higher trophic levels are made up of **carnivores,** animals that eat other animals. Some animals eat at any trophic level, and so are called **omnivores.** A small number of species, including whales, polar bears, eagles, and humans, are **top predators,** meaning that they feed at an ecosystem's top, or highest, trophic level.

To help visualize these trophic levels, recall the story at the beginning of this chapter: The goldenrod plant that produced the leaf in the field is a primary producer; the grasshopper that ate the leaf is an herbivore occupying the second trophic level; the pheasant that ate the grasshopper is a carnivore at the third trophic level; and because pheasants are hunted in certain areas for human consumption, humans could be considered the top predator. A series of organisms deriving their energy and materials from one trophic level to another in this type of simple, linear progression is called a **food chain.** However, in most ecosystems, energy paths are not simple and linear because many animals feed at more than one trophic level; they comprise complex networks of interconnected food chains called a **food web.**

Decomposition and Waste Heat

All organisms are intimately involved in recycling matter, especially water and nutrients. Some, however, have a specialized ability to break down **detritus** (dead organisms, organic matter, and animal wastes) into inorganic matter. These so-called detritus feeders obtain their energy by decomposing nonliving organic matter into simpler chemicals and nutrients, which are then recycled. Important detritus feeders are bacteria, fungi, insects, worms, and termites.

It is important to remember that, although matter can be recycled, energy cannot. As energy performs useful work, part of it is always "lost" as waste heat. For example, the decomposition of detritus into inorganic matter releases waste heat that cannot be recycled in nature. Waste heat is also released by other organisms as a result of digestion, growth, repair, and reproduction. Thus, as food energy moves to higher trophic levels within a biotic pyramid, energy is lost as waste heat at each and every step (Figure I.7). This applies to simple food chains as well as to more complex food webs. One of the important consequences of energy loss within a biotic pyramid is that less food energy is available to consumers at higher trophic levels. Applied to humans, this means many more people can be fed at a lower trophic level (consuming wheat, corn, rice, potatoes, and other plant crops) than at higher levels (consuming beef, pork, chicken, and fish).

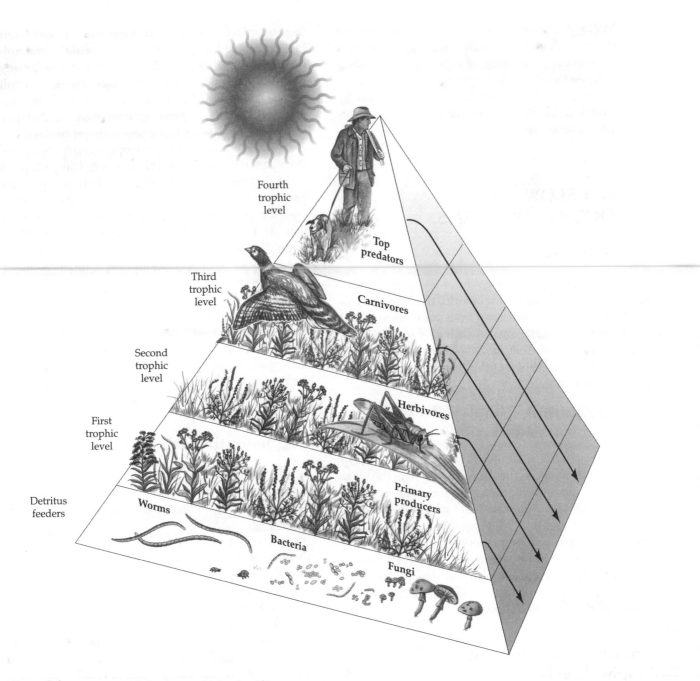

Figure I.6 Biotic pyramid showing trophic levels within an ecosystem.

BIOGEOCHEMICAL CYCLES IN ECOSYSTEMS

Unlike energy, which flows "one-way" through food webs and is ultimately spent as waste heat, the chemical elements in minerals and nutrients are recycled. That is, their overall quantities remain the same and they continue to be reused, even though their chemical forms may change. The processes whereby chemicals are continually recycled through interactions of organisms with their environment are called **biogeochemical cycles**. An example, the hydrologic cycle, mentioned earlier in this chapter, is explained further in Chapter 3, while the nitrogen cycle is covered in Chapter 8. In this chapter, we will take a brief look at biogeochemical cycles involving carbon and oxygen, which are fundamental to photosynthesis and many other ecological processes.

The Carbon and Oxygen Cycles

Like many other chemical elements, carbon and oxygen atoms continuously move through ecosystems on cyclical journeys referred to as carbon and oxygen

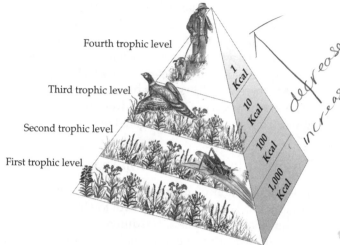

Fourth trophic level

Third trophic level

Second trophic level

First trophic level

Figure I.7   Energy pyramid within an ecosystem. Energy is lost as waste heat as plant and/or animal matter is consumed.

and animal life, energy stored in lipids and fats is continuously released through *aerobic respiration* and in the process, carbon and oxygen combine to form carbon dioxide which is discharged to the atmosphere.

Aerobic respiration is a cellular-level biochemical process occurring in most plants and animals whereby energy stored in organic matter is released through metabolism. Oxygen gas is consumed and carbon dioxide gas is formed. Although respiration can occur in the absence of oxygen (anaerobic respiration), oxygen-based (aerobic) respiration is more common.

As the following chemical equation shows, respiration can be regarded as the opposite of photosynthesis:

$$C_6H_{12}O_6 + 6O_2 \rightarrow 6CO_2 + 6H_2O$$

In other words, organic matter (represented by one molecule of the simple sugar glucose) combines with six molecules of oxygen (O_2) to form six molecules of carbon dioxide (CO_2) and six molecules of water (H_2O).

Through respiration, carbon atoms in organic matter (food) are transformed into carbon dioxide, released to the atmosphere, and then recycled, perhaps in a totally different ecosystem. But the next step in the journey may not again be photosynthesis. The carbon dioxide released through respiration could, for example, dissolve in a rain drop making it acidic as carbonic acid is formed; acidic rainfall assists in the weathering of rock minerals changing some of them into water-soluble nutrients such as calcium bicarbonate.

Biogeochemical cycles such as those we have just discussed are examples of nature's complex, dynamic tapestry reflecting how the natural world works.

biogeochemical cycles. At times during these cycles, given carbon and oxygen atoms may be part of a mineral like calcium carbonate. At other times, they may be present in carbon dioxide gas, and at still other times they can be found in organic matter such as sugars, carbohydrates, and proteins.

Earlier, we saw that chlorophyll in green plants traps solar energy and acts as a catalyst causing carbon atoms in carbon dioxide gas to chemically combine with oxygen atoms in water molecules to form organic compounds such as sugar, an important nutrient. Herbivores feeding on plants convert sugar to lipids and proteins and, in this way, carbon and oxygen become part of muscle, nerve, and bone tissue. In both plant

REGIONAL PERSPECTIVES

Environmental Protection in Canada

Canada's identity is shaped in large part by the abundance and diversity of its wilderness and wildlife. But, as in many other parts of the world, this rich legacy is being eroded by human actions such as deforestation, habitat destruction, environmental air and water pollution, inadequate management of toxic substances, nonsustainable agricultural practices, and losses of plant and animal biodiversity. Canada also faces a number of global problems, including:

- Increasing exposure to solar ultraviolet radiation due to periodically declining levels of stratospheric ozone.

- Effects of global warming and related climate-change phenomena linked to rising level of atmospheric carbon dioxide and other greenhouse gases.

Marine pollution that occurred due to the 1970 sinking of the oil tanker *Arrow* and the barge *Irving Whale* led Canadian authorities to establish the Department of the Environment (1971), which later became Environment Canada. In 1988, the Canadian

Environmental Protection Act (CEPA) was passed, establishing broadly based legislation and regulations to deal with the most pressing environmental issues. Under CEPA, a number of important actions were taken, including:

- Reduction of lead and benzene levels in gasoline fuels.

- Virtual elimination of dioxins and furans, highly toxic chemicals, from pulp and paper-mill effluent discharges to coastal waters.

- Significant reduction in the use of chlorofluoro-carbons (CFCs) as refrigerant chemicals. Canada's phase-out of CFCs and other ozone-depleting substances took place faster than required under the Montreal Protocol, an agreement among many industrial nations to discontinue CFC production.

- Elimination of polychlorinated biphenyls (PCBs) from new electrical transformers, capacitors, and other electrical equipment.

- Establishment of a National Pollutant Release Inventory to track releases of toxic chemicals.

In 2000, responsibility and authority for environmental protection in Canada will expand under the new CEPA, which focuses on pollution prevention rather than simply pollution control. It mandates:

- Effective management of 23,000 toxic substances.

- Citizens' rights to sue the government if significant environmental harm occurs as a result of failure to enforce CEPA regulations.

- Creation of a National Pollutants Release Inventory linked to an Internet-based Environmental Registry of CEPA laws and regulations.

- Setting up a National Advisory Committee with representatives of federal, provincial, territorial, and aboriginal governments to ensure effective actions in environmental protection.

- Implementation of internationally agreed-upon conventions dealing with the control of trans-boundary movement of hazardous wastes.

- Expanded powers of enforcement to control air, water, and land pollution.

Canada was the first industrial nation to ratify the United Nations Convention on Biological Diversity in 1992. Soon afterward, in 1995, Canada published its Canadian Biodiversity Strategy, a plan endorsed by both federal and provincial environmental, parks, wildlife, and forestry ministers. The 1998 Endangered Species Recovery Fund, an outgrowth of this strategy, focuses on Canadian species at risk and continues to be supported by the Canadian Wildlife Service, a division of Environment Canada, and the World Wildlife Fund Canada. Successful efforts thus far include protecting high-priority wildlife species such as the peregrine falcon, piping plover, trumpeter swan, and beluga whale. In 1999, there were 339 species on Canada's national list of species at risk. The Canadian government is currently at work developing an Endangered Species Act.

Environmental Protection in Germany

The reunification of East and West Germany in 1990 brought together two societies with totally different environmental philosophies. The differences were perhaps best reflected in their energy policies. West Germany had a highly diversified and largely privately owned energy supply system with high standards for energy efficiency, while East Germany had a very centralized, mainly state-owned system dependent on lignite, or "brown coal," a highly polluting energy source.

Since then, considerable privatization of East Germany's energy industry has occurred, contributing to significant reductions in air pollution, especially levels of atmospheric sulfur dioxide, carbon monoxide, and smokestack particulate emissions from power plants and factories. Coal power, particularly "hard coal," which is far less polluting than lignite, continues to play a major role in Germany's mix of energy sources. The German government subsidizes its use and has managed to diminish the air pollution problems caused by its combustion by mandating flue-gas desulfurization and greater use of low-sulfur coal. Nevertheless, the government is determined to reduce coal subsidies by 2005.

Nuclear power continues to play a major role in Germany. In 1999, 32 percent of the country's electric power was generated by 20 reactors, but there is considerable ongoing debate about their future. Power generated by reactors is "clean" in terms of carbon dioxide emissions. However, many citizens and government leaders would like to abandon nuclear power, and environmental activists stand opposed to expanding it, pressing instead for investment in alternative energy sources.

A restructuring of Germany's electric power marketplace is currently in progress and is favorable to independent suppliers of renewable power. In 1999, for example, Germany announced its six-year 100,000 Solar Roofs program, which provides interest-free loans with 10-year flexible payback plans to people who buy and install rooftop PV power systems, which are solar cells capable of converting light directly into electricity. Up to 40 percent of the cost of the average PV rooftop unit is government subsidized (see Chapter 11).

Wind power is one of the most important alternative energy developments in Germany. In 1999, Germany had a total of 2,875 MW of installed wind power, making it the world leader in wind-generated electric power production. Its commitment to wind energy began shortly after the Chernobyl nuclear disaster in the Ukraine in 1986. Now, in northern Germany, 15 percent of electric power is produced by wind turbines, and the government plans to shut down a nearby nuclear power plant (see Chapter 11).

Today, climate change issues loom large on Germany's environmental protection agenda ever since its government agreed to an 8 percent reduction in carbon dioxide emission levels at the 1997 Framework Convention on Climate Change in Kyoto, Japan (see Chapter 12). To accomplish this, Germany plans to reduce industrial carbon dioxide emissions by at least 25 percent by 2005, a very ambitious goal.

Environmental Protection in Sweden

Sweden's Environmental Protection Agency (SEPA), established in 1967, is the country's lead organization for protecting and sustaining the environment. But even before SEPA, Sweden was leading the way in Europe toward enlightened environmental policies and practices.

Sweden was the first European country to establish a system of national parks. Nine were opened to the public in 1909, and today a total of 26 national parks span Sweden, from its majestic forest ecosystems in the south to its impressive mountains in the north. The parks encompass 6,524 square kilometers ($\approx$ 652,400 hectares or 1,612,000 acres), 1.5 percent of the country's total area. In addition, more than 2,000 nature preserves covering close to 30,000 square kilometers ($\approx$ 3,000,000 hectares or 7,413,000 acres) have been set aside, and three more national parks are being planned.

Environmental problem-solving in Sweden is largely based on environmental taxes called "green" taxes. For example, consumers pay higher gasoline and diesel fuel taxes according to the fuel's sulfur content. Consumers are also assessed charges based on motor vehicle nitrogen oxide emissions. This approach is leading to a cleaner environment. A 30 percent reduction in sulfur dioxide emissions was achieved between 1989 and 1995. During that same period, nitrogen oxide emissions were cut in half in areas where applicable green taxes were imposed.

Taxing carbon dioxide emissions is another part of energy policy in Sweden. Recent studies show that by 1994 this policy had brought about an 11 percent reduction in the country's total greenhouse gas emissions. It is believed that the carbon dioxide tax will ultimately lead to less dependence on fossil fuels, which currently supply about 10 percent of the country's electric power. In the late 1990s, 53 percent of the electric power generated in Sweden came from nuclear sources and 37 percent from hydroelectric sources. The future of nuclear power in Sweden continues to be highly controversial (see Chapter 10).

In 1999, Sweden's Parliament, or Riksdag, adopted 15 environmental policy objectives to be shared by all sectors of society and achieved in large part by 2020. The goal is to achieve a sustainable environment for future generations. Here are the 15 objectives:

1. Clean air
2. A balanced marine environment, including sustainable coastal areas and archipelagos
3. A nontoxic environment
4. High-quality groundwater
5. Sustainable lakes and watercourses
6. Flourishing wetlands
7. No eutrophication
8. Natural acidification only
9. Sustainable forests
10. A varied agricultural landscape
11. A magnificent mountain environment
12. A good urban environment
13. A safe radiation environment
14. A protective ozone layer
15. Limited influence on climate change

KEY TERMS

abiotic factors	consumers
aerobic respiration	detritus
allelopathy	ecology
aquifer	ecosystem
autotrophs	environmental ethics
biodiversity	environmental policies
biogeochemical cycles	environmental science
biome	food chain
biosphere	food web
biotic factors	freshwater
carnivores	greenhouse effect
chaparral	habitat
commensalism	herbivores
community	heterotrophs
competition	hydrologic cycle
conservation	mutualism

natural resources
neritic
niche
non-point sources
omnivores
pelagic
photosynthesis
point sources
pollution
population
prairie

predation
preservation
primary producers
primary production
scientific method
stewardship
sustainable
taxonomists
top predators
trophic levels

construction, forestry, agriculture, home building, and industry.) What problems could develop within this ecosystem in the future? How could these problems be prevented or managed?

DISCUSSION QUESTIONS

1. Aldo Leopold wrote, "When we see land as a community to which we belong, we may begin to use it with love and respect." What exactly do you suppose he meant by those words?

2. John Muir felt that absolute preservation of wild and scenic environments was the only way to save them for future generations. Do you agree, or is it really going too far to set aside large natural areas from ever being developed?

3. What do you think about Gifford Pinchot's definition of conservation: "the use of the earth for the good of man"? Could this approach to conservation generate a greater appreciation for the value of natural systems?

4. In 1972, most uses of the chemical pesticide DDT were banned in the United States and subsequently abandoned in other countries as well. But malaria is now on the increase, especially in sub-Saharan African nations, where millions of children are dying from it yearly. Since widescale use of DDT could control malaria in these nations, would you argue for continuing or banning its use in Africa? Explain your answer.

5. Choose an ecosystem close to where you live and describe it in terms of its plant and animal life. Is the ecosystem disturbed or undisturbed? How can you tell? (Consider such activities as road

INTERNET WEBSITES

Visit our website at http://www.mhhe.com/environmentalscience/ for specific resources and Internet links on the following topics:

Biomes of the world

Major biomes

Conservation around the world

Images of Walden Pond in Massachusetts

National Wildlife Federation

Biodiversity Conservation Information System

Ecology and biodiversity

Canada's Endangered Species Recovery Fund

SUGGESTED READINGS

Callicott, J. Baird, and Michael P. Nelson, eds. 1998. *The great new wilderness debate*. Athens and London: The University of Georgia Press.

Carson, Rachel. 1962. *Silent spring*. Boston: Houghton Mifflin.

Leopold, Aldo. 1949. *A sand county almanac*. New York: Oxford University Press, Inc. [Note: 11th printing, Ballantine Books Edition, July 1977.]

Lyndon, Shanley, J. 1957. *The making of Walden*, with the text of the first edition. Chicago: University of Chicago Press.

Miller, Debbie. 2000. *Midnight wilderness*, 10th anniversary edition. Anchorage, AL: Northwest Books.

Weiss, Edith Brown. 1988. *In fairness to future generations: International law, common patrimony, and intergenerational equity*. Tokyo, Japan: United Nations University; Dobbs Ferry, NY: Transnational Publishers.

PART I

Understanding and Protecting Natural Environments

PhotoDisc, Inc./Vol. 16

Chapter 1
Arctic Wilderness and Wildlife

Chapter 2
Tropical Forests and Biodiversity

CHAPTER 1

Arctic Wilderness and Wildlife

PhotoDisc, Inc./Vol. 44

Gazing beyond the open tundra plain, I stare at the shimmering endless pack ice, curving toward the North Pole some fourteen hundred miles away. Beneath a dome of larkspur-blue sky, on this final fringe of tundra, with our civilized world far behind us, a tremendous feeling of wildness surges through me. We are truly on the edge of the most remote wilderness area remaining in the United States. We are on top of the world.

Debbie S. Miller, *Midnight Wilderness: Journeys in Alaska's Arctic National Wildlife Refuge*

The major theme of this chapter is wilderness—how and why Alaska's Arctic National Wildlife Refuge was set aside and protected as a wilderness area. We will study the essence of wilderness, its value in providing critical habitats for untold numbers of plant and animal species, and its significance to humans.

The Arctic National Wildlife Refuge (ANWR) is a pristine wilderness and wildlife sanctuary established in northeastern Alaska by the United States. The refuge protects a rich diversity of arctic and subarctic habitats for fish and other wildlife. It also provides compelling insights into natural ecosystems and how they work. Creation of the refuge followed a struggle of opposing ideas: protecting wilderness and wildlife versus developing petroleum and natural gas resources. The struggle isn't over yet, and difficult decisions that affect the future of the refuge lie ahead. We will begin this study by describing the tundra ecosystem and how energy from the sun makes the tundra—and all ecosystems—work.

BACKGROUND

Humans Discover the Arctic Tundra

Anthropologists believe that Asian people first migrated to the Arctic about 12,000 BP (before the present time) during the last North American glaciation, although recent studies suggest these migrations may have occurred even earlier, perhaps as long ago as 20,000 BP. Sea level at that time was about 90 meters ($\approx$ 300 feet) lower than it is now, allowing a land bridge to form across the Bering Strait. Other people arrived later, perhaps 5,000 BP, via skin-covered skiffs, wooden boats, and ice floes.

Aborigines of the far North have long been called Eskimos, but many arctic people of Alaska, Canada, and Greenland prefer the term Inuit, meaning "real people." Some of the natives indigenous to northeastern Alaska, the northern Yukon, and Canada's Northwest Territories are Athabaskan Indians and are called the Gwich'in, or "caribou people."

Alaska's Inuits settled arctic coastal areas north of the Brooks Range and at Kaktovik on Barter Island (Figure 1.1). The Gwich'in, on the other hand, established villages such as Old Crow, Arctic Village, and Inuvik, areas they called The Birthing Place. The outside world knew little about the Arctic, its wildlife, its native peoples, or their way of life until whalers, prospectors, and hunters began to relate their tales of exploration and adventure.

Olaus and Margaret Murie were among the first to explore northeastern Alaska. Working for the U.S. Biological Survey in the 1920s, they studied caribou herds migrating from their winter habitats south of the Porcupine River, across lofty Brooks Range passes, to coastal-plain summer calving grounds, flat treeless tundra bordering the Beaufort Sea. They observed caribou

INSIGHT 1.1

Lichens—A Specialized Relationship

Lichens are plant species composed of algae and fungi living together in a way that benefits both species, a relationship called **symbiosis.** Fungi provide protective habitat that retains moisture, while photosynthetic algae produce food needed by both species. This particular symbiotic connection is an example of **mutualism,** in which each species not only benefits from the association but actually depends on it.

Some lichens contain **cyanobacteria** (blue-green algae), which can fix atmospheric nitrogen gas—that is, they cause nitrogen to combine chemically with oxygen to form useful nutrients such as ammonia.

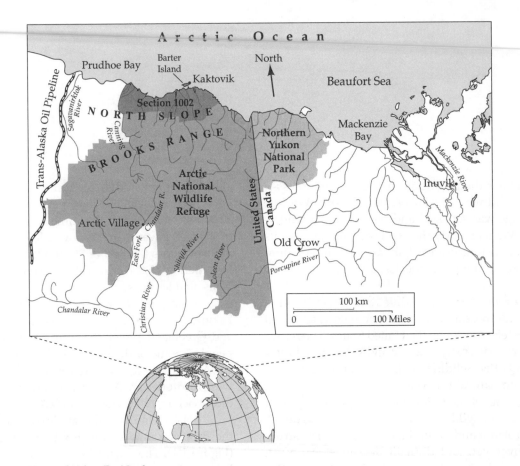

Figure 1.1   The Alaskan and Yukon Far North.
(Source: Map based on maps provided by the U.S. Fish and Wildlife Service, Anchorage, Alaska.)

giving birth to their young and foraging on tundra grasses, mosses, and lichens (Insight 1.1).

Bob Marshall, founder of The Wilderness Society, was another early explorer. Between 1929 and 1939, Marshall made four expeditions to the Arctic. His book, *Arctic Village,* gave the outside world insight into the wild magnificence and timeless beauty of arctic wilderness. His writings evoked compelling images of what he called the "Great Divide," the Central Brooks Range, with its ice-laden mountain peaks, massive gla-

ciers, deep canyons, wild, untamed rivers, and splendid waterfalls. He described the summer tundra as a diversity of mosses, ferns, grasses, countless species of brilliant flowering plants, dwarf willow trees, and small and great rivers. And he wrote about the native people who lived there. Marshall believed that preserving wilderness was important not only to protect plant and animal wildlife but also to save unique environments for posterity, as places where the human spirit could be renewed and enlivened.

INSIGHT 1.2
Permafrost

Permafrost lies below most tundra soils. It remains frozen all year long, even when overlying tundra thaws during arctic summers. Permafrost lends physical stability to the tundra above it, while the tundra's insulating effect keeps permafrost from melting. But if tundra soils are disturbed by such activities as road building, oil drilling, or pipeline construction, its insulating barrier can be compromised, and the underlying permafrost can melt. This melting causes tundra soils to thaw and settle.

Permafrost exerts a continuous cooling effect on overlying tundra, in many cases stunting the growth of trees such as willow, birch, and juniper. Low temperatures in tundra soils diminish the rates of decomposition of **organic** matter. This is beneficial because it results in less nutrient release and runoff. There is also a global benefit: Permafrost cooling limits the amount of the greenhouse gas *methane* that is generated in arctic wetlands compared to that produced naturally in temperate and tropical wetlands. (See Chapter 12.)

Figure 1.2   Alaska's coastal-plain summer tundra wetlands.
(Photo courtesy of U.S. Fish and Wildlife Service, Anchorage, Alaska.)

Characteristics of the Tundra

Tundra is a perpetually cold, almost treeless ecosystem. Its landscapes—soils, low hills, braided rivers, and cobbled beaches—typically receive less than 250 millimeters (≈ 10 inches) of annual precipitation, making the ecosystem a cold desert. Temperatures fall as low as −50 °C (−58 °F), freezing the soil and forcing many species to migrate south to warmer habitats. Permanently frozen ground, called **permafrost,** lies a few inches to a few feet beneath most tundra soils. In the tundra of northern Alaska, permafrost penetrates to depths as great as 650 meters (≈ 2,100 feet) (Insight 1.2). The onset of summer causes frozen rivers to melt and tundra soils to thaw, but the underlying permafrost prevents meltwaters from penetrating the soil. As a result, thousands of shallow lakes, ponds, and pools form, transforming the landscape (Figure 1.2). Some of these wetlands become mosquito hatcheries, an important part of life on the tundra, as we will see later.

The tundra of northern Alaska is cold because the earth's North Pole is tilted away from the sun most of the year. That is, the earth's axis, an imaginary line passing through the North and South Poles, is tilted with respect to its orbit around the sun (Figure 1.3). But for approximately six weeks, from late May to mid-July, the north polar region is slanted toward the sun, and the sun appears to circle the horizon. Then, regions north of the Arctic Circle experience 24 hours of sunlight every day, becoming the "Land of the Midnight Sun." During this relatively short six-week time period, solar irradiation increases as the sun's elevation above the horizon reaches its maximum on about June 20.

Solar energy is crucial to life in almost all ecosystems, including the tundra of the far North. In the following section, we will explore how energy from the sun flows through the tundra ecosystem.

Energy Flow Through the Tundra Ecosystem

Owing to its distance from the sun, the earth receives only a small fraction of the sun's total energy output. This fraction, called the **solar constant,** averages 1,370 watts per square meter or about 20,000 Calories/square meter/minute above the earth's atmosphere. (One **Calorie** [Cal] is the energy needed to raise the temperature of 1,000 grams of water by one degree Celsius.) The atmosphere reflects and absorbs about

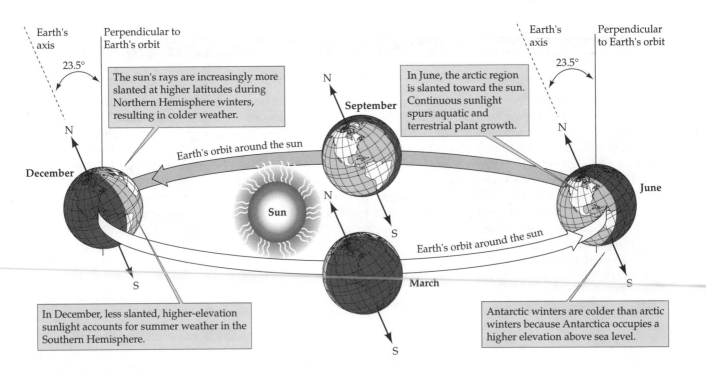

Figure 1.3 The earth's axis is tilted about 23.5° from the plane of the planet's orbit around the sun. The tilt influences the planet's climate and weather and is responsible for the Arctic's 24-hour photoperiod during its brief summers.

half this energy. The other half, about 10,000 Calories/square meter/minute, reaches the planet's surface at the equator. However, less energy is generally received at higher latitudes. This can be calculated using Lambert's Sine Law:

$$I = I_o \sin \beta$$

where I = sunlight intensity at any higher latitude

I_o = sunlight intensity at the equator, about 10,000 Cal/m²/min, and

β = the sun's elevation at noon above the horizon at a given latitude

Since the maximum value of β is 90° at the equator, the solar intensity there would be:

$$I = (10,000) \times (\sin 90°) = 10,000 \text{ Cal/m}^2/\text{min}$$

where the sine of 90° = 1.00. This is the typical noontime solar intensity above equatorial forests. If we assume 10 hours of sunlight daily, a tropical forest would receive about 6,000,000 Cal/m²/day.

Arctic tundra and the Beaufort Sea lie at about 70° N latitude. There, the summer sun's maximum elevation, β, is 43.5°. During arctic summers, solar intensity would be:

$$I = (10,000) \times (\sin 43.5°) = 6,900 \text{ Cal/m}^2/\text{min}$$

Given 24 hours of daily summer sunlight, the tundra receives as much as 9,900,000 Cal/m²/day—*65 percent more solar energy than a tropical forest receives each day!* This explains why tundra plant life and Beaufort Sea phytoplankton exhibit luxuriant growth during north polar summers.

Primary Production

Through photosynthesis, green plants capture energy from the sun and convert **inorganic** sources of carbon, such as carbon dioxide, into **organic** compounds, such as carbohydrates, proteins, lipids, and sugars. This process is called primary production. The twenty-four-hour-long photoperiods on the tundra accelerate photosynthesis and plant growth. Thus, the primary production process is carried out by an impressive variety of arctic plants: mosses, lichens, sedges, cotton grass, alpine bearberry, sweet pea, forget-me-nots, Lapland rosebay, bluebells, asters, lupines, fireweed, willow, birch, juniper, and louseworts.

The rate at which solar energy is captured during photosynthesis is called **gross primary production.** However, not all of this energy is used to make plant biomass. Some of it is needed for plant metabolism, tissue repair, and reproduction. Thus, the rate at which solar energy is converted into plant biomass, thereby making food available to organisms at higher trophic levels, is known as **net primary production.** On the

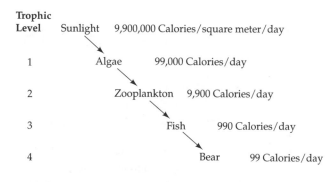

Trophic
Level   Sunlight   9,900,000 Calories/square meter/day

1   Algae   99,000 Calories/day

2   Zooplankton   9,900 Calories/day

3   Fish   990 Calories/day

4   Bear   99 Calories/day

Figure 1.4   Energy flow at a 1-square-meter Beaufort Sea study site.

average, net primary production converts about 1 percent of available sunlight energy into plant biomass. This is called the *One Percent Rule*.

The **One Percent Rule** is the concept that on the average about 1 percent of sunlight energy is captured during photosynthesis and stored in plant biomass. For example, consider net primary production in a 1-square-meter area of the Beaufort Sea, the body of water off the coast of northeastern Alaska. During the arctic summer, this area receives about 9,900,000 Calories of solar energy daily. The One Percent Rule indicates that algae capture about 99,000 Calories of solar energy each day (1 percent of 9,900,000) and incorporate it into algal biomass (Figure 1.4).

Food Webs and Food Chains

A food web is a biological network of feeding pathways through which organisms obtain the energy and materials they need to live. Figure 1.5 illustrates a simplified Beaufort Sea food web. Surface-water phytoplankton, positioned to utilize solar energy multiply, producing new plant matter through photosynthesis. Nearby zooplankton graze on phytoplankton and manufacture animal protein, a food source for arctic cod, arctic char, baleen whales, and other marine fauna. Food web energy-pathways in other ecosystems are similar in complexity and diversity to these Beaufort Sea feeding patterns.

Tundra food webs, however, are simpler than those in the Beaufort Sea and are sometimes described simply as food chains. Let's compare feeding pathways on the arctic tundra with those in the Beaufort Sea in greater detail.

Arctic Tundra Trophic Levels   Figure 1.6 illustrates food production and food consumption in tundra and Beaufort Sea ecosystems. Here, primary production of food and its consumption are shown as compartments in biotic pyramids. As noted earlier, the tundra's primary producers are mainly grasses, lichens, sedges, and an almost unbelievable profusion of flowering plants.

In striking contrast to the diversity of **first trophic level** plant life, relatively few species of herbivores occupy the tundra's **second trophic level.** They are mainly caribou, moose, sheep, lemming, snowshoe hare, and the arctic ground squirrel (Figure 1.7). Their eating preferences, which are simple, are typified by caribou grazing on summer tundra grasses.

Even fewer species of carnivores occupy the tundra's third and higher trophic levels, primarily arctic foxes, snowy owls, weasels, wolves, and bears. Their feeding habits are also simple. Some carnivores eat at the second trophic level by consuming herbivores. For example, arctic foxes and snowy owls chiefly eat lemming. Other carnivores obtain their energy by devouring fellow carnivores. Grizzly bears, on the other hand, are omnivores; they eat just about anything at any trophic level.

Uncomplicated feeding patterns such as those on the tundra can therefore be described in terms of simple food chains. For example, a typical food chain might reflect an arctic wolf consuming a caribou which had obtained its food by eating tundra grasses. Because short food chains convey food energy more efficiently than complex food webs, solar energy on the tundra is conserved and even large carnivores like bears can thrive there.

Beaufort Sea Trophic Levels   Primary production in the Beaufort Sea is carried out principally by marine phytoplankton called diatoms (die'-a-toms), which are single-celled algae Examples of marine diatoms are shown in Figure 1.8. They make up a major part of the first trophic level of the Beaufort Sea biotic pyramid (see Figure 1.6). Primary production occurs from the water surface to depths of 100 meters ($\approx$ 330 feet). This upper layer of the sea is called the **photic zone,** the region where enough light can penetrate to support photosynthesis.

While net primary production is generally limited to capturing about 1 percent of available sunlight, a more efficient energy transfer occurs at the second trophic level where, for example, zooplankton consume algae (see Figure 1.6). Here, about 10 percent of algal biomass is converted into zooplankton protein. This illustrates the *Ten Percent Rule.*

The **Ten Percent Rule** applies to every food web energy-transfer step following net primary production. Consider the earlier example in which algae captured 99,000 Calories of solar energy. Zooplankton grazing on these algae will capture 9,900 Calories daily (10 percent); fish feeding on zooplankton will capture 990 Calories daily (10 percent); and a bear eating fish will capture 99 Calories daily (10 percent) (Insight 1.3; see Figure 1.4).

INSIGHT 1.3
Energy Efficiency

Energy is the ability to perform useful work. In environmental systems, energy is required by living organisms for respiration, digestion, motion, growth, repair, and reproduction.

Only a small percentage of energy is captured by organisms in a food web, but overall no energy is lost. This fact illustrates the **First Law of Thermodynamics**, a fundamental scientific principle, which teaches that energy cannot be created or destroyed. It is always conserved. What may appear to be lost energy can be accounted for as waste heat—energy of limited value incapable of performing useful

work. For example, when an ordinary electric lamp converts electricity into light, considerable waste heat is also produced. The light and heat energy add up to the electric energy consumed.

The **Second Law of Thermodynamics**, which deals with energy efficiency, teaches that no process is 100 percent energy efficient. Every time energy is used, a fraction of it appears as waste heat. The less waste heat produced, the more energy-efficient a particular process is. The one percent and ten percent rules reflect this fact. Other examples of energy efficiency are given in Chapter 11.

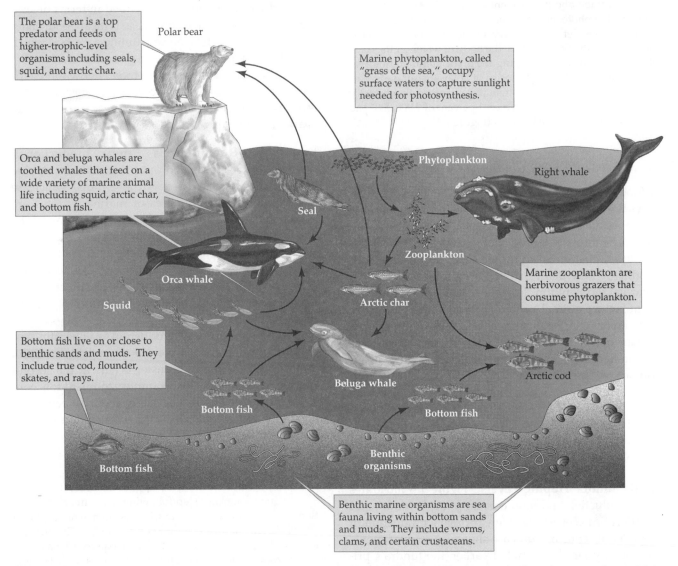

The polar bear is a top predator and feeds on higher-trophic-level organisms including seals, squid, and arctic char.

Polar bear

Marine phytoplankton, called "grass of the sea," occupy surface waters to capture sunlight needed for photosynthesis.

Orca and beluga whales are toothed whales that feed on a wide variety of marine animal life including squid, arctic char, and bottom fish.

Phytoplankton

Right whale

Seal

Zooplankton

Orca whale

Marine zooplankton are herbivorous grazers that consume phytoplankton.

Squid

Arctic char

Bottom fish live on or close to benthic sands and muds. They include true cod, flounder, skates, and rays.

Beluga whale

Arctic cod

Bottom fish

Bottom fish

Bottom fish

Benthic organisms

Benthic marine organisms are sea fauna living within bottom sands and muds. They include worms, clams, and certain crustaceans.

Figure 1.5   Examples of Beaufort Sea plankton and nekton, with a simplified food web diagram showing the flow of energy and materials in this ecosystem.

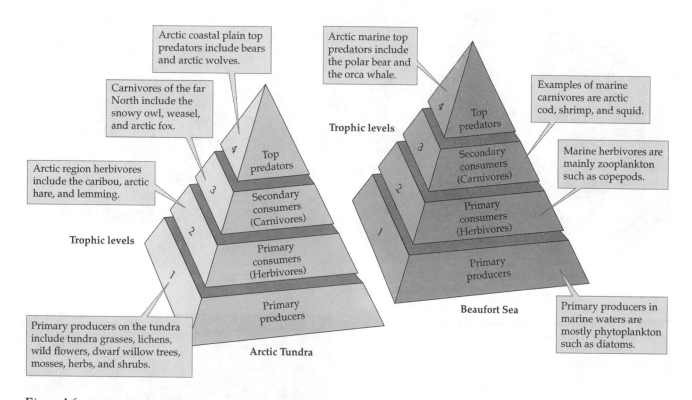

Arctic coastal plain top predators include bears and arctic wolves.

Carnivores of the far North include the snowy owl, weasel, and arctic fox.

Arctic region herbivores include the caribou, arctic hare, and lemming.

Trophic levels

Primary producers on the tundra include tundra grasses, lichens, wild flowers, dwarf willow trees, mosses, herbs, and shrubs.

Top predators

Secondary consumers (Carnivores)

Primary consumers (Herbivores)

Primary producers

Arctic Tundra

Arctic marine top predators include the polar bear and the orca whale.

Trophic levels

Top predators

Secondary consumers (Carnivores)

Primary consumers (Herbivores)

Primary producers

Beaufort Sea

Examples of marine carnivores are arctic cod, shrimp, and squid.

Marine herbivores are mainly zooplankton such as copepods.

Primary producers in marine waters are mostly phytoplankton such as diatoms.

Figure 1.6   Biotic pyramids showing trophic levels on the arctic tundra and in the Beaufort Sea.

Figure 1.7   Arctic ground squirrel.
(Photo courtesy of the U.S. Fish and Wildlife Service, Anchorage, Alaska.)

Beaufort Sea zooplankton consist mainly of **copepods**, small crustaceans that function as herbivores grazing on algae. Copepods typically range in length from 1 to 5 millimeters. Examples of marine zooplankton are shown in Figure 1.9.

Euphausiids are large, shrimp-like zooplankton up to 5 centimeters long. They function as both herbivores and carnivores because they eat both algae and copepods. In antarctic waters, krill, one type of euphausiid, is a principal food source for baleen whales (Figure 1.10).

Biodiversity on the Tundra

The tundra environment is cold and barren most of the year, but dramatic changes occur during the brief summers when continuous sunlight stimulates prolific plant growth. The cotton grass, mosses, lichens, sedges, and flowers that dominate the tundra support a wide variety of animal life. Nearby, in the Beaufort Sea, continuous sunlight triggers blooms of marine algae that support animal life in the sea, including whales that migrate there every year. Many species live on the tundra year long, while others are summer visitors. Fish and wildlife play key roles in the culture and economy of the far North, particularly the caribou, sea birds and waterfowl, and marine animals that inhabit the Beaufort Sea.

Porcupine Caribou

In the Arctic, many native people embrace a traditional subsistence way of life that depends on hunting, fishing, trapping, and whaling. For them, protecting caribou is as much a cultural issue as a wildlife issue. The Gwich'in, who inhabit 15 or more villages across

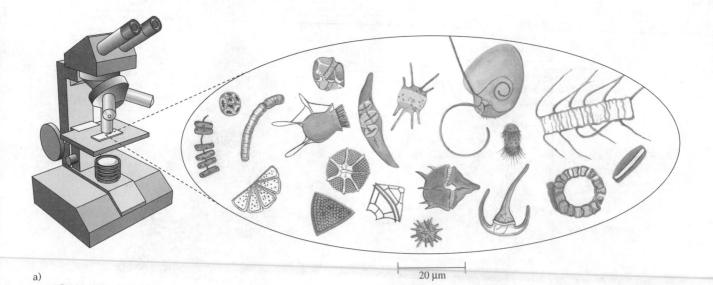

a)

20 µm

Figure 1.8 Typical marine phytoplankton—single-celled, chlorophyll-containing algae that manufacture plant biomass from inorganic aquatic nutrients through photosynthesis.
(*a*) Hand-drawn representation.
(*b*) Photomicrograph.
(*b: D.P. Wilson/Science Source/Photo Researchers, Inc.*)

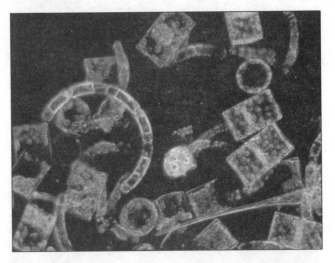

b)

northeastern Alaska and northwestern Canada, have depended on caribou for food, clothing, and tools for thousands of years. In his book *Earth and the Great Weather: The Brooks Range*, Kenneth Brower wrote,

> Caribou meant everything to them that buffalo meant to the Plains Indians. From the neck, spine, and from the backs of the caribou's legs, [they] got sinew for sewing. From the bone and antler they fashioned utensils, tools, and weapons; from the hides they made tents, bags, kayaks. The caribou's skin was perfect for winter clothing. It was light, flexible, and good insulation. . . . Caribou skin warmed [them] outside, and caribou fat warmed them inside.

It is estimated that more than 800,000 caribou inhabit Alaska. They live and perpetually migrate as 25 distinct herds. Arctic caribou, relatives of the European reindeer, are known as barren-ground caribou. Porcupine caribou comprise approximately 180,000 of these magnificent wild animals (Figure 1.11). They are named for the Porcupine River, which they cross twice each year—in the spring on their way to coastal tundra (or North Slope) calving grounds and in the fall when they return to coniferous forest winter habitats south of Alaska's Brooks Range and the Yukon's Ogilvie Mountains. There, protected from intense cold by black-and-white spruce, poplar, birch, and aspen, caribou feed on a diversity of food sources, including lichen.

Different herds occupy different winter habitats. For example, the Alaska Peninsula herd retreats to forests within the state's southwestern peninsula, the Western Arctic herd over-winters primarily south of the Arctic Circle on the Seward Peninsula, and the Central Arctic herd finds winter protection on interior-facing slopes of the Brooks Range south of Prudhoe Bay.

Perhaps it is the lengthening hours of daylight in March that signal the start of caribou migrations. Small herds, more or less isolated from one another, begin their long trek toward the North Slope. Cows (female caribou) and yearlings (last year's new calves) lead the way. Bulls (male caribou) follow later. By late May, sometimes early June, they have crossed the Arctic's mountains and arrived on the tundra. Each herd gathers at a somewhat different location there.

Porcupine caribou prefer calving grounds on a narrow, 240-kilometer (≈ 150-mile)-long stretch of tundra spanning the United States–Canada border. Soon after arriving, pregnant caribou (often as many as 90 percent of the cows) calve, each producing one offspring. Calving occurs on grassy tussocks and sedge

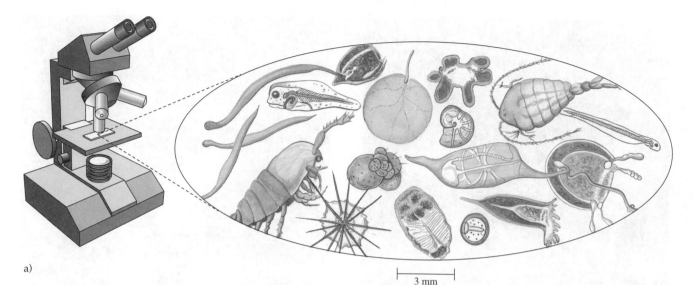

a)

3 mm

b)

Figure 1.9 Typical marine zooplankton—primary consumers that convert phytoplankton plant matter into protein, an important nutrient in all food webs.
(*a*) Hand-drawn representation.
(*b*) Photomicrograph.
(*b: D.P. Wilson/David & Eric Hosking/Science Source/Photo Researchers, Inc.*)

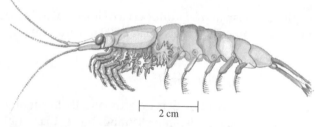

2 cm

Figure 1.10 Krill, *Euphausia superba*, is a type of zooplankton known scientifically as the euphausiids. Krill are an important marine food source, especially in Antarctic waters.

meadows near low-rolling foothills. Each year approximately 40,000 Porcupine caribou calves are born within a 10-day period. They thrive on the rich milk their mothers produce as they graze on tundra plant life.

About two weeks after calving, the bulls join the caribou cows on the tundra. For a time, they forage peacefully. Then, seemingly out of nowhere, an onslaught by tundra mosquitoes begins. Billions of black-flies and deerflies hatch from larvae in the countless shallow bogs and wetland ponds that dot the tundra landscape. The male mosquitoes are important pollinators of tundra plants, but the females are carnivores. They feed on caribou, drawing up to a quart of blood a week from each one. Caribou blood, the mosquitoes' energy source, is needed to produce eggs—the promise of next year's mosquito onslaught.

Caribou seek relief from the relentless mosquito attacks by wading into shallow Beaufort Sea waters

and heading into the wind. When the mosquito wars end, the caribou return to their coastal plain, meandering to forage and recover from weight loss. Toward the end of summer, all the caribou herds converge on the tundra, a gathering known as the "great aggregation." For a short time, the tundra teems with hundreds of thousands of caribou in one of nature's most extraordinary exhibitions of life.

By mid-August, their summer sojourn on the tundra over, caribou herds disperse toward their respective wintering grounds. The Porcupine herd heads east along the Beaufort Sea coastal plain and then south, part of the herd traversing the Brooks Range and part passing through Canada's Ogilvie Mountains. They ford the Porcupine River in late September and retrace ancient migration routes into coniferous forests. The time of the caribou **rut** is at hand; their mating season begins, and the circle of life continues. Earth-satellite tracking data of Porcupine caribou collected

Figure 1.11 A porcupine caribou herd on the coastal plain of the Arctic National Wildlife Refuge.
(Photo courtesy of the U.S. Fish and Wildlife Service, Anchorage, Alaska.)

Figure 1.12 Golden plover nesting on ANWR's coastal tundra grasses.
(Photo courtesy of the U.S. Fish and Wildlife Service, Anchorage, Alaska.)

by the U.S. Fish and Wildlife Service show that a single caribou commonly travels up to 2,700 miles a year.

Sea Birds and Waterfowl

The Canadian Wildlife Service estimates that as many as two million birds, representing more than 140 species, forage, nest, and raise their young on the summer tundra. These include arctic tern, Lapland longspur, willow ptarmigan, snowy owl, snow bunting, peeping sandpiper, savannah sparrow, whistling swan, snow goose, mallard, Pacific loon, pintail duck, golden eagle, rough-legged hawk, golden plover, arctic peregrine falcon, and American peregrine falcon.

It is amazing how far many of these birds migrate each year to reach the summer tundra. They retrace ancestral pathways hundreds, even thousands, of miles long. Arctic tern, for example, fly more than 16,000 kilometers (≈ 10,000 miles) from their antarctic wintering habitat to feed and breed in the far North. Theirs is the longest migration of any bird. As diving birds, tern are second- and third-trophic-level consumers because they eat zooplankton and fish.

Golden plovers (Figure 1.12) fly to the tundra from Argentina, 12,800 kilometers (≈ 8,000 miles) away, to nest and reproduce. When their eggs hatch, plover chicks, which cannot fly at first, feed on multitudes of tundra mosquitoes easily within their reach—the same mosquitoes that cause so much distress to so many caribou. These mosquitoes are a vital food resource for thousands of shorebirds and their chicks. Young birds thrive, mature quickly, and by the end of the arctic summer are able to make the long journey to Argentina.

Biodiversity in the Beaufort Sea

As previously mentioned, in addition to tundra and coniferous forests, northern Alaska includes another important ecosystem: the Beaufort Sea (see Figure 1.1). Warmed by summer sunlight and northward-flowing Pacific Ocean currents, the sea abounds with life, including *plankton* and *nekton*.

Plankton and Nekton Plankton are very small aquatic plants and animals. They are drifters whose motions are controlled mainly by water currents, winds, and waves. Plankton are of two kinds:

- **Phytoplankton** Single or multicellular, free-floating, chlorophyll-containing aquatic plant life found in fresh and marine waters. They are also called algae. Since phytoplankton are photosynthetic plants, they occupy surface or near-surface waters in order to capture solar energy. Phytoplankton are sometimes called the "grass of the sea."

- **Zooplankton** Small, free-floating, aquatic animals present in both fresh and marine waters. They are called "grazers" because they feed on phytoplankton. Since zooplankton are animals, they are able to convert plant matter into animal protein, thereby becoming a major food source for higher animal life.

  Nekton comprise the group of aquatic animals that are larger than plankton. Rather than being mere drifters, like plankton, nekton can swim against water currents. Nekton include adult fish such as various species of Arctic salmon, whitefish, flounder, and char as well as whales, seals, and sea turtles. Shrimp are on the borderline—they are purposeful swimmers if water currents are not too strong but often merely drift in strong currents.

  While it may appear that Beaufort Sea plankton and nekton are totally isolated from the arctic tundra, life in the sea is connected to life on the tundra. For example, polar bears depend on tundra denning sites as places where females can give birth to their cubs, but at the same time they depend on Beaufort Sea arctic char, shrimp, and squid for food.

- **Whales** Six whale species inhabit Beaufort Sea waters. Three are toothed whales: orca, narwhal, and beluga. They typically consume shrimp, crabs, squid, flounder, halibut, and salmon. As many as 5,000 beluga congregate, feed, and calve in the shallow waters of Mackenzie Bay each year (see Figure 1.1). Three species of baleen whales are also there: California gray, bowhead, and right whales. Gray whales migrate from coastal waters off Mexico and California some 16,000 kilometers (≈ 10,000 miles) away. When summer ends, they return to these warmer waters to have their young.

  In the 1800s, California gray whales were intensively hunted for whale oil to be used in lamps for household lighting. Unfortunately, these whales are easy to sight and capture because they occupy near-shore waters, so eventually gray whales, along with blue, humpback, right, and bowheads, became threatened species. Killing gray whales was banned in 1930, but effective protection didn't occur until the Marine Mammal Protection Act was passed in 1972. This act prevents U.S. citizens from disturbing and hunting marine mammals anywhere in the world. However, native Alaskans are allowed to hunt whales as they always have as long as they stay within catch limits specified by the International Whaling Commission and the Alaska Eskimo Whaling Commission.

CASE STUDY: ARCTIC NATIONAL WILDLIFE REFUGE

Bob Marshall and others who explored the Arctic wilderness came to understand that wilderness is more than majestic landscapes, imposing rugged mountains, and wild rivers (Figure 1.13). It is Nature's exhibition of plant and animal life within awe-inspiring, pristine ecosystems—environments that evoke a spiritual response from humans. They were convinced that such special places should be forever preserved for future generations. Their sentiments were echoed by many Alaskans as well as by The Wilderness Society and the Sierra Club. In 1938, Marshall published a proposal to set aside an arctic wilderness area. He wrote:

> Because the unique recreational value of Alaska lies in its frontier character, it would seem desirable to establish a really sizable area, free from roads and industries, where frontier conditions will be preserved. . . . I would like to recommend that all of Alaska north of the Yukon River, with the exception of a small area immediately adjacent to Nome, should be zoned as a region where the federal government will contribute no funds for road building and permit no leases for industrial development. . . . Alaska is unique among all recreational areas belonging to the United States because Alaska is yet largely a wilderness. In the name of a balanced use of American resources, let's keep northern Alaska largely a wilderness!

Creation of the Arctic Wildlife Range

The first step toward preserving the arctic wilderness was taken in 1960 when a large, pristine region in Alaska was set aside as the Arctic Wildlife Range. The range was created by President Dwight Eisenhower through the National Wildlife Refuge System. It encompassed about 3.6 million hectares (≈ 8.9 million acres), extending eastward from the Canning River to the Canadian Border and southward from the Beaufort Sea to interior coniferous forests south of the Brooks Range (see Figure 1.1). The range encompassed:

- Coastal-plain tundra whose luxurious summer vegetation created key habitat and breeding areas for caribou, musk oxen (Insight 1.4), and other arctic wildlife.
- Snow and glacier-capped mountains south of arctic coastal tundra, home of the arctic fox, wolf, wolverine, and Dall sheep (Figure 1.14).

INSIGHT 1.4
Arctic Region Musk Oxen

Musk oxen were once very common in polar regions of the far North, including Alaska, Canada, Greenland, and Asia. They are well-adapted to harsh winters, but in the early nineteenth century their numbers declined as a result of overharvesting for their meat, hides, and fur. By 1929, no musk oxen remained in Alaska, and relatively few existed elsewhere.

With musk oxen extinct in Alaska, the Territorial Legislature asked the U.S. Congress to help import a breeding stock from Greenland. In 1930, 34 animals were brought to Alaska. The species is now well-established and protected within ANWR.

Figure 1.13   The majestic Brooks Range, a part of North America's Continental Divide, is the backbone of the Arctic National Wildlife Refuge. *(Photo courtesy of the U.S. Fish and Wildlife Service, Anchorage, Alaska.)*

- Boreal conifer forest south of the Brooks Range, home of the arctic moose and wintering grounds for many caribou herds.
- Spectacular migratory avian species, summer visitors to the far North that include arctic tern, peregrine falcon, and golden plover.
- Numerous freshwater rivers, streams, lakes, and marshlands that serve as fish and waterfowl habitat.

Creating the Arctic Wildlife Range gave tangible meaning to the idea of protecting the region's wildlife and wilderness. The main reasons for establishing the range were:

1. To preserve a large, self-sustaining arctic ecosystem.
2. To protect natural breeding and feeding grounds of fish and wildlife species that inhabit or migrate to this region.
3. To shelter distinctive wildlife, especially caribou herds, for posterity.

Creation of the Arctic National Wildlife Refuge (ANWR)

Even though certain land areas were protected in the Arctic Wildlife Range, it was known that developments

Figure 1.14 Dall sheep in the Arctic National Wildlife Refuge. *(Photo courtesy of the U.S. Fish and Wildlife Service, Anchorage, Alaska.)*

within the range might someday occur, especially oil exploration, drilling, and pipeline construction for transporting petroleum and natural gas. Conservationists argued that permanent habitat and wildlife protection was needed. This was achievable only through legislation that could designate "wilderness" areas. These considerations ultimately led to the creation of the Arctic National Wildlife Refuge and the establishment of wilderness areas within it.

Thus, in 1980 the size of the Arctic Range more than doubled with the creation of the Arctic National Wildlife Refuge, made possible when President Jimmy Carter signed into law Alaska's National Interest Lands Conservation Act, called the Alaska Lands Act. ANWR is a 7.9-million-hectare ($\approx$ 19.5-million-acre) sanctuary in northeastern Alaska, a region larger than the states of New Hampshire, Vermont, and Massachusetts combined (see Figure 1.1). About 41 percent of the refuge is designated as wilderness under the 1964 Wilderness Act, an area defined for the most part by the boundaries of the original Arctic Wildlife Range. Many consider this 3.2-million-hectare ($\approx$ 8-million-

acre) wilderness area within ANWR the crown jewels of the American wilderness.

As a legally defined land-use category, ANWR's wilderness area is off-limits to the construction of roads, buildings, and pipelines. It is also off-limits to most vehicles, to timber harvesting, and to mining activities, including oil exploration and drilling. Permanent settlement within ANWR is not allowed except where it existed before the creation of the refuge as in the case of Inuits living at Kaktovic on Barter Island. Arctic Village, one of the communities the Gwich'in occupy, lies outside of ANWR on the south slope of the Brooks Range. Activities allowed in ANWR are mostly limited to hiking, camping, sport fishing, nonmotorized boating, and similar recreational pursuits. Aircraft are allowed to fly in and out of the refuge.

In 1984, the Canadian government established Northern Yukon Park (NYP), 1.2 million hectares ($\approx$ 3 million acres) of protected lands adjacent to ANWR. Together, ANWR and NYP make up the largest block of protected wild habitat on earth. Three years later, in 1987, the United States and Canada signed the Porcupine Caribou Herd Treaty, an agreement to preserve one of the world's greatest remaining assemblages of wild animals.

The Effects of ANWR

Establishing ANWR meant different things to different people. To some, it staked a claim that they hope will forever protect aboriginal arctic wildlife and safeguard the region's unparalleled scenic beauty. To others, it recognized and helped perpetuate the history, culture, and lifestyle of native peoples of the North, which depend on hunting and fishing. Since a part of the refuge is now legally designated as wilderness, many see this as an awakening to the importance of wilderness. Four major objectives underlie the creation of ANWR:

1. To conserve the region's fish and wildlife, maintain its diversity, and protect its natural habitats.

2. To enable native people to continue their subsistence way of life.

3. To safeguard the region's environmental resources, including its air and water quality.

4. To comply with the Canada–United States Porcupine Herd Treaty through which both nations are committed to protecting the Porcupine caribou herd.

Expanding the Arctic Wildlife Range to today's arctic refuge was based on a compromise between conservationists and the petroleum industry. The compromise set aside part of ANWR, Section 1002 (ten-oh-two), with the understanding that oil and gas developments might be allowed there in the future. The compromise came about because earlier petroleum explorations had indicated that oil deposits were likely to lie beneath this part of the refuge.

Section 1002: A Many-Sided Issue

Ever since the Wilderness Act was passed in 1964 with the aim of preserving natural areas, considerable controversy has resulted because areas set aside through the act are off-limits to any future development. Prior to the Wilderness Act, certain primitive areas under U.S. Forest Service jurisdiction had been shielded from development as if they were legally protected wilderness. The Forest Service operates under a multiple-use philosophy that protects some areas while allowing timber production, mining, animal grazing, and recreational pursuits in other areas.

One such controversial area is Section 1002, a 0.6-million-hectare (≈ 1.5-million-acre) stretch of coastal tundra about 30 miles wide and 100 miles long bordering the Beaufort Sea (see Figure 1.1). Section 1002 is seen as both a magnificent, primeval, distinctive environment and a potentially rich source of petroleum. Only the U.S. Congress has the authority to decide whether oil and gas developments will be allowed in this part of the refuge or whether Section 1002 will become a part of ANWR's protected wilderness.

Some environmental experts believe that serious environmental problems will result if Section 1002 is opened to oil and gas exploration, production, and transport. Others see little ecological danger in doing this and believe that needed energy resources could be developed. It would seem that the wisdom of Solomon is needed to decide how best to manage this part of the refuge.

Potential Environmental Problems

By 1977, many successful Prudhoe Bay oil wells were in operation, the Trans Alaska Pipeline System (TAPS) had been completed, and close to 1.7 million barrels of oil were being pumped daily to the city of Valdez on the southern coast of Alaska, off-loaded to oil tankers, and shipped to West Coast oil refineries (see Chapter 9). Many studies have been carried out to assess the environmental impacts of the petroleum industry in Alaska, particularly at Prudhoe Bay.

The Institute of Arctic and Alpine Research in Anchorage estimates that only 5 percent of the tundra near Prudhoe Bay has undergone petroleum development. Caribou migrate cautiously across roadways

Figure 1.15 Porcupine caribou meandering beneath the Trans Alaska Pipeline constructed, in part, above the arctic tundra. *(Photo courtesy of the U.S. Fish and Wildlife Service, Anchorage, Alaska.)*

and under oil pipelines in these highly developed oil fields (Figure 1.15). Grizzly bears and arctic foxes are also seen, and smaller animals such as arctic hare, ground squirrels, and lemming are abundant. Thousands of waterfowl and sea birds nest and forage on tundra near Prudhoe Bay, including Canada geese, snow geese, tundra swans, brant geese, plovers, and several duck species. Arctic char, whitefish, and grayling are prevalent, and whales continue to frequent nearby coastal waters. There is little evidence that petroleum developments at Prudhoe Bay have had harmful effects on wildlife.

One of the most interesting observations made at Prudhoe Bay is the large increase in the size of the Central Arctic caribou herd since the 1960s. The herd, whose grazing and calving grounds are close to Prudhoe Bay, expanded from an estimated 3,000 to 23,500 animals since oil developments began. No one is sure why the increase occurred. There is a lack of caribou population data before North Slope development, and some biologists have suggested that fluctuations in arctic wildlife populations are not unusual. Others point out that the state of Alaska supports the killing of wolves, which are natural predators of caribou and whose numbers have now fallen sharply as a result. Oil industry scientists have stated that Prudhoe Bay petroleum developments have obviously not affected the herd adversely.

However, considerable damage to tundra vegetation is evident at Prudhoe Bay. Many oil spills have occurred due to drilling, pumping, and oil storage operations. Cleanup operations have proven effective in some cases, and the oil industry is sponsoring long-term revegetation studies of abandoned gravel-based airstrips, roads, and drilling pads. The goal is to learn how to rehabilitate these sites by colonizing local tundra grass and shrub species.

Biologists studying Section 1002 ecology have outlined a number of potential risks to arctic wildlife if the decision is made to open the area to petroleum operations:

1. Porcupine caribou calve in Section 1002. Coastal-plain tundra here is only 48 kilometers (≈ 30 miles) wide, much narrower than at Prudhoe Bay. Presently, predator scarcity near caribou calving sites helps conserve the herd's population size. But if oil development is allowed, the herd will be driven toward the foothills to the south where they will be exposed to increased predation by wolves.

2. Caribou seek relief from summer mosquito attacks by venturing into Beaufort Sea coastal waters. This avenue of escape could be cut off by oil industry developments in Section 1002.

3. The Section 1002 coastal plain is the most significant onshore polar bear denning habitat in the United States. It is utilized by scores of polar bears each year, including many who journey there from other regions.

4. Oil developments on the Kenai National Moose Range have incurred significant environmental impacts, especially alteration of moose habitats. Similar impacts could occur in Section 1002.

5. An oil spill in Section 1002 could affect Beaufort Sea aquatic wildlife habitats where extremely cold conditions diminish rates of oil evaporation and weathering, thereby increasing environmental damage.

Potential Energy Benefits

The need for the United States to have uninterrupted oil and natural gas supplies for continued development and national security seems unarguable to most people. The importance of becoming less dependent on imported foreign oil and diminishing the nation's balance-of-trade deficit is also important. For example, petroleum imports in 1996 accounted for about 75 percent of the U.S. trade deficit that year. Arguments for switching from dependence on fossil fuels to increased energy conservation and renewable energy sources such as solar and wind power are also justified (see Part IV). One of the questions we face is whether oil and gas reserves of economic and strategic value should be exploited even when the environmental risks appear high.

Most developed economies, including the United States, Canada, European Union, and Japan, depend heavily on petroleum and the fuels produced from it, such as gasoline, aviation fuel, diesel oil, and kerosene.

A transition to sustainable, renewable energy is on the horizon, but it is occurring very slowly. In the year 2000, projections indicate that 38.7 percent of total U.S. energy needs will depend on oil, which will be consumed at a rate of about 19 million barrels per day (see Figure 9.11). More than half (≈ 55 percent) of the oil will be imported; the rest will be produced domestically, with 25 percent of domestic production occurring at Prudhoe Bay. Proven oil reserves at Prudhoe Bay are declining and are projected to run out by about 2020.

The Alaska Lands Act required the U.S. Secretary of the Interior to conduct geological studies of Section 1002 and recommend whether the area should be opened to leasing for oil and gas exploration, drilling, and development. The studies assessed the size of petroleum deposits that might exist under this part of ANWR's coastal plain. Based on limited investigations, the study results indicated that there is a 19 percent probability of finding a 3-billion-barrel oil deposit and a 5 percent probability of finding a 9-billion-barrel oil deposit under Section 1002. These numbers can be compared with the 12 billion barrels of proven reserves shown to have existed initially at Prudhoe Bay.

If the 19 percent probability of a 3-billion-barrel oil deposit is correct, and if this oil were pumped through the Trans Alaska Pipeline at 1.7 million barrels per day, ANWR's oil would last about five years. Some conservationists ask whether a 19 percent probability of developing an oil field that could supply 10 percent of the nation's needs for less than five years is worth the environmental risk. On the other hand, there is a 5 percent chance of discovering a much larger oil field—perhaps 9 billion barrels. A field this large could supply oil at 1.7 million barrels per day (10 percent of U.S. needs) for about 14 years.

The Secretary of the Interior and those advising him completed their studies in 1986 and recommended to the U.S. Congress that Section 1002 of the Arctic Refuge be opened to oil development. Alaska's governor and legislature also support oil development in Section 1002. Both President Ronald Reagan in 1987 and President George Bush in 1990 asked Congress to open Section 1002 to oil and gas leasing, but the Congress did not support these recommendations.

During the summer of 1995, the Alaska Federation of Natives Board, the state's largest native organization, voted 19 to 9 in favor of opening Section 1002. It is easy to see why. North Slope Inuits have benefited greatly from oil industry lease and tax revenues collected as a result of Prudhoe Bay operations. This money has enabled them to build new homes, roads, schools, streetlights, clinics, satellite TV receivers, and a power plant. However, other native groups oppose opening 1002. The Gwich'in, for example, believe that to do so would disrupt wildlife, especially the Porcupine caribou herd upon which their way of life depends. They fear that once oil development occurs, Porcupine caribou will be affected, and oil pollution on the coastal plain and in the Beaufort Sea will be inevitable.

The petroleum industry also argues for opening Section 1002 of ANWR to oil and gas drilling for two reasons. First, petroleum and natural gas are needed to offset American dependence on foreign sources. Second, oil field drilling techniques have been greatly improved since the early days of Prudhoe Bay development so that less than 1 percent of the Section 1002 area would be affected by wells, roads, pumping stations, and pipelines, causing little adverse impact on the tundra ecosystem. Here are the main reasons the oil industry believes oil and gas development in Section 1002 will not harm the environment:

1. Arctic petroleum operations, especially transporting drilling rigs and personnel, can be shifted to winter seasons when tundra soils are frozen. Equipment and personnel can be transported on ice roads, and drilling can be carried out on ice pads. A shift to winter-season operations in Section 1002 would have minimal impact on the tundra.

2. Constructing ice roads and ice-based drilling rigs instead of gravel roads and gravel drilling pads (all of which have to be six feet deep for insulation to protect underlying permafrost) is much less harmful to tundra and minimizes the requirement for gravel.

3. Drilling rigs have been redesigned to use underground directional-drilling technology that allows a cluster of wells to be drilled from one ice pad, each drilling trajectory slanted at a different angle to tap widely separated, deep underground oil targets. The new technique results in a smaller environmental footprint.

4. Today's drilling pad is smaller because, prior to 1994, spent muds used in drilling operations were stored in reserve pits built as part of gravel pads. New technology allows spent muds to be injected into wells drilled thousands of feet below the tundra. A typical Prudhoe Bay drilling pad constructed in the 1970s occupied about 18 hectares (≈ 44 acres), while pads built since 1994 cover as little as 3.2 hectares (≈ 8 acres)—82 percent less area.

However, despite these improvements in petroleum industry technology, a 1987 environmental impact statement (EIS) prepared by the U.S. Department of Interior reported that the following infrastructure would be needed to fully develop oil fields in Section 1002:

- A major pipeline 100 miles long connecting Section 1002 wells to the Trans Alaska Pipeline, which commences at Prudhoe Bay.
- 120 miles of main roads and 160 miles of spur roads.
- Two large, permanent airfields and two smaller airfields.
- 50 to 60 drilling pads.
- 10 to 15 gravel-mining sites.

Only the U.S. Congress has the authority to decide whether oil development will take place in ANWR's Section 1002. The decision has not yet been made, and Section 1002 remains closed to petroleum exploration and development.

REGIONAL PERSPECTIVES

More than 480 wildlife ranges and refuges have been set aside in the United States under the National Wildlife Refuge System, a program established through the Endangered Species Preservation Act of 1966. Some were created to protect big game species and waterfowl for hunting and trapping. (Cattle grazing, timber cutting, sport and commercial fishing, exploration for oil and gas, drilling, and oil transport are allowed in wildlife ranges if approved by the U.S. Secretary of the Interior.) Other sanctuaries for threatened and endangered species are managed by the U.S. Department of the Interior and the U.S. Fish and Wildlife Service.

In addition to these areas, 155 national forests and 20 grasslands are protected through the U.S. Forest Service, and 360 regions of special significance are protected as U.S. National Parks. In 1972, the U.S. Congress authorized a National Marine Sanctuaries Program promoting the creation of ocean-related parks administered by the National Oceanic and Atmospheric Administration (NOAA).

The United States and Canada have a long history of wildlife protection. But other industrial nations have also set aside and protected wildlife habitats, recognizing that they are of special biological significance. A few developing nations are adopting similar programs, the most noteworthy of which is Costa Rica (see Chapter 2). Overall, thousands of forest, grassland, and wetland areas have been established as national parks, national forests, wildlife reserves, marine sanctuaries, biological preserves, and wildlife refuges. Here are some North American regional perspectives on protected wilderness and wildlife.

United States

Alaska: Kenai National Moose Range

An example of a wildlife range is Alaska's Kenai National Moose Range, an 809,000-hectare (≈ 2-million-acre) area adjacent to Cook Inlet on the Kenai Peninsula. Established in 1941, the range protects Kenai moose breeding and feeding grounds. Oil deposits were long known to exist here, but the range was initially off-limits to oil exploration, drilling, and pipeline construction. In 1958, the Secretary of the Interior opened sections of the range to oil development. Within seven years, 30,000 barrels of oil were being produced and transported daily via a 22-mile pipeline to a Standard Oil Company refinery on Cook Inlet.

Oil development on the Kenai Moose Range has eliminated many of the region's spruce-fir forests due to construction of seismic trails, roads, pipelines, and pumping stations. Native trees have not regrown, but they have been replaced by willow, alder, birch, and aspen, creating a far different habitat. In this case, the legal status "range" did not prevent developers from exploiting subsurface oil and gas resources. The only way to prevent developments such as this is to designate an area as "wilderness." This was not possible in the United States until passage of the Wilderness Act in 1964. Since then, many areas within U.S. parks, forests, ranges, and refuges have been granted wilderness status.

Florida: Everglades National Park

President Harry S. Truman dedicated the Everglades National Park in 1947. In part, he said:

> Here are no lofty peaks seeking the sky, no mighty glaciers or rushing streams wearing away the uplifted land. Here is land, tranquil in its quiet beauty, serving not as the source of water, but as the receiver of it. To its natural abundance we owe the spectacular plant and animal life that distinguishes this place from all others in our country.

Marjory Stoneman Douglas described this unique ecosystem in her book *The Everglades: River of Grass*. Indeed, the Everglades are a shallow sea of slow-moving water about 2 meters (≈ 7 feet) deep and 80 kilometers (≈ 50 miles) wide. They begin as freshwater runoff from Lake Okeechobee and then flow south 161 kilometers (≈ 100 miles), emptying into Florida Bay and the Gulf of Mexico. The ecosystem embraces almost a million hectares (≈ 3,500 square miles) of broad sloughs and saw grass, tree islands, mossy-covered live oaks and water oaks, dwarf cypress and tall freshwater cypress, strangler fig trees, ferns and vines, mangrove forests, and tropical hardwoods called hammocks.

Hidden among the saw grass are coral snakes, striped skunks, brown scorpions, tree snails, twilight hawk moths, brown deer, panthers, great barred owls, otters, gray herons, egrets, and wading birds. The Everglades are known also as wintering and staging areas for the Cape May warbler, peregrine falcon, bobolink, snail-kite bird, and tree swallow. Today, native fish such as Florida's largemouth bass are under pressure as tilapia and oscars introduced from Africa and South America compete for nesting sites.

The lure of Florida's semitropical warm weather began to attract visitors and retirees in the 1920s. Soon, a wave of rising population engulfed South Florida, and by the late 1940s parts of the Everglades were being dredged for flood control and to create land for vegetable and sugarcane agriculture, cattle grazing, and home building. Freshwater was diverted from the Everglades by means of canals, dikes, and pipelines to supply irrigation water for farms and drinking water for municipalities.

In the 1970s, researchers began to document precipitous declines among bird species once common to the Everglades. Fewer than 10 percent of some species remained, compared to numbers recorded a hundred years earlier. They included great egrets, snowy egrets, and wood storks. The wood stork stands over three feet tall and has a five-foot wing spread. It was placed on the endangered species list after its numbers fell from more than 6,000 nesting birds in the 1960s to only 500 in 1984.

By the 1990s, over 50 percent of Everglade wetlands had disappeared and 14 wildlife species had been declared threatened or endangered, including the American crocodile, southern bald eagle, and loggerhead turtle. Diverting water from the Everglades had altered naturally occurring wet and dry cycles essential to the ecosystem's peculiar biology.

In 1994, then-Governor Lawton Chiles established a 42-member commission made up of scientists, engineers, and officials from federal and state agencies to study South Florida's ecological problems and develop plans to restore the Everglades. The commission's final report was written into the U.S. Congress' Water Resources Development Act of 1996, which committed funds to undertake restoration of the Everglades.

Many restoration projects are now underway to save the Everglades, and it is hoped that the creation of Everglades National Park will ultimately lead to preservation of a truly unique American ecosystem.

Hawaii: Volcanoes National Park

Two of the world's most spectacular volcanoes are found on Hawaii, the largest of the Hawaiian Islands: Mauna Loa, 4,169 meters ($\approx$ 13,677 feet) high, and Kilauea, the world's most active volcano, standing 1,246 meters ($\approx$ 4,090 feet) high. Both volcanoes serve as centerpieces for Hawaii's Volcanoes National Park created in 1916.

The high-altitude wilderness on Mauna Loa is characterized by volcanic cinder cones, barren lava rocks, and rugged landscapes. Kilauea's ongoing eruptions and molten lava flows are known worldwide. Less well-known are Kilauea's tree fern forests, which display striking transition zones among early successional, pioneer-organism stages and more mature stands of climax tree species.

Canada

Canada's national parks system began in 1885 when Banff National Park was established in Alberta's Canadian Rocky Mountains. There are currently 38 national parks and park reserves in Canada—representative natural areas of Canadian significance. They are protected and managed to sustain unimpaired wild ecosystems for public education and appreciation both now and for future generations. Examples of Canada's parks include:

Banff National Park

Canada's oldest national park, Banff, began after three Canadian Pacific Railway workers found a hot-springs cave on the eastern slopes of Alberta's Rocky Mountains. The park has since grown in size and significance to include 6,641 square kilometers (2,564 square miles) of majestic mountains, rock cliffs, awe-inspiring glaciers, waterfalls, primeval forests, strikingly beautiful meadows, and pristine rivers.

Banff National Park includes Lake Minnewanka and three glacier-fed lakes: Lake Louise, Moraine Lake, and Peyto Lake. Also included are the headwaters of the North Saskatchewan and Bow Rivers as well as extensive wetlands known as the Vermilion Lakes. Outstanding photographic imagery of the Banff region is accessible through the website given at the end of the chapter.

The Banff area is known for its abundant wildlife, including wood bison, elk, deer, bighorn sheep, mountain goats, moose, black bears, grizzly bears, and wolves.

St. Lawrence Islands National Park

Ontario's St. Lawrence Islands Park, authorized in 1914, is Canada's smallest national park and the first to be established east of the Rocky Mountains. The park is composed of 111 granite islands and islets strewn like stepping stones across 80 kilometers ($\approx$ 50 miles) of the St. Lawrence River's headwaters. Lake Ontario to the west, owing to its great size, moderates the park's climate to the extent that several plant and animal species, including the Pitch Pine, Black Rat snake, and Least Bittern (a wading bird) exhibit the most northern limits of their ranges on these islands.

The park's islands show the influence of North America's most recent Ice Age, the Laurentide glaciation, which grew to its maximum size close to 18,000 years ago. Exposed bedrock, scraped smooth by moving glaciers, is still evident because soil formation here is very slow.

Mexico

There are more than 90 designated protected lands and aquatic areas in Mexico. They are identified as national parks, national monuments, marine parks, biosphere reserves, urban parks, and special ecological zones. Mexico's national parks exist mainly to conserve the country's cultural heritage. They also serve as recreational sites and help support tourism. Other protected areas, such as biosphere reserves, are noted for their biological diversity and were set aside to protect threatened or endangered plant and animal species. Ten of Mexico's several reserves are part of an international research, conservation, and training program known as "Man and the Biosphere" organized in 1970 by the United Nations Educational, Scientific, and Cultural Organization (UNESCO). One of the most interesting reserves is El Cielo Biosphere Reserve.

El Cielo Biosphere Reserve

Lying at elevations that span 200 to 2,290 meters ($\approx$ 650 to 7,500 feet) above sea level, El Cielo Biosphere Reserve covers 144,426 hectares ($\approx$ 356,872 acres) of the

Sierra Madre Oriental mountains. It was established in 1985 by the Mexican state of Tamaulipas in northeastern Mexico and identified as a biosphere reserve of international significance in 1987 by UNESCO. The reserve is about 240 kilometers ($\approx$ 150 miles) north by northeast of Mexico City, not very far from the Texas–Mexico border.

The reserve encloses four distinct ecosystems: dwarf oak and heath forest, pine-oak forest, tropical forest at higher elevations, and an elegant cloud forest. Photographs of the reserve's tropical forest are available through the website given at the end of this chapter.

El Cielo's unusually high biodiversity is due to the reserve's climatic transition zone between North and Central America. In particular, its tropical forest is known for a great diversity of avian species, whose names suggest compelling sounds and images: Amethyst-throated Hummingbird, Melodious Blackbird, Yellow-winged Tanager, Blue Mockingbird, Bronzed Cowbird, Common Raven, Yellow-green Vireo, Green Jay, Olive Sparrow, Vermilion Flycatcher, Singing Quail, Gray Hawk, Flame-colored Tanager, and Acorn Woodpecker.

KEY TERMS

Calorie	organic
copepods	permafrost
cyanobacteria	photic zone
First Law of Thermodynamics	phytoplankton
first trophic level	plankton
gross primary production	rut
inorganic	Second Law of Thermodynamics
lichens	second trophic level
mutualism	solar constant
nekton	symbiosis
net primary production	Ten Percent Rule
One Percent Rule	tundra
	zooplankton

DISCUSSION QUESTIONS

1. What does wilderness mean to you? Have you ever experienced a wilderness area? What values do you think are the most important in deciding to set aside a wilderness area?

2. If the U.S. Congress decides to open ANWR's Section 1002 to oil and gas development, do you believe this decision could affect the entire refuge ecosystem? Explain your thoughts. How significantly do you think the Porcupine caribou herd would be affected?

3. The oil industry argues that by using new technology it can undertake arctic oil and gas development without significant harm to tundra habitat or wildlife species. Are these arguments convincing to you? Explain why you feel the way you do.

4. Should discoveries of oil and/or natural gas, especially really big ones, be exploited no matter what the environmental costs? What if your country's national security depends heavily on oil?

INDEPENDENT PROJECT

This chapter emphasizes wilderness values. Complete one of the following independent projects:

Project A

Spend some time, perhaps an hour, in the most undisturbed wilderness ecosystem you can find. Make sure the area you have chosen is safe. Take a clipboard and jot down the various kinds of plants, shrubs, and trees you see even if you can't identify them by name. How many different species of animal life can you observe? In what ways is this ecosystem important to them? Is there evidence of human-caused disturbance? Are the disturbances significant? What connection, if any, do you sense or feel with this natural system? How is it different from what you experience in your daily life? Should this particular environment be protected? Why or why not?

Project B

Obtain a copy of the book *Midnight Wilderness: Journeys in Alaska's Arctic National Wildlife Refuge*, by Debbie S. Miller (Alaska Northwest Books). Read this book and describe your insights and reactions to an event Miller wrote about that is most significant to you. What particular wilderness values are reflected? Should wildernesses be legally protected for all time? Why or why not?

INTERNET WEBSITES

Visit our website at http://www.mhhe.com/environmentalscience/ for specific resources and Internet links on the following topics:

Arctic perspectives

Arctic National Wildlife Refuge

Arctic National Wildlife Refuge:
    Caribou–The Porcupine herd

Biology of wolves in Alaska

Arctic communities in Northern Canada

Alaska Department of Fish and Game—
    Wildlife Notebook Series

U.S. Fish and Wildlife—Arctic National
    Wildlife Refuge Site

Sounds in different biomes

Banff National Park—Photos of the Banff region

Mexico's El Cielo Biosphere Reserve—
    Photos of the rain forest

SUGGESTED READINGS

Brower, Kenneth. 1971. *Earth and the great weather: The Brooks Range.* San Francisco, CA: Friends of the Earth.

Callicott, J. Baird, and Michael P. Nelson, eds. 1998. *The great new wilderness debate.* Athens, GA: The University of Georgia Press.

Chadwick, Douglas. 1979. Our wildest wilderness: So empty, yet so full. *National Geographic* 156(6): 737–769

Douglas, Marjory Stoneman. 1997. *The Everglades: River of grass.* Sarasota, FL: Pineapple Press, Inc.

Kauffmann, John M. 1992. *Alaska's Brooks Range: The ultimate mountains.* Seattle, WA: The Mountaineers.

Mairson, Alan. 1994. The Everglades. *National Geographic* (April):2–35.

Marshall, Robert. 1970. *Alaska wilderness: Exploring the Central Brooks Range.* 2d ed. Berkeley, CA: University of California Press.

Matthews, Down. 1992. On the trail. *Destination Discovery* (December): 42–46.

McConnell, Grant. 1972. The failures and successes of organized conservation. P. 49 in *Environment and Americans: The problem of priorities*, edited by Roderick Nash. New York: Holt, Rinehart and Winston.

McQuaid, Kim. 1995. Arctic time. *Wilderness* (Spring): 15–27.

Miller, Debbie S. 1990. *Midnight wilderness: Journeys in Alaska's Arctic National Wildlife Refuge.* San Francisco: Sierra Club Books.

Pielov, E. C. 1994. *A naturalist's guide to the Arctic.* Chicago, IL: The University of Chicago Press.

Smith, Richard G. 1973. Alaskan oil and national wildlife ranges. *Association of Pacific Coast Geographers Yearbook* 35: 75–85.

Smith, Zachary A. 1992. *The environmental policy paradox.* Englewood Cliffs, NJ: Prentice Hall.

U.S. Department of the Interior, Fish and Wildlife Service. 1988. *Arctic National Wildlife Refuge: Comprehensive conservation plan, final summary.*

U.S. Geological Service. 1998. National Wildlife Refuge 1002 Area, Petroleum Assessment. http://energy.usgs.gov/factsheets/ANWR/ANWR.html

Watkins, T. H. 1995. Beyond mile zero. *Wilderness* (Spring): 9–14; 30–31.

Williams, Gregg. 1994. In the midnight hour—Summering in Alaska's Arctic. *EcoTraveler* (July/August):30–39.

CHAPTER 2

Tropical Forests and Biodiversity

PhotoDisc, Inc./Vol. 16

Its lands are most beautiful and filled with trees of a thousand different kinds—and tall!

They seem to touch the sky. I am told that they never lose their foliage, as I can

understand, for I saw them as green and as lovely as they are in Spain in May.

Christopher Columbus, upon landing in
Costa Rica after his fourth and final voyage
to the New World in 1502

Tropical forests are not just stands of trees—they are ecological balance wheels where photosynthesis is carried out on a grand scale, consuming carbon dioxide, producing food, and adding oxygen to the earth's atmosphere. They provide critical habitats to unknown numbers of biotic populations, and they create products important to human welfare. Tropical forests also reflect spiritual and cultural values that have long nourished people everywhere. Ecologists and citizens alike are concerned about the continuing loss of forest lands and habitats, especially tropical forests.

Until recently, Costa Rica suffered one of the highest rates of deforestation in the world. But in the 1980s, environmental policies were adopted that limited the conversion of forests into cattle pastures and croplands. In addition, national parks, biological reserves, and wilderness areas were established to protect distinctive wild places. As a result, Costa Rica has placed a greater percentage of land off-limits to development than any other country. The focus of this chapter is tropical forests in general and Costa Rica's forests in particular. We will lay the groundwork for that discussion by first examining the nature of tropical forests.

BACKGROUND

Characteristics of Tropical Forests

Remote-sensing satellite surveys carried out in 1990 and 1995 by the World Conservation Monitoring Centre, a division of the United Nations Food and Agriculture Organization (UNFAO), indicated that tropical forests cover approximately 1.79 billion hectares ($\approx$ 4.42 billion acres) and account for about one-third of all forested land. The 1990 survey classified tropical forests as lowland and upland formations; as "closed" and "open" forests; and as tropical rain forests, tropical moist forests, tropical dry forests, tree plantations, and unclassified forests (Table 2.1).

Closed forests are distinguished by continuous, unbroken, dense vegetation that commonly blocks 90 percent or more of overhead sunlight from reaching the forest floor. Because of the limited light, the forest floor is mostly bare except for fallen branches, dwarf saplings, ferns, club mosses, leaf litter, and scattered shrubs. Nevertheless, mature trees prosper in closed forests because most of their leaves are in high canopies exposed to daily sunlight. Closed forests are also known for their almost daily torrential rainstorms, high temperatures, humidity, and striking biodiversity of plant and animal species. The two most important types of tropical closed forests are *rain forests* and *moist forests*.

Open forests have considerably less canopy cover than closed forests, and as a result, as little as 50 percent of the sunlight is commonly shaded out. The more open structure is due to limited rainfall that occurs only seasonally. Common types of open tropical forests include *dry forests* and *tree plantations*.

Rain Forests

Tropical rain forests account for about 8 percent of the world's total land area. Worldwide, three tropical rain forest formations are recognized:

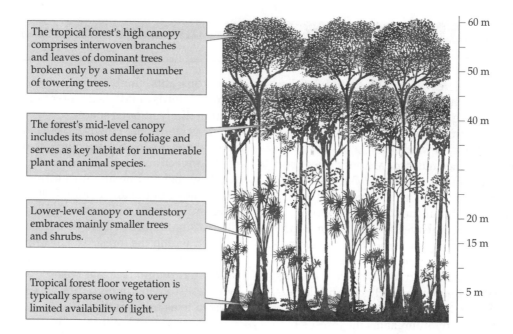

The tropical forest's high canopy comprises interwoven branches and leaves of dominant trees broken only by a smaller number of towering trees.

The forest's mid-level canopy includes its most dense foliage and serves as key habitat for innumerable plant and animal species.

Lower-level canopy or understory embraces mainly smaller trees and shrubs.

Tropical forest floor vegetation is typically sparse owing to very limited availability of light.

Figure 2.1   Cross section of a three-tiered canopy rain forest showing a high-level canopy, a middle-level canopy, and a lower-level canopy.

TABLE 2.1   Tropical Forests

TYPE OF FOREST	AREA (BILLION HECTARES)
Lowland forests	
Closed forests	
Tropical rain forests	0.72
Tropical moist forests	0.59
Open forests	
Tropical dry forests	0.24
Tree plantations	0.04
Upland forests	
Unclassified forests*	0.20
TOTAL TROPICAL FORESTS	1.79

SOURCE: Data from World Conservation Monitoring Centre, UN Food and Agriculture Organization (FAO), 1990.

*Upland forests often cannot be studied or classified by satellite imagery due to almost continuous cloud cover.

1. The American Rain Forest Formation, including parts of Brazil, Peru, Ecuador, Colombia, Panama, Costa Rica, Nicaragua, Honduras, and Guatemala.

2. The African Rain Forest Formation, including parts of Congo, Gabon, Equatorial Guinea, Cameroon, and Nigeria.

3. The Indo-Malayan Rain Forest Formation, including parts of Malaysia, the Philippines, Vietnam, Myanmar (formerly Burma), and Papua, New Guinea.

The rate of precipitation in tropical rain forests is commonly 2,200 millimeters (≈ 87 inches) a year, but it can be as high as 3,000 millimeters (≈ 120 inches) a year. Temperatures often rise to 35 °C (≈ 95 °F) during the day and fall to about 18 °C (≈ 64 °F) at night. Relative humidity varies between 90 percent and 100 percent. Tropical vegetation flourishes in this heat and humidity, creating lush, green ecosystems whose trees typically display a layering of tree canopies (Figure 2.1).

Rain forest canopies are noted for their **epiphytes,** plants attached to other plant life. Epiphytes include lichens, mosses, ferns, orchids, liverworts, lianas (woody climbing vines), and colorful bromeliads, part of the pineapple family. Epiphytes extract nutrients from rainwater and from decomposing plant and animal matter, not from the trees themselves. Their symbiotic relationship to tropical trees is an example of **commensalism,** a cooperative biological relationship between two species whereby one species benefits while the other neither benefits nor is harmed. Epiphytes conserve forest moisture and provide habitat and food sources for canopy-dwelling birds, monkeys, sloths, flying squirrels, rats, mice, bats, and insects. At the end of their life cycle, they fall to the forest floor where

INSIGHT 2.1

A Day in the Life of a Tropical Rain Forest

It is morning. The forest is hot and humid. Patches of mist and fog that formed during the cool night drift slowly among the trees. Everything is saturated with moisture—soils, vegetation, saplings, tree trunks. There is no wind.

Dominating the landscape are huge trees up to 2 meters (≈ 6.6 feet) in diameter, 6.3 meters (≈ 21 feet) in circumference, and 60 meters (≈ 200 feet) in height. Their lowest branches may stand more than 30 meters (≈ 100 feet) above the forest floor. A profuse display of plant life is everywhere: pineapple-like bromeliads (colorful flowering plants living on tree trunks and branches), red and yellow orchids, twisting and dangling vines, bright colorful flowers, dark green mosses, and graceful ferns.

Sunlight brightly illuminates the forest's high canopies, but it is dark in the forest's interior. Sounds of raucous animals fill the air: Howler monkeys scream; the great potoo cries out as it has during the night; nocturnal bats return to underground nests and hollow trees. Brightly colored inhabitants flash against shadowy backgrounds: morpho butterflies, toucans, red-lored parrots with striking red and green plumage, scarlet macaws, orange-chinned parrots, pale-billed woodpeckers. Some animals are less conspicuous: crickets, cicadas, iguana lizards, bees, boa constrictors, and two of the most greatly feared snakes—the bushmaster and the fer-de-lance.

Most of the rain that fell the previous day has been absorbed by forest vegetation and tree roots interlacing the upper few centimeters of soil. Tropical trees also have roots in their high canopies, called **crown roots,** where they capture nutrients from decomposing plant and animal life. Rainwater is absorbed by canopy bromeliads, orchids, lichen, mosses, and other plant life as well.

It is noon. Forest sounds have subsided. One might hear only the rustling of leaves high above. It is the hottest time of the day, and temperatures near the forest floor average about 27 °C (≈ 80 °F). A high humidity (between 90 percent and 100 percent) prevails due to the almost-closed, windless, environment. Honey bears lurk in tree branches.

Numerous ants climb up and down tree trunks. Hummingbirds and helicopter damselflies flit through dimly lit open spaces.

It is afternoon. Howls, cries, and mating calls are renewed. Rising masses of moist air create powerful **convection** currents above the forest. Storm clouds gather and break into a torrential rainstorm just as they did the day before. It is not uncommon for as much as 130 millimeters (≈ 5 inches) of rain to fall in a single storm. About 50 percent of the rain is recycled within the forest. The rest runs off through streams and rivers. Precipitation taken up by roots is drawn through trees and released as water vapor by pores (stomata) in high-canopy leaves, a process referred to as **transpiration.**

Tropical wind and rainstorms can shatter old and decaying trees, causing them to crash to the ground and allow a shaft of light (a light gap) to illuminate the forest floor. Light triggers rapid growth among forest-floor saplings, many of which, while perhaps 50 years old, are only 25 millimeters (≈ 1 inch) in diameter and 2 meters (≈ 6.6 feet) high. Generally, only one of these young trees will grow fast enough to close the light gap in the forest canopy.

It is evening. Different noises are now heard. Donald Perry, author of *Life above the Jungle Floor*, describes the forest's night sounds as "a jungle chorus of katydids, crickets, frogs, owls, and a great many other animals sounding territorial warnings and singing songs of love." Rain forests are sometimes called jungles, a word that brings to mind images of densely interwoven vines, impassable trees, and thick foliage. But the interior of most rain forests isn't like this. Trees are well-spaced. Only stunted and sparse vegetation grow on the forest floor. There is no jungle here, mainly because of insufficient light. Often, less than 10 percent of overhead sunlight filters through the foliage.

However, real jungles do exist. They are found along rivers that cut through tropical forests. The rivers create an **edge effect,** a corridor that allows sunlight to penetrate the forest for a distance, stimulating intensive forest-floor vegetative growth.

decomposers, agents of decay such as bacteria, fungi, termites, beetles, and worms, release their nutrients to soils. Insight 2.1 describes a typical day in a rain forest.

Upon arriving in the New World, Columbus was told that rain forest trees were always green—they

never lose their foliage. The report was partly right. Rain forest trees are often called "evergreen," but few of them are true evergreen trees (conifers). In Costa Rica, for example, there are no conifers. The trees are mostly broad-leaved, deciduous hardwoods that con-

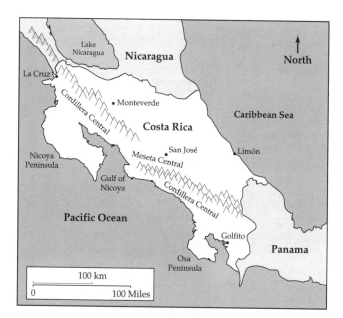

Figure 2.2 Costa Rica lies between Nicaragua to the north and Panama to the south. Its eastern coastline borders the Caribbean Sea; its western coastline, the Pacific Ocean.

TABLE 2.2 Tropical Forest Tree Plantations

COUNTRY	AREA (MILLION HECTARES)
China	33.8
India	14.6
Indonesia	6.1
Brazil	4.9
Vietnam	1.5
Republic of Korea	1.4
Chile	1.0
TOTAL	63.3

SOURCE: Data from UN Food and Agriculture Organization (UNFAO), 1995.

tinuously grow new leaves to replace older ones that die and fall to the forest floor. Thus, the forest appears to be evergreen year-around.

Tropical Moist Forests

Tropical moist forests, another important type of closed forest, generally receive an average of 2,000 millimeters (≈ 80 inches) of annual rainfall. Rainstorms are common, and the periods of little or no rainfall are limited to three months or less. Moist forests typically border the edges of rain forests. In

Costa Rica, they are found in southwestern Pacific lowlands on the Osa Peninsula (Figure 2.2). There, as in rain forests, broad-leaved hardwood trees predominate, and leaf generation and leaf fall take place evenly throughout the year, making moist forests appear evergreen.

Tropical Dry Forests

Tropical dry forests include woodlands and scrublands where annual rainfall is limited to about 1,000 millimeters (≈ 40 inches). A pronounced dry season prevails, often lasting six to eight months. Dry forests are more open in structure and less densely wooded than rain forests and moist forests. They are not evergreen year-around. During dry seasons, trees lose most of their leaves to conserve water. Also, since they experience a pronounced dry season, tropical dry forests are easily burned down. The tropical dry forests that once dominated Costa Rica's northwestern lowlands and much of Central America have been burned down to create pasturelands for cattle ranching, a big business in Latin America today. Unfortunately, tropical dry forests are now rare in Central America; it is estimated that 98 percent of Costa Rica's Pacific northwest tropical dry forests are gone.

Tree Plantations

Increasingly, both developed and developing countries are establishing plantations where fast-growing trees are grown to produce timber, plywood, woodchips for pulp and papermaking, fuelwood, and charcoal. In several regions of Asia-Oceania, mainly Indonesia, Malaysia, and Thailand, forest plantations include rubber, coconut, and oil palm trees.

In 1995, tree plantations accounted for just over 81 million hectares (≈ 200 million acres) worldwide. While this was only 2.3 percent of total forest cover, it was double what existed in 1980. The largest tree plantations are in developing nations (Table 2.2). In most cases, they have been established in formerly nonforested areas such as grasslands, savannas, and scrub woodlands. For example, Brazil is developing large tree plantations in eastern Amazonia where pine trees will be raised and harvested to make charcoal for use in iron smelters. Other Brazilian plantations are planned to meet needs for fuelwood. In India, plantation-grown poplar trees are being used to manufacture matches, plywood, and other wood products.

Soil Quality

The recurring rainstorms in tropical forests cause soil erosion and nutrient leaching so that tropical forest

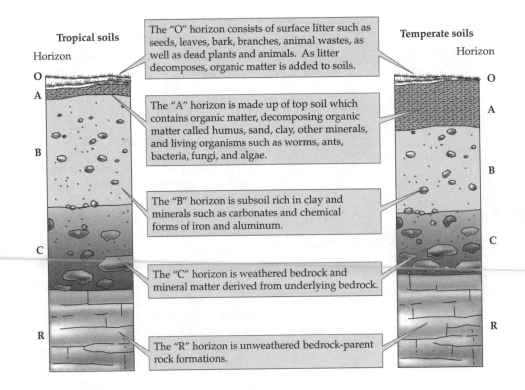

Figure 2.3 Tropical forest soil horizons compared with temperate soil horizons.

The "O" horizon consists of surface litter such as seeds, leaves, bark, branches, animal wastes, as well as dead plants and animals. As litter decomposes, organic matter is added to soils.

The "A" horizon is made up of top soil which contains organic matter, decomposing organic matter called humus, sand, clay, other minerals, and living organisms such as worms, ants, bacteria, fungi, and algae.

The "B" horizon is subsoil rich in clay and minerals such as carbonates and chemical forms of iron and aluminum.

The "C" horizon is weathered bedrock and mineral matter derived from underlying bedrock.

The "R" horizon is unweathered bedrock-parent rock formations.

soils are usually impoverished in terms of organic matter and nutrients. The typical layers, or **horizons,** in tropical forest soil are compared with temperate soils in Figure 2.3. Woody debris, leaves, seeds, bark, flowers, and animal wastes define the layer known as the organic litter horizon, or O horizon. Decomposers concentrate in the O horizon. As agents of decay, they decompose organic matter, thereby releasing nutrients to the A horizon, a thin layer of topsoil just below the O horizon. Soil nutrients released to the A horizon are assimilated by complex root systems that interlace both the O and A horizons. Clay, which underlies the A horizon, makes up the B horizon and serves as a source of essential minerals needed by forest vegetation. The B horizon, together with rock minerals below it in the C horizon, is high in aluminum and iron oxide content. High temperatures in the forest interior, together with high humidity, promote bedrock weathering, which adds metal oxides to tropical soils and gives them a yellow to red color. Weathered rock materials and minerals characterize the C horizon while unweathered bedrock defines the R horizon.

Some tropical soils are surprisingly fertile. For example, soils of volcanic origin like those in parts of Costa Rica's Central Plateau are excellent for growing coffee. In addition, soils enriched by seasonal flooding support major banana plantations along Caribbean coastal lowlands south of Limón and on Pacific southwestern lowlands near Golfito (see Figure 2.2).

Tropical forest deforestation, when it occurs, exposes soils to erosion. To make matters worse, soil erosion can be accelerated, depending on agricultural practices (Table 2.3). Without tree canopies, intense sunlight literally bakes the forest floor, transforming normally spongy, rainfall-absorbent soils into a bricklike pavement that retards vegetative growth and hinders agricultural activity.

The Role of Trees in a Tropical Forest

Tropical forest soils are generally nutrient poor because tropical trees capture and assimilate nutrients rapidly and efficiently through intricate root networks within and on top of the forest floor. As a result, the forest's storehouse of nutrients is conserved mainly in its trees, not in its soils.

Nutrient recycling in tropical forests depends on decomposing organic matter in fallen trees, tree trunks, bark, branches, and leaf litter, followed by a rapid uptake of nutrients as they are released. For example, it is estimated that half of Costa Rica's tropical forest trees are **legumes.** They are therefore able to convert atmospheric nitrogen gas into plant nutrients. The process is similar to that occurring in other legumes such as

DEGREE OF FOREST DISTURBANCE	SOIL EROSION (METRIC TONS/HECTARE/YEAR)
Undisturbed tropical forest	0.01
Localized, primitive, forest agriculture	0.02
Manually cleared forests; no tilling	0.40
Mechanically cleared forests; little tilling	4
Machine-tillage	16

TABLE 2.3.   Typical Rates of Soil Erosion in Tropical Regions

SOURCE: Data from R. Lai, 1981.

alfalfa, peas, beans, and clover. Legumes colonize nitrogen-fixing Rhizobium bacteria that grow in nodules on their roots. The bacteria facilitate the biochemical transformation of atmospheric nitrogen, which is not a plant nutrient, into chemical forms such as ammonia and nitrate, which are plant nutrients.

The noted naturalist E. O. Wilson of Harvard University suggests that while tropical forests cover about 8 percent of the planet's land area, they are the primary habitat for at least half, and possibly as much as 80 percent, of all the world's species of flora and fauna. This makes tropical forests the most biodiverse of all the planet's biomes. This is because the biological complexity of equatorial forests creates incomparably numerous and varied habitats capable of accommodating untold numbers of organisms.

In 1995, 450 species of trees were identified in a one-hectare patch ($\approx$ 2.5 acres) of Brazil's Amazonian Rain Forest just east of Brasilia, the country's capital. This is the most highly diverse, terrestrial ecosystem known. Tree diversity in other one-hectare areas of Brazilian rain forest is typically about 225 species. Undisturbed one-half hectare ($\approx$ 1.2-acre) areas of tropical forest in Colombia are habitat for about 200 different plant species. By comparison, similar areas in New England forests support about 20 plant species.

In tropical forests, trees of the same species are often widely dispersed. Two trees of the same species may be well over 800 meters ($\approx$ ½ mile) apart simply because of the very large number of different tree species present. Physical separation between individual trees of the same species explains why insect damage and plant diseases are minimized in biodiverse forests. Some ecologists, including E. O. Wilson, believe that biodiversity contributes to ecological stability and that ecosystems with fewer species tend to be less stable in the long term. To the extent that this is true, losses in biological complexity may well signal diminished stability and resilience in ecosystems.

Overview of Deforestation

In earlier centuries, many countries clear-cut large tracts of their forested lands. For example, almost all of the British Isles was once covered by trees. However, by the end of the eighteenth century, at least 95 percent of those forests had been cut down, making it necessary to import timber for building ships, homes, and furniture. Similar declines occurred elsewhere in parts of Europe and North America.

Deforestation on a major scale has been occurring in both developed and developing countries since about 1700 when it is estimated that forests covered as much as 4.5 billion hectares ($\approx$ 11 billion acres) of the planet. If this estimate is correct, total forest cover has declined by about 1.0 billion hectares ($\approx$ 2.5 billion acres) or 23 percent during the last 300 years ($\approx$ 0.08 percent per year).

More recent rates of deforestation are even higher. Data from UNFAO show that between 1990 and 1995, forest cover declined by 56.3 million hectares (139 million acres). This represents an average worldwide rate of decline of 11.3 million hectares ($\approx$ 27.9 million acres) or 0.32 percent a year, four times greater than earlier rates.

By all odds, however, the greatest rates of deforestation are occurring today in tropical forests, where approximately 1 percent of existing cover is being lost yearly. Tropical forest deforestation is occurring primarily in developing countries, most of which occupy regions near the equator. The main driving force behind tropical deforestation in many tropical countries is land conversion. Another important driving force is slash-and-burn agriculture, also known as shifting cultivation. Of somewhat lesser importance is fuelwood gathering.

Land Conversion

The practice of converting forest land into public and private ventures is called **land conversion.** In Southeast Asia, for example, forests are being exploited for deciduous hardwoods like ebony, teak, and mahogany, which are in high demand because of their fine-grained elegance and durability. Central and South American forests are being cleared for cattle ranching, urban development, permanent agriculture, fuelwood production, and hydroelectric power developments. In Costa Rica, approximately two-thirds of deforestation has been caused by land conversion. Worldwide, land conversion is the principal means whereby poor nations raise money to pay off foreign debts and finance new development projects.

Highway construction, a type of land conversion, is a way to capitalize on otherwise inaccessible forest land. Brazil's 3,400-mile-long Trans Amazon Highway, built in 1978, has opened the rain forest to hardwood tree harvesting, woodchip fuel production, and urban development. New cities like Rondonia are being carved out of the forest.

Slash-and-Burn Agriculture

An ancient custom in tropical lands is **slash-and-burn agriculture,** or **shifting cultivation.** In this practice, trees are cut down, allowed to dry, and then burned to create nutrient-rich ashes, which are mixed with soils in the cleared patch of forest. The soil is then used to grow subsistence crops, such as corn, beans, cassava, cacao, and cotton, and pastureland grasses for grazing cattle. But although cutting and burning trees releases nutrients into the soil, the resulting fertility is short-lived, lasting only two to three years. Sooner or later, the people have to abandon a soil-impoverished agricultural area, move on, and clear another part of the forest. Abandoned forest lands are generally allowed to lie fallow for three or more decades before they are farmed again.

Fuelwood Gathering

In many developing societies, native people depend on **fuelwood,** mostly dead tree branches, to heat their dwellings and cook their meals. Searching the forest for fallen branches is an everyday task. Fuelwood is the primary source of domestic energy for 40 percent of the world's population. But this practice disrupts forest ecology by eliminating a nutrient source—decaying wood. When dead branches cannot be found, live branches are cut, which ultimately kills trees and accelerates deforestation.

While considerable fuelwood is used in individual households, in some developing countries it is converted into charcoal by heating wood in clay pots while excluding as much air as possible. About 70 percent of the wood's heat energy is lost during this process. But because it is lighter than wood, charcoal is more easily transported to urban areas. Charcoal is also used as an industrial energy source—for example, in iron-making. It takes four tons of wood to make one ton of charcoal—enough to fire an iron smelter for five minutes!

Table 2.4 provides some insight into how much fuelwood energy was used in 1992 in several different countries.

Effects of Deforestation

Although deforestation is often carried out for valid economic or cultural reasons, it has adverse effects on the environment. Deforestation alters forest habitats, disrupts plant and animal populations, diminishes biodiversity, endangers wildlife survival, and affects global climate patterns.

For example, recurring drought in India is blamed on extensive deforestation. Scientists estimate that forest cover must be increased fourfold if the country is to regain adequate rainfall. And in Ethiopia, recent drought and famine appear to be linked to deforestation in the country's western highlands. As forested lands there were cleared, rainfall declined, dry periods occurred more frequently, and agriculture failed. There are three reasons for what happened in Ethiopia:

1. Trees pump groundwater from deep soils into the atmosphere. When trees are cut down, less moisture is pumped into the air and less precipitation occurs.

2. Deforestation opens forest ecosystems, allowing humid air to escape, thereby diminishing the forest's moisture reserve. Less water is recycled in drier forests. The overall result is less cloud cover and less rain.

3. Loss of trees and tree foliage exposes the forest floor to the full impact of tropical rains, promotes soil erosion, and accelerates water export from the forest to nearby rivers. Less cycling of water takes place, depending on the scale of clearcutting, land slope, topography, and severity and frequency of rainstorms.

TABLE 2.4   Percent of Total Energy Derived from Wood in Selected Countries

COUNTRY	PERCENT ENERGY	COUNTRY	PERCENT ENERGY
Nepal	94	Nicaragua	50
Tanzania	92	Indonesia	50
Nigeria	82	India	33
Sudan	74	Costa Rica	33
Kenya	71	China	25
Paraguay	64	Brazil	20

SOURCE: Data from *Vital Signs.* Worldwatch Institute, 1992.

Chances for Recovery

Sometimes nature is resilient and once-forested lands recover. In the eastern United States, for example, forest regrowth following clear-cuts is often rapid; some commercially valuable trees regrow in about 40 years. But at other times, nature is unforgiving. Tropical forests cut during the construction of the Pan American Highway have yet to recover; grasses and shrubs are still the only dominant plants there. Here are some of the reasons tropical forests may not regenerate even after long periods of time:

- Nutrients released by cutting and burning trees are quickly transported out of the forest by rainstorms.

- Tropical forest tree reproduction depends on insects, birds, and small mammals for cross-pollination and seed dispersal. These populations cannot survive in forest habitats that have been drastically altered.

- Tropical tree seedlings tolerate only a narrow range of humidity and light. They often do not survive in totally open areas.

- While ungerminated seeds (a seed bank) are known to survive for long periods of time, tree seeds may fail to germinate in areas disturbed by human activity. Soil compaction, for example, may prevent air and water from reaching seeds.

Sometimes tropical forest regrowth occurs through a series of changes called **ecological succession** (Insight 2.2). When the Panama Canal was constructed (1904–15), forests on both sides of the canal were cleared. Today, a diverse forest has grown up there, although some plant and tree species are different from those that existed before. Forest recovery has also been observed using earth-orbiting satellites.

For example, native tropical trees are being reestablished in abandoned pastureland in the Amazonian Rain Forest. In these two cases, reforestation is creating diverse forest ecosystems. This was possible because clear-cut areas were not very large and seed banks were close by.

CASE STUDY: COSTA RICA'S TROPICAL FORESTS

Costa Rica (meaning "rich coast" in Spanish) lies between Nicaragua and Panama within the isthmus of Central America (see Figure 2.2). In 1821, Costa Rica became a democratic republic after declaring its independence from Spain. Two-thirds of its population lives in the Meseta Central (Central Plateau), where a somewhat idyllic climate prevails. The year-round average temperature is about 21 °C (≈ 70 °F). Coffee, the top crop and number one export, is grown on nearby mountain slopes. Other important export crops include bananas, cocoa, beef, and sugarcane. The recent emergence of tourism in Costa Rica has overshadowed the country's traditional industries.

Costa Rica has a total land area of 50,060 km² (≈ 5,000,000 hectares) and is comprised of several geographic regions. They are defined by the Cordillera Central (Central Mountains) that stretch from northwest to southeast, often exceeding 2,500 meters (≈ 8,200 feet) in elevation.

Costa Rica, despite its small size (comparable to West Virginia), is known for its spectacular plant and animal diversity. For example, there are more species of birds and trees there than in all of North America. Terry Erwin, a biologist at the Smithsonian Institution in Washington, D.C., estimates that possibly 10 million insect species inhabit rain forest high canopies in Costa

INSIGHT 2.2
Ecological Succession

Succession is a series of distinct changes in plant and animal communities that takes place over time in a particular ecosystem. Such changes generally lead to more complex communities.

Primary succession depends first on rebuilding soils lost through destructive events such as volcanic explosions and lava flows. Very slow processes, including weathering of rocks and minerals and decomposition of organic matter, must take place to rebuild soils before vegetation can recur.

Secondary succession refers to recolonization of a particular landscape by shrubs, trees, and animal life following a disturbance such as deforestation or a forest fire. It is a natural process of restoration that rebuilds on soils that remain intact and are capable of supporting plant life.

Both primary and secondary succession occur in stages that are marked by changing flora and fauna. Opportunistic **pioneer organisms,** including grasses, ferns, mosses, and lichens, are well represented during early successional stages. More complex plant life appears later, followed over time by a mature, self-regulating, steady-state community of living things called the **climax ecosystem.** In many cases, a climax system resembles the original ecosystem before any disturbances occurred.

While distinct stages of plant growth appear during succession, the emergence of each stage is thought to be independent of prior communities. Vegetative and tree emergence is a function of differing rates of seed propagation, germination, and growth of each species present.

Rica. This number is *twenty times* greater than earlier estimates of the total number of insect species on the entire planet! A single tropical tree in Costa Rica may well have more than 50 different epiphytic plant species growing on it, including algae, lichens, ferns, vines, mosses, bromeliads, and exotic flowering plants like orchids. Similar temperate-region trees support fewer than 10 epiphyte species. More than 100 species of bats have been identified in Costa Rica. In comparison, only 40 bat species are known in all of North America. Conservative estimates of species biodiversity in Costa Rica are given in Table 2.5.

Characteristics of Costa Rican Forests

In the early 1500s, when Costa Rica became a Spanish colony, tropical forests covered virtually the entire territory. But, as in many tropical countries, the country's hardwood trees have been cut and exported for over a hundred years. They are one of Costa Rica's chief economic assets. In the 1800s, the establishment of coffee plantations added to deforestation in the Central Plateau where fertile volcanic soils and an equable climate prevail. In the 1900s, additional forest lands gave way to banana plantations in the country's Atlantic and Pacific lowlands as well as cattle-grazing pastures in what were northwestern dry forests. Today, most of Costa Rica's original forest cover has been lost. What is left are smaller, scattered forests, remnants of earlier luxuriant tropical ecosystems.

TABLE 2.5   Estimates of Numbers of Plant and Animal Species in Costa Rica

Insects, spiders	365,000
Other invertebrates	85,000
Bacteria, viruses	35,000
Vascular plants	10,000
Fungi	2,500
Vertebrates	1,500
Fish	1,013
Birds	845
Reptiles	218
Mammals	205
Amphibians	160
TOTAL	≈ 500,000

SOURCE: Data from Instituto Nacional de Biodiversidad de Costa Rica (INBio), 1993.

In Costa Rica's remaining rain forests, high canopies tower 50 to 60 meters (≈ 160 to 200 feet) above the forest floor (see Figure 2.1). The broad crowns and interwoven branches of tall, dominant trees create an almost unbroken carpet of vegetative cover high above the forest floor. A smaller number of towering trees, such as the ceiba or silk-tassel tree, stand as high as 75 meters (≈ 250 feet), punctuating the high canopy.

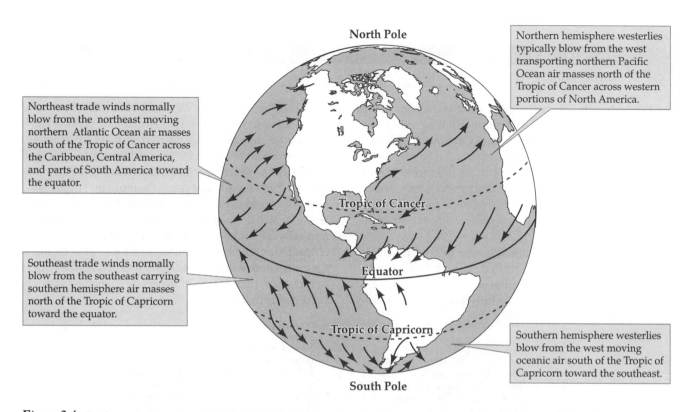

Northern hemisphere westerlies typically blow from the west transporting northern Pacific Ocean air masses north of the Tropic of Cancer across western portions of North America.

Northeast trade winds normally blow from the northeast moving northern Atlantic Ocean air masses south of the Tropic of Cancer across the Caribbean, Central America, and parts of South America toward the equator.

Southeast trade winds normally blow from the southeast carrying southern hemisphere air masses north of the Tropic of Capricorn toward the equator.

Southern hemisphere westerlies blow from the west moving oceanic air south of the Tropic of Capricorn toward the southeast.

Figure 2.4 Earth's prevailing winds. Note that in the tropics the principal winds are easterlies—that is, they generally blow from east to west.

Mid-level trees occupy a tier about 20 to 40 meters (≈ 65 to 130 feet) high. Below them is a low-level canopy or understory comprised of small trees, including tree ferns about 5 to 15 meters (≈ 16 to 50 feet) high.

Natural events acting over millions of years shaped Costa Rica's forests. These events included prevailing winds, copious precipitation, high humidity, tropical temperatures, and volcanic action. Trade winds ("easterlies" blowing from east to west) transport warm, moisture-laden Caribbean air to the country's eastern lowlands and mountains (Figure 2.4). As moist air moves inland, it is lifted by rising landscapes to higher elevations where it expands and cools, causing precipitation. This results in almost daily rainstorms. Rainfall produced in this way is called **orographic precipitation** (Insight 2.3).

Onslaughts of rain in Costa Rica's eastern lowlands and mountains result in somewhat drier air being transported to the country's interior. Here, precipitation is less frequent, and in some places occurs only during a two- to three-month rainy season and averages less than 1,000 millimeters (≈ 40 inches) a year. Tropical dry forests are best adapted to these conditions.

Factors Leading to Deforestation

Until recently, rates of deforestation in Costa Rica were among the highest in the world. Clear-cutting started

there as early as the 1600s. By 1900, 10 percent of the country's forests had been cleared, and by 1950, 44 percent of them were gone. By the 1990s, 80 percent of the country's forest cover had been cleared, and what had once been tropical forests became cattle ranches, farms, villages, tree plantations, water reservoirs, highways, and commercial enterprises. The 1900s also marked the beginning of major foreign investments in Costa Rican agriculture. Low-interest loans from the World Bank and other international agencies propelled the transformation of dry forests in the country's northwest into pastureland. Cattle ranching became big business, and Costa Rica became one of Latin America's leading exporters of fast-food hamburger meat.

Besides land conversion, other factors led to Costa Rica's deforestation. Because the country is a developing nation with widespread poverty, its native people, poor peasant farmers called *campesinos,* have long depended on slash-and-burn agriculture, which accounts for about one-third of Costa Rica's deforestation. Also, in Costa Rica, fuelwood accounts for one-third of the total energy consumed. The energy crisis there is sometimes not having enough fuelwood.

Costa Rica Reverses the Trend

In the 1960s, the government of Costa Rica recognized that prevailing rates of deforestation, if allowed to

INSIGHT 2.3
Orographic Precipitation

While North America's prevailing westerlies drive air masses eastward over mountain ranges like the Cascades in Washington and Oregon, Caribbean easterlies carry moist air westward across Central America. In both regions, moist air is lifted to higher elevations, where it expands due to lower atmospheric pressure. Expansion of an air mass lowers its temperature, a physical process known as **adiabatic cooling.**

When air cools, it is less able to hold moisture and becomes super-saturated. Under these conditions, orographic precipitation consisting of rain and snow occurs on the **windward** side of moun-

tains (Figure 2.A). Since moisture is squeezed out of rising air masses on the windward side, relatively dry air reaches the other side—the **leeward** side—forming a rain shadow. In Washington State, for example, the western side of the Cascade Mountains is a green, lush region that receives about 800 millimeters (≈ 32 inches) of annual precipitation in the lowlands and up to 2,500 millimeters (≈ 100 inches) annually at higher elevations. Dominant trees include Douglas fir, cedar, and hemlock. East of the Cascades, a dry Ponderosa-pine desert prevails, averaging less than 250 millimeters (≈ 10 inches) of annual precipitation.

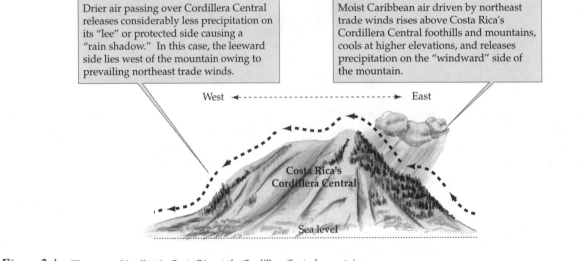

Drier air passing over Cordillera Central releases considerably less precipitation on its "lee" or protected side causing a "rain shadow." In this case, the leeward side lies west of the mountain owing to prevailing northeast trade winds.

Moist Caribbean air driven by northeast trade winds rises above Costa Rica's Cordillera Central foothills and mountains, cools at higher elevations, and releases precipitation on the "windward" side of the mountain.

West ← - - - - - - - - - - - - - - - - - - → East

Costa Rica's Cordillera Central

Sea level

Figure 2.A   *The orographic effect in Costa Rica at the Cordillera Central mountain range.*

continue, would result in catastrophic losses of the country's forests. By the early 1970s, a rising tide of national concern led to the creation of the first Congress on Natural Resources to consider the problem. The Congress identified new national priorities aimed at saving the nation's remaining tropical forests. By 1986, a high-ranking cabinet-level ministry for environmental affairs was established, the Ministry of Natural Resources, Energy and Mines. Costa Rica's government determined that environmental affairs would be given high priority in all government decision-making.

In 1987, Costa Rica launched its National Conservation Strategy for Sustainable Development (ECODES) to enhance resource conservation and sustainable development. The government committed

itself to a new way of life where decision-making at the highest levels integrated economics, demography, industry, mining, water resources, agriculture, energy, and technology. This was a sharp departure from life as usual in Costa Rica where laws and regulations often reflected special interests. The country's success in adopting its new environmental strategy was due in part to its high literacy rate (93 percent), the highest in Central America.

Reforestation

An example of sustainable development is Costa Rica's Forestry Action Program, which fosters reforestation in national forests and commercial tree plantations. In

1. Santa Rosa National Park
2. Rincón de la Vieja National Park
3. Ostional National Wildlife Refuge
4. Las Baulas Natinal Park
5. Cabo Blanco Strict Nature Reserve
6. Barra Honda National Park
7. Palo Verde National Park
8. Guayabo, Negritos, and Pájaros Islands Biological Reserves
9. Peñas Blancas Wildlife Refuge
10. Carara Biological Reserve
11. Manuel Antonio National Park
12. Caño Island Biolaogical Reserve
13. Corcovado National Park
14. Golfito Biological Reserve
15. La Amistad International Park
16. Chirripó National Park
17. Hitoy-Cerere Biological Reserve
18. Gandoca-Manzanillo National Wildlife Refuge
19. Cahuita National Park
20. Tortuguero National Park
21. Barra del Colorado National Wildlife Refuge
22. Braulio Carrillo National Park
23. Poás Volcano National Park
24. Irazú Volcano National Park
25. Guayabo National Monument
26. Tapantí Wildlife Refuge
27. Cocos Island National Park
28. Monteverde Cloud Forest Biological Reserve
29. Caño Negro Wildlife Refuge

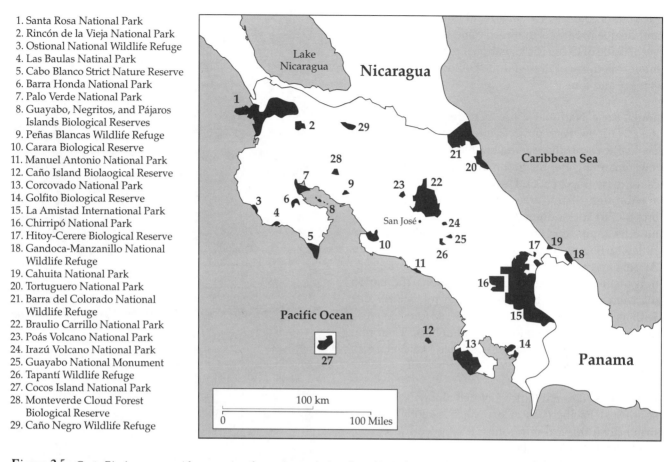

Figure 2.5 Costa Rica's concern with preserving the environment is reflected in its large number of national parks, nature reserves, wildlife refuges, and biological reserves.
(Source: Data from INFO Costa Rica, San Juan, Costa Rica, 1997.)

some areas, fast-growing trees like pine are used to produce pulpwood for paper-making and fuelwood for native people. In other areas, timber such as teakwood for furniture construction is grown. While only 62 hectares (≈ 150 acres) were reforested under this program in 1970, 2,700 hectares (≈ 6,700 acres) a year were being planted by the 1980s. The Forestry Action Program is lowering rates of deforestation in Costa Rica.

Creating National Parks, Biological Reserves, and Wildlife Refuges

Many forest soils are poorly suited to agriculture, but they can be preserved as parks, wilderness, and biological reserves. Costa Rica has become a world leader in protecting tropical forests, distinctive wildlife areas, and their rich biological diversity. This was accomplished by making natural resource protection a top political priority and by setting aside a network of protected environments. This achievement reflects the understanding of Costa Rica's government that the country's future is linked to the health and well-being

of its natural ecosystems. Figure 2.5 identifies Costa Rica's principal national parks, wildlife refuges, and biological reserves.

The first land area to be set aside, Poás Volcano National Park, is near San José in the Central Valley. This park is noted for its active volcano (Volcán Poás), its extraordinary variety of bird species, and its striking beauty. By 1998, 12 percent of Costa Rica's land was protected in 46 national parks. An additional 16 percent was protected as Indian lands, biological reserves, wildlife refuges, and corridors. By sheltering these areas and placing them off-limits to development, Costa Rica has protected 28 percent of its total land area. This is a larger percentage of protected territory than any other country in the world.

Environmental organizations from all over the world are at work in Costa Rica conducting studies and research in tropical biology and resource conservation. Journalists and filmmakers are making the country's extraordinary wildlife and landscapes the subjects of their work. There is even an ecology channel on television helping to reinforce the development of an environmental ethic. The University of

Rica offers courses in wetland and forest management and degree programs in forestry and ecotourism.

Three of Costa Rica's national parks are of special interest because of their unique habitats, wildlife, vegetation, and forests.

Santa Rosa National Park   Created in 1971, Santa Rosa National Park projects into the Pacific Ocean in Costa Rica's northwest region. One of the country's few remaining fragments of tropical dry forest is located there (Figure 2.6). During the dry season, January through March, trees and vegetation are mostly brown. The park's animal life includes peccaries, coatimundis, tapirs, and coyotes. Sea turtles, like the olive ridley turtle, congregate along ocean beaches between August and December, and as many as a quarter million turtles nest there every year. Among the park's many tree species, the naked Indian tree stands out with its red, peeling bark. Tropical moist forested areas are also found in Santa Rosa National Park. Their tall, evergreen-appearing, deciduous trees stand close to the ocean shoreline.

Corcovado National Park   Located along Costa Rica's southwest Pacific coast, Corcovado National Park is primarily a tropical rain forest. It is noted for its lush biodiversity. At least 6,000 different insects, 500 tropical trees, 367 birds, 140 mammals, and 117 reptiles and amphibians have been identified there. The silk-cotton tree, one of the park's giant trees, stands as high as 70 meters (≈ 230 feet) and is supported by thick root buttresses. The park's animal species, some of them endangered elsewhere, include the jaguar, American crocodile, scarlet macaw, caiman, and Baird's tapir. White-faced monkeys are commonly seen (Figure 2.7).

Gold was discovered in this region in the 1930s, setting off a gold rush. Mercury metal, used to extract gold from stream sediments, became a toxic contaminant in streams and rivers, making the fish unsafe to eat and the water dangerous to drink. By 1986, local and even international opinion had forced Costa Rica to ban this kind of exploitation. Strict new controls and penalties for unauthorized deforestation, logging, and mining were imposed. Gold mining in the country's national parks is now illegal.

Tortuguero National Park   Lying along Costa Rica's northeastern Caribbean coastline, Tortuguero National Park is one of the country's best-known parks. Tortuguero's beaches are famous nesting sites for green sea turtles. These rather large turtles can be 1 meter (≈ 3.3 feet) long and weigh up to 200 kilograms (≈ 440 pounds). Females come ashore at night, from August through October, to dig nests and lay eggs.

Figure 2.6 Dry tropical forest in Costa Rica's Santa Rosa National Park.
(Photo courtesy of Kathryn E. Smith.)

Figure 2.7 White-faced monkeys in Costa Rica's Corcovado National Park.
(Photo courtesy of Kathryn E. Smith.)

Tortuguero means "turtle catcher" in Spanish. For hundreds of years, indigenous peoples included turtles in their diet. While a sustainable balance had been struck, it didn't last. Commercial trade in sea turtles developed, and tons of turtles and turtle eggs were sold around the world. Conservation efforts, especially the creation of national parks, have succeeded in protecting the turtles.

Engaging in Debt-for-Nature Swaps

Another way that ecologically valuable resources can be protected is through a practice known as **debt-for-nature swaps,** which hinges on two factors:

1. Many developing nations are heavily in debt to foreign governments and nongovernmental organizations (NGOs), such as the World Bank, in return for development projects like dams, power plants, highways, and food assistance programs.

2. Conservation groups, such as the Nature Conservancy, Conservation International, and the World Wildlife Fund, are eager to save the world's tropical forest diversity, most of it in developing nations.

Here's how a debt-for-nature swap works: A nonprofit conservation group or a lending nation buys part of a country's debt at a discount. The debt is forgiven in exchange for an agreed-upon commitment to conserve particular forest lands, to create a wildlife reserve, or to fund tropical studies.

Bolivia participated in the first debt-for-nature swap in 1987. The United States forgave a large developmental assistance debt in return for Bolivia's agreement to fund $20 million of conservation-related projects. By 1992, similar swaps resulted in $80 million of Costa Rican foreign debt being bought by conservation groups for $12.5 million. In exchange, Costa Rica committed $43 million toward conservation projects. In this way, the country was able to overcome some of its financial debts and at the same time fund important environmental initiatives.

Promoting Ecotourism

Environmental education has become a major theme in a kind of tourism called **ecotourism.** Visitors not only visit the tropical forest to appreciate its climate and scenery, but also to learn about conservation and wise uses of natural resources. In 1994, ecotourism became Costa Rica's primary source of foreign income. Since then, hotels have been booked far in advance of Christmas and Easter holidays as vacationers, hikers, and tourists from various parts of the world make plans to visit the country's national parks and biological preserves.

Similar activities have occurred in Belize, where the population of about 200,000 is a mixture of English, Scottish, African, Mayan, and Mexican ancestry. Today, tourists and vacationers often outnumber the country's natives. They visit Belize's jungle paradise to hike, snorkel, study Mayan ruins, and experience a tropical rain forest.

But ecotourism can be damaging to the environment when hordes of people crowd primitive trails through biologically sensitive areas. Like Costa Rica, Belize's economic future is tied to tropical forest protection. As a result, a new national goal has emerged: to keep at least 80 percent of the land as tropical forest and make Belize the best-preserved country in Central America. Clearly, this goal parallels Costa Rica's model for forest protection.

Fostering Sustainable Forest Products

Natural products yet to be extracted from forest flora and fauna may in the long run prove to be more valuable than a forest's standing timber or the cattle-grazing pasturelands that can be created by cutting them down. New ways of thinking about forests are needed. In some cases, ways can be found to extract forest products economically and sustainably. Perhaps the best-known example is rubber-tree tapping. Natural rubber is made from latex tapped from tropical forest rubber trees. This was once the main source of rubber to make tires for cars, trucks, and bicycles. Today, many people believe we no longer need natural rubber—that synthetic rubber has taken its place. But this is not so. The demand for natural rubber is increasing in the United States because tires formulated with both natural and synthetic rubber have superior heat resistance to tires made with synthetic rubber alone.

In 1990, Conservation International in Washington, D.C., initiated an interesting sustainable-development project in Ecuador's rain forest. Round, ivory-like nuts from tagua trees were collected by poor peasants and sold to industries that make buttons for clothing manufacturers. By 1993, more than 1,000 native people were employed, and more than $1.5 million had been generated. The industry is based on a forest product, not on the trees themselves. The trees are left undisturbed. This project provides an example of how indigenous people can prosper from the richness of a forest ecosystem without destroying it.

Studies conducted in Belize by Robert Mendelsohn of Yale University and Michael Balick of the New York Botanical Garden have shown that harvesting medicinal and other natural products from rain forests

INSIGHT 2.4

Natural Products for Medicinal Purposes

One of the first medicines from the rain forest was an extract from the bark of the cinchona tree. Natives called this tree the "fever tree" because drinking its bark extract quelled the recurring fevers caused by malaria, making it the first effective remedy for malaria. The two tree-bark chemicals later identified as active anti-fever agents were quinine and quinidine, and both are still used today to treat malaria.

Other examples of early medication from the forest are chemical extracts from wintergreen leaves and the bark of willow and sweet birch trees. The extracts contain salicylates, compounds related to aspirin. Later, aspirin itself was made from them.

Current research is uncovering a number of tropical forest chemicals that demonstrate antibiotic and anticancer properties. One of the most important of these is taxol, an extract of the Pacific yew tree, which is used in treating ovarian cancer. Other important new drugs include vinblastine and vincristine, extracts of Madagascar periwinkle. Both are used in conjunction with other drugs to cure childhood leukemia. Biologists and medical scientists believe that many yet-to-be-discovered therapies, treatments, and cures will be found in biodiverse regions of the planet such as rain forests.

In the United States, 25 percent of all prescription drugs and 60 percent of all nonprescription remedies are based on chemical ingredients extracted from plant and animal natural products.

Included in the long list of medicines derived from plants and animals are:

Amoxycillin	Broad-spectrum antibiotic
Aspirin	Painkiller and headache treatment
Cortisone	Anti-inflammatory agent
Curare	Powerful muscle relaxant used in major surgery
Cyclosporin	Organ transplant anti-rejection agent
Digitalis	Cardiovascular regulator
Erythromycin	Broad-spectrum antibiotic
Ortho Novum	Oral contraceptive
Penicillin	Broad-spectrum antibiotic
Quinine	Effective in the treatment of malaria
Reserpine	Tranquilizer and blood pressure control agent
Streptomycin	Broad-spectrum antibiotic
Taxol	Treatment of ovarian cancer
Taxotere	A derivative of taxol used to treat lung cancer
Tetracyclines	Broad-spectrum antibiotics
Vasotec	Anti-hypertension drug
Vinblastine	Treatment of childhood leukemia
Vincristine	Treatment of childhood leukemia

is economically sustainable. They estimate that the value of rain forests used for harvesting plant products is $1,347 per acre. By comparison, the asset value of tropical pine-wood plantations is $1,289 per acre and that of cropland is $137 per acre.

Establishing INBio

One new initiative aimed at saving the earth's forests and at the same time using them in sustainable ways is the founding of a nonprofit organization called INBio (Instituto Nacional de Biodiversidad de Costa Rica). INBio is committed to research, conservation, and discovering sustainable, nondestructive ways to use tropical forests. The idea was instituted by two scientists, Daniel Janzen, a tropical biologist at the University of Pennsylvania, and Rodrigo Gamez, a plant virologist at the University of Costa Rica, and it gained the support of Costa Rica's government, becoming established in 1989.

The purpose of INBio is to investigate tropical forest biodiversity and learn how to tap the economic potential of the country's immense biotic wealth. Working with INBio, taxonomists supported by the Costa Rican government have started identifying and cataloging the country's vast biodiversity, particularly its plants, insects, and microorganisms. Parataxonomists, who are skilled in helping taxonomists identify and classify plant and animal species, assist with this enormous task. As a result, Costa Rica is becoming a world center for research and education in tropical biology.

The Merck–INBio Agreement Traditional medicine among native peoples the world over has long depended on natural plant and animal products (Insight 2.4). Today, considerable pharmaceutical research is based on finding potential sources of new medicinal agents of plant and animal origin. Much of this search is taking place in equatorial forests where the greatest diversity of living things exists, so clearly the success of ventures aimed at finding new drugs and therapies depends on conserving the biodiversity of these natural systems.

In 1991, INBio and Merck & Co. Inc., the world's largest pharmaceutical company, signed an innovative agreement allowing Merck to evaluate plant and insect samples from Costa Rica's forests for potential new drugs. One reason is that four of Merck's newest drugs were derived from natural products: the antibiotics mefoxin and primaxin, the antiparasitic agent ivomec used in veterinary medicine, and the cholesterol-lowering agent mevacor. Here is the introduction to the agreement between INBio and Merck:

> Conservation of biodiversity is a critical component of economic growth. Today, conserving biodiversity means seeking an effective balance between progress through development and assuring the long-term sustainability of the world's biotic wealth.

Selecting plants, insects, and environmental samples to be investigated is done jointly by Merck and INBio scientists. Extracting natural products of potential interest is taking place at two Costa Rican universities: National University and the University of Costa Rica. Testing and evaluating extracts for possible new pharmaceutical applications is carried out at Merck's laboratory facilities in Spain and in the United States. Under this agreement, Merck has provided INBio with $1 million plus $130,000 of laboratory equipment. The project is helping train many Costa Ricans to collect, screen, and classify tropical plant and animal life. Royalties on products Merck develops through this agreement are shared with INBio and used to support conservation efforts.

The agreement between INBio and Merck is also helping Costa Rica build a scientific and technical infrastructure—field stations, research laboratories, and facilities to evaluate potential new drugs. Educational programs to train scientists and technicians are being established, and industrial organizations will help produce and market new products. Other countries, including Mexico, Indonesia, Nepal, and Nicaragua, are planning to set up organizations similar to INBio.

REGIONAL PERSPECTIVES

Costa Rica is not the only place where landmark decisions are being made to protect pristine forest environments. Indeed, many countries are taking action to preserve their forested lands (Insight 2.5). Planning is also underway aimed at using protected ecosystems to benefit local people by saving their home territory, sustaining fish and wildlife populations, and maintaining cultural, economic, and aesthetic values.

Clayoquot Sound, Canada: Temperate-Zone Rain Forests

Not all rain forests stand in tropical regions. Clayoquot Sound, which occupies a part of the western edge of Canada's Vancouver Island, is surrounded by coastal, temperate-zone rain forests. The sound receives most of its freshwater runoff from these ancient, forested watersheds—runoff that flows through wild, spectacular, salmon-spawning rivers. The sound is the backdrop for Meares Island, the towns of Tofino and

Ucluelet, the home of the Tla-o-qui-aht, Ahousaht, and Hesquiaht First Nations, and ecosystems long known for their old-growth rain forests.

For decades, major logging industries in British Columbia have been clear-cutting Clayoquot Sound forests. The largest of these industries include MacMillan Bloedel Limited and International Forest Products. The deforestation they have caused has fragmented the sound's pristine watersheds and reduced old-growth forested areas by one-third.

In the 1980s, Tofino environmentalists created an organization called Friends of Clayoquot Sound to prevent logging on Meares Island, their source of drinking water. Others organized to protect shellfish beds, salmon-rearing sites, and unique coastal watershed areas from the onslaught of accelerated clear-cutting. Repeated failures to control industrial clear-cutting led to acts of civil disobedience; hundreds were arrested. In 1993, after the government of British Columbia announced a plan to preserve only one-third of Clayoquot Sound and allow the remaining two-thirds to be clear-cut, thousands of people from across Canada

INSIGHT 2.5
Locations of the Earth's Forests

Today, our best estimates of forest cover and different types of forests are obtained by remote sensing satellites in orbit around the earth equipped with, for example, aerial cameras, balloon-borne sensors, or satellite instrumentation. Based on this information, the United Nations Food and Agriculture Organization (UNFAO) estimated in 1995 that forested lands covered 3.45 billion hectares ($\approx$ 8.52 billion acres)—close to 27 percent—of the world's total land area not including Greenland and Antarctica.

Some of these forests are natural, old-growth, undisturbed ecosystems; some are fragmented, disturbed environments; and some are tree plantations. The five countries having the greatest forested areas today are: the Russian Federation, which includes the former republics of the Soviet Union, 22 percent; Brazil, 16 percent; Canada, 7 percent; the United States, 6 percent; and China, 4 percent.

Three distinctly different forest biomes are commonly identified: boreal coniferous forests (also called taiga), which circle the earth south of north polar tundra regions; temperate deciduous forests made up largely of broad-leaved, hardwood trees; and tropical forests, which occupy the planet's equatorial zone 23.5 °N and 23.5 °S of the equator. In addition, forest plantations, where trees are raised much like an agricultural crop, are becoming increasingly important.

converged on the sound to protest. More than 900 were arrested; it was the largest action of civil disobedience in Canadian history.

Action to save Clayoquot Sound began in 1994 when Spiegel, a German publishing company, stopped buying MacMillan Bloedel pulp to make its paper. America's Sierra Club decided to publish its magazine *Sierra* using paper from sources outside British Columbia. Greenpeace International convinced the Scott Paper Company to cancel its pulp contract with MacMillan Bloedel. These and similar actions forced the B.C. government to establish a moratorium on logging, especially in pristine areas of Clayoquot Sound. But tension in Clayoquot Sound continued as decisions and plans aimed at forest protection were either disregarded or ignored.

In 1996, First Nations representatives requested a time of peace in Clayoquot Sound to seek a reasoned, permanent solution to the problems affecting the region. One idea gained support: establishing Clayoquot Sound as a U.N. Biosphere Reserve. Biosphere reserves were first established in 1976 under the aegis of the United Nations' Man and the Biosphere Programme of UNESCO (United Nations Educational, Scientific, and Cultural Organization). Biosphere reserves have three principal goals:

1. To conserve ecosystem, species, and genetic biodiversity.

2. To foster development that is socially, culturally, and ecologically sustainable.

3. To support the creation of environmental education, training, and demonstration projects.

By the end of 1996, the World Conservation Union, an alliance of more than 130 governments including British Columbia, unanimously supported designating Clayoquot Sound as a U.N. Biosphere Reserve. The proposal was approved by the U.N. in January, 2000 and Clayoquot Sound was added to the 325 biosphere reserves that already exist in 83 countries.

Puerto Rico: The Caribbean National Forest

Puerto Rico's Caribbean National Forest, located in the rugged Sierra de Luquillo 40 kilometers ($\approx$ 25 miles) southeast of San Juan, is the smallest, yet the most highly biodiverse national park in the United States. It is also North America's only tropical rain forest. Rainfall of up to 6,100 millimeters ($\approx$ 240 inches) annually has created an ancient evergreen forest, one of the oldest forest reserves in the Western Hemisphere. More than 240 native tree species have been identified, 88 of them quite rare, growing among mosses, vines, epiphytes, orchids, and giant ferns. A cloud forest dominates mountain peaks that stand 760 meters ($\approx$ 2,500 feet) above sea level.

Forest fauna include more than 50 species of endangered Puerto Rican parrots and 16 species of tree frogs. Major population declines among two species of tree frogs, web-footed and mottled coquis frogs, have been noted. Like other declines among frog species around the world, the cause is unknown.

Mexico: Rain Forests in Chiapas State

Thick tropical rain forests once stretched from hot lowlands through the foothills of the sierra to steep and misty mountaintops in the southeastern corner of Mexico's Chiapas State. Today, the forest is being cleared for agriculture and timber, and dusty cattle towns have become cities. Mayan settlements have multiplied, and once-common jaguars, peccaries, and scarlet macaws are gone. Rain forests, which at one time occupied close to 3 million hectares (≈ 7.4 million acres) in Chiapas, have been reduced to less than 600,000 hectares (≈ 1.5 million acres). They are now endangered, and their impressive biodiversity is imperiled.

Pressure on rain forest resources is increasing. Political rebels have become forest squatters, poor farmers who grow corn, beans, and a few other crops. After their relatively infertile soils are depleted of nutrients, they are forced to move on to claim a new place in the forest, cut down trees, and continue their subsistence way of life. The land is free to whomever will clear the forest and till the soil.

Some in Chiapas believe that creating alternative lifestyles and occupations can save the rain forest. For example, there are plans to encourage ecotourism, develop coffee plantations, employ contour planting techniques, adopt no-till cultivation, and teach environmental education. Another new occupation in Chiapas is collecting butterflies, which are then bought by Mexican biologists for biodiversity cataloging and supplied as specimens to museums worldwide. While it is a controversial project, it provides insight into the daunting dilemma of trying to learn how to live on rain forest products without destroying the rain forest itself.

KEY TERMS

adiabatic cooling

climax ecosystem

closed forests

commensalism

convection

crown roots

debt-for-nature swaps

decomposers

deforestation

ecological succession

ecotourism

edge effect

epiphytes

fuelwood

land conversion

leeward

legumes

open forests

orographic precipitation

pioneer organisms

primary succession

secondary succession

slash-and-burn agriculture (shifting cultivation)

soil horizons

succession

transpiration

windward

DISCUSSION QUESTIONS

1. Developed nations like the United States believe that lesser-developed nations, especially in the tropics, should be discouraged from clear-cutting their forests. Are developed nations justified in taking this view? Should they have the right to force lesser-developed nations to protect their forests? Discuss this issue.

2. Many who have studied slash-and-burn agriculture suggest that more of the world's poor people should own their own land. Generally, they do not. Would larger areas of tropical forests be conserved if poor peoples owned and controlled their own land? Discuss your view.

3. In the Pacific Northwest, controversy prevails regarding protection of an endangered wildlife species, the Northern Spotted Owl. The owl requires large areas of old-growth forest for its reproduction and survival, so logging has been limited in most forests. As a result, many loggers are out of work, and entire communities have been hard hit economically. What are your views regarding protecting forests and endangered wildlife versus protecting forest-related jobs?

4. Even though two-thirds of its total land area is covered by forests, Japan does not allow wide-scale cutting of its own trees. Instead, Japan is a major importer of timber. Comment on this.

5. Do you believe that ecotourism should be encouraged in places like Costa Rica? What dangers to the environment do you see as a result of unregulated ecotourism? Discuss this issue.

INDEPENDENT PROJECT

Forests are often clear-cut. However, there is potential for their recovery. Regrowth of forests usually follows well-defined successional stages. Choose a forest in your region and describe its mature or climax tree species. Whether it has ever been clear-cut or not, find out what vegetative species would colonize first following a clearcut (the pioneer organisms). Then describe the successional pattern likely to occur. How long would it take to reestablish the climax forest you are describing?

INTERNET WEBSITES

Visit our web site at http://www.mhhe.com/environmentalscience/ for specific resources and Internet links on the following topics:

Costa Rica/Photographic files

Web "SuperSite" detailing Costa Rican national parks and other protected areas

Costa Rican wildlife

Costa Rican wildlife and biodiversity

Map of Costa Rica's protected areas

Rain forest preservation in Costa Rica

Rain forest habitats

Foreign aid and conservation of tropical forests: An action plan for change

Rain forest plants

What is biodiversity?

Canada's Biodiversity Convention

SUGGESTED READINGS

Abramovitz, Janet N. 1998. Sustaining the world's forests. Pp. 21–40 in *State of the world 1998*, edited by L. Brown, et al. New York: W.W. Norton & Company.

Balick, Michael J., Elaine Elisabetsky, and Sarah A. Laird, eds. 1996. *Medicinal resources of the tropical forest: Biodiversity and its importance to human health*. New York: Columbia University Press.

Biodiversity. 1992. *Science* 256:1142–43.

Boza, Mario A. 1993. Conservation in action: Past, present, and future of the national park system of Costa Rica. *Conservation Biology* 7(2):239–47.

Carpenter, Betsy and Bob Holmes. 1992. Living with nature. *U.S. News and World Report* (November 30): 60–67.

Durning, Alan T. 1993. Saving the forests: What will it take? *Worldwatch Paper 117*. Washington, D.C.: Worldwatch Institute.

Gamez, Rodrigo, and Alvaro Uglade. 1988. Costa Rica's national park system and the preservation of biological diversity: Linking conservation with socio-economic development. Pp. 131–43 in *Tropical rainforests: Diversity and conservation*, edited by F. Almeda and C. M. Pringle. San Francisco: California Academy of Sciences.

Janzen, Daniel H. 1983. *Costa Rican natural history*. Chicago: University of Chicago Press.

Kellman, Martin, and Rosanne Tackaberry. 1997. *Tropical environments*. London; New York: Routledge.

Mendelsohn, R. O., C. Peters, and A. Gentry. 1989. Valuation of an Amazonian rainforest. *Nature* 339:655–56.

Park, Chris C. 1992. *Tropical rainforests*. London; New York: Routledge.

Perry, Donald. 1986. *Life above the forest floor: A biologist explores a strange and hidden treetop world*. New York: Simon and Schuster.

Stein, Bruce A., and Stephanie R. Flack. 1997. Conservation priorities: The state of U.S. plants and animals. *Environment* 39(4): 8–11; 34–39.

Wilson, E. O. 1992. *The diversity of life*. Cambridge: Harvard University Press.

Wilson, E. O. 1993. The new environmentalism. *Nature Conservancy* (March/April): 38

PART II

Water and Air

Fundamental Resources

PhotoDisc, Inc./Vol. 6

Chapter 3
Industrial Revolution and Pollution

Chapter 4
Estuaries, Seagrass Meadows, and
Saltwater Ecosystems

Chapter 5
Air Pollution
History, Legislation, and Trends

CHAPTER 3

Industrial Revolution and Pollution

PhotoDisc, Inc./Vol. 44

There is a phenomenal resiliency in the mechanisms of the earth. A river or lake is
almost never dead. If you give it the slightest chance by stopping pollutants from going
into it, then nature usually comes back. When we deal gently with the earth–even when
we have thoroughly damaged it–we can repair our friendship with it.

René Dubos

From the earliest times, civilizations have risen near rivers, bays, and estuaries. Aquatic environments like these provide highways for people and commerce, habitat for fish and wildlife, water supplies for crop irrigation, sources of drinking water, and places to dispose of wastes. For these reasons, many of the world's great cities are located near water—cities like New York, Boston, Paris, Cairo, Montreal, Lisbon, and London.

The River Thames is England's most important waterway. Its role in British history, its importance to London, and its decline and subsequent restoration make it a model of ecological problem-solving. We will begin this chapter by describing the Thames estuary and explaining the hydrologic cycle of water, which is vital to the functioning of all the earth's ecosystems.

BACKGROUND

London and the River Thames

The Romans settled along the Thames, about 64 kilometers (≈ 40 miles) upriver from the North Sea, in 43 A.D. Their fortifications mark the site of present-day London, and evidence of their four centuries of occupation is still visible in the roads and aqueducts they built. Later, Germanic peoples, Scandinavians, Danes, and Normans forged what ultimately became England. But it was the River Thames that most influenced London's settlement and development. At first,

the river's main value was its strategic military location, but before long, commercial fishing dominated the scene and London's ports gained worldwide importance.

The River Thames begins as a small stream in the Cotswold Hills of southwestern England. It flows generally eastward 346 kilometers (≈ 215 miles) alongside such cities as Oxford, Wallingford, Reading, Sunbury, and London before emptying into the North Sea (Figure 3.1). As the river meanders through the countryside, it receives *runoff* from innumerable streams and small rivers that drain various parts of its *watershed*.

A **watershed** is a bowl-shaped catchment basin where **runoff** precipitation (mainly rain and snow melt) collects. Some of the precipitation evaporates, some is stored in soils and vegetation, some infiltrates soils and recharges groundwaters, and the rest runs off to nearby surface waters such as streams, rivers, lakes, estuaries, and oceans. Runoff transports dissolved and suspended matter, and can cause soil erosion, which adds soils, nutrients, mineral matter, and pollutants to receiving waters when it occurs. The physical features of a watershed, including the contours of the landscape, trees and vegetation, and types of soils, determine the direction, rate of flow, and chemistry of runoff waters. The Thames watershed occupies an area of about 1,295,000 hectares (≈ 5,000 square miles).

The Freshwater Thames

Runoff in the Thames watershed collects in numerous tributaries that connect with the River Thames. The

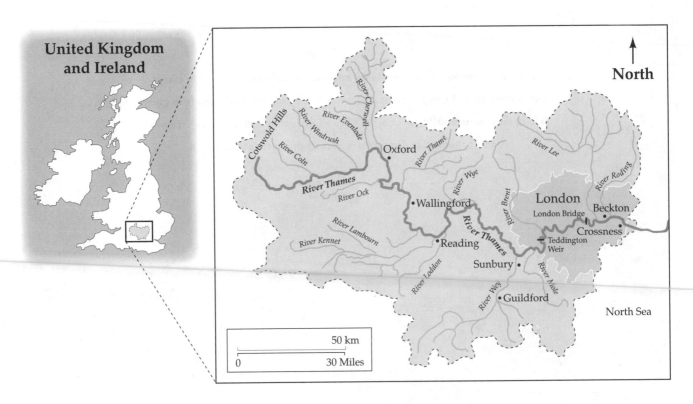

Figure 3.1 The River Thames Watershed.

Thames flows through a number of weirs or low-lying dams, that regulate water flow (Insight 3.1). Near London, the river's flow is controlled by Teddington Weir (Figure 3.2); above the weir, the Thames is a *freshwater* river where waterfowl dwell and people engage in sports fishing and other recreational pursuits. Parts of the river serve as navigational waterways and as sources of drinking water.

Freshwater is non-salty-tasting water that is relatively free of dissolved salts and minerals. Freshwater is required by most land-based plants and is used as drinking water for animals and humans. It is also necessary for crop irrigation, livestock watering, food processing, and many industrial operations. It is important to note that "freshwater" does not necessarily imply safe drinking water. Safe drinking water (called **potable water**) must not only be freshwater, it must also be largely free of disease-causing microorganisms and chemical substances that could cause illness.

The Tidal Thames

The Thames is really two rivers, because below the Teddington Weir it mixes with waters from the North Sea and becomes a tidal *estuary*. The tidal Thames, also called the "tideway," flows from Teddington Weir to the North Sea, a distance of about 100 kilometers (≈ 62 miles).

Figure 3.2 Teddington Weir on the River Thames.
(Photo courtesy of the Environment Agency, London, England.)

An **estuary** is defined by oceanographer Donald W. Pritchard as a " . . . semi-enclosed body of water which has free connection with the open sea and within which seawater is measurably diluted by freshwater from land drainage." It is a relatively shallow basin where watershed runoff mixes with seawater that pulses up and down the estuary with recurring tides.

Estuaries are generally highly productive biological ecosystems. They provide critical habitat for a

INSIGHT 3.1

Weirs and Locks on the Thames

A **weir** is a low-lying dam designed to regulate river water flow using a series of sluice gates (controllable barriers). During periods of low flow, sluice gates are lowered to restrict water flow and maintain adequate water levels. This facilitates navigation, enhances recreational uses, and maintains drinking-water reservoirs. Without weirs, the Thames would become extremely shallow during summer periods of low water flow, making navigation almost impossible. During times of flooding, sluice gates are raised to allow water to flow to the sea as rapidly as possible.

Weirs on the Thames create a series of steps in river elevation that become barriers to ships and recreational vessels. To solve this problem, locks have been built to bypass weirs. Locks have gates at each end that can be raised or lowered to control the water levels within them. A lock is filled by closing its downstream gate and opening the upstream gate. A vessel within the lock is thereby raised to the upstream river level and can navigate freely into it. The reverse procedure allows a vessel to navigate downstream.

Figure 3.3 The River Thames as it flows through metropolitan London.

(Photo courtesy of the Environment Agency, London, England.)

large diversity of fish and wildlife. They also serve as aquatic nurseries where young finfish and shellfish find protection and sources of food. This explains why the tidal Thames became such an important commercial fishery, once yielding bountiful catches of salmon, rainbow trout, smelt, lampern, shad, eel, cod, sole, bass, and flounder.

The tidal Thames is the better known part of the river since it flows through metropolitan London and has long served as a much traveled navigational channel carrying people and goods between London and distant ports in other countries.(Figure 3.3). Distances on the river are often measured in terms of kilometers upriver and downriver from London Bridge (Figure 3.4). Among its multiple uses, this part of the Thames was a waste disposal site where sewage, industrial wastes, and urban runoff were regularly discharged without any treatment. For a long time, the tideway served as London's main source of drinking water.

The Hydrologic Cycle

The earth is called the "water planet" because water is so prevalent and life depends on it. Water moves through the earth's ecosystems in a process called the **hydrologic cycle,** which includes evaporation, atmospheric transport, precipitation, runoff, infiltration, and re-evaporation (Figure 3.5).

Evaporation occurs when solar energy increases molecular motion in aquatic environments, causing water molecules to escape into the air. Water in plants is transported through tree trunks, branches, stems, leaves, and needles, and released to the air through

INSIGHT 3.2

Aquifers

Water infiltration through watershed soils maintains groundwater resources in porous sand, gravel, or rock formations. Some of these serve as **aquifers,** underground sources of freshwater. Because aquifers can be tapped for drinking water, agricultural irrigation, and industrial needs, they are among our most important water resources. For example, most of the agricultural productivity of America's mid-continental prairie lands, which stretch from southern South Dakota to northwestern Texas, depends on water pumped from the Ogallala aquifer, the largest groundwater resource in North America. In the United States overall, about 50 percent of all drinking water is drawn from wells

drilled in shallow aquifers less than 30 meters (≈ 100 feet) deep.

Generally speaking, there are two kinds of aquifers: shallow, unconfined (water table) aquifers and relatively deep, confined aquifers. Water in unconfined aquifers is repeatedly recharged (renewed) by precipitation falling on the ground above. This makes those aquifers vulnerable to contamination by hazardous materials such as gasoline, diesel oil, solvents, and pesticides. Confined aquifers are better protected because they lie below layers of clay or rock and are usually recharged by precipitation falling at distant sites.

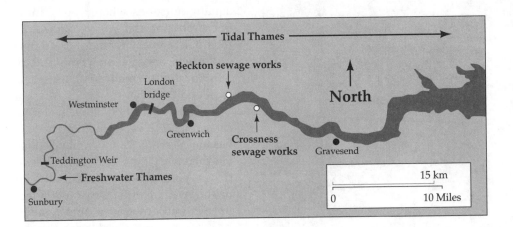

Figure 3.4 The Tidal Thames as it flows eastward from Teddington Weir, through London, and to the North Sea.

plant pores called stomata. This process of **transpiration,** combined with water evaporation from plant surfaces, is called **evapo-transpiration.**

Evaporation purifies water by separating it from contaminants, including bacteria, viruses, soils, salts, minerals, and pollutants. Water temperature and relative humidity determine rates of water evaporation, and wind influences the direction and extent of water vapor transport. As water vapor rises, clouds may form at varying elevations in the earth's lower atmosphere (the troposphere). Cloud formation requires the presence of very small atmospheric particles called **condensation nuclei.** These particles originate from volcanoes, sea-salt spray, forest fires, and wind-driven dust, attracting moisture and forming small cloud

droplets. Precipitation as rain, sleet, snow, or hail is triggered when cloud droplets increase in size.

In England, as elsewhere, precipitation occurs across many landscapes but is distributed unevenly. For example, northwestern Scotland experiences more than 2,500 millimeters (≈ 100 inches) of precipitation yearly. London typically receives much less—about 630 millimeters (≈ 25 inches). Rain and melted snow run off to receiving waters, infiltrate soils, are taken up in trees and vegetation, and evaporate. Infiltration means that precipitation moves vertically through soils, where it is absorbed, taken up by vegetation, or added to groundwater aquifers (Insight 3.2).

The approximate distribution of the earth's water is shown in Table 3.1. Ocean waters and estuaries cover

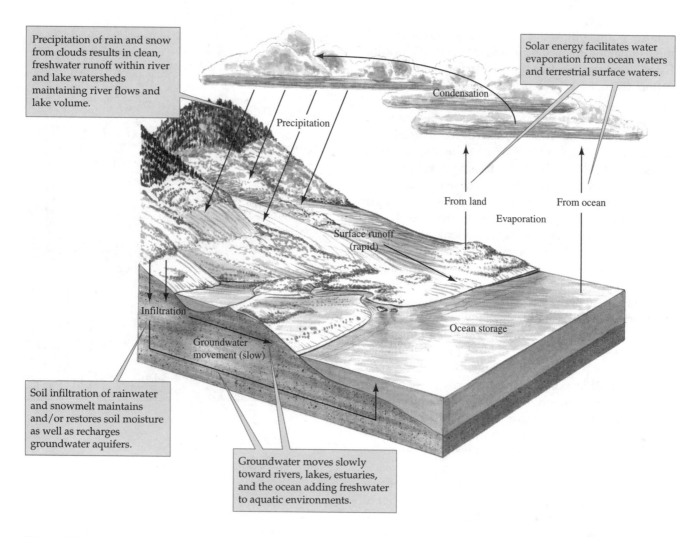

Figure 3.5 The hydrologic cycle.

TABLE 3.1 Distribution of Water on Earth	
Aquatic System	Percent of Total Water
Seawater and estuaries	97
Glaciers, ice fields, and ice caps	2
Freshwaters	1*

SOURCE: Data from *The Water Encyclopedia*, Second Edition, 1990.

*The 1 percent of freshwater is distributed as follows: groundwater aquifers, 97 percent; surface waters, 3 percent.

2 percent. Freshwaters add up to only 1 percent of the total. Approximately 97 percent of the planet's freshwater is stored in groundwater aquifers. This means that only 0.03 percent of the earth's water supply is available as surface freshwater.

CASE STUDY: ENGLAND'S RIVER THAMES

How the River Became Polluted

Industrial expansion and population increase led to the pollution of the Thames estuary. Historians connect the rise of England's Industrial Revolution with the development of the steam engine by Thomas Savery in 1698, Thomas Newcomen in 1705, and James Watt in 1770. One of the earliest uses of the steam engine, which generated power by burning coal to boil water,

about 71 percent of the earth's surface and account for 97 percent of the planet's total water supply. Glaciers and other frozen water resources account for another

was to pump out water that had seeped into coal mines, making them unsafe and difficult to work in. Later, steam engines transformed factory work so that cotton, linen, and woolen fabrics could be woven on steam-engine-powered looms instead of by hand. Britain's textile industry arose from these beginnings.

The revolution could be seen in commercial fishing, shipping, shipbuilding, glass factories, and iron works. Coal gas, produced by heating coal in the absence of air (a process called pyrolysis), illuminated homes, streets, and factories. Iron-making, which used coal to chemically reduce iron ore to iron metal, was improved by using steam engines to force air into furnaces to create high temperatures. Steam engines were used to make furniture, soap, explosives, ammunition, and clothing. In 1856, Henry Bessemer developed a blast furnace to make steel, an innovation of major significance to industry and technology worldwide.

The Industrial Revolution flourished in London largely because of port and shipping facilities on the River Thames. London's population doubled between 1700 and 1800, making it the largest city in the world, with about one million people. It then doubled again by 1850 (Figure 3.6), transforming London from a pastoral, agrarian village into an industrial city. As multitudes crowded into congested, unsanitary slums to take advantage of job opportunities, animal wastes and human excrement became a common sight in streets and alleys. Disposing of these wastes was left to nature. The next rainfall washed most of them into nearby streams that flowed into the Thames. London was becoming a city of crowded, dingy tenements, narrow dirt streets, filth, wretchedness, and squalor. Pollution visibly degraded the Thames, eradicating fish populations and triggering epidemic diseases like cholera. Most people had never experienced conditions this appalling before.

Cesspools, Street Sewers, and River Pollution

In the early 1800s, it was common practice to dispose of human wastes in nearby cesspools, brick-lined cisterns or tanks built inside houses or outside, below street level. It is estimated that there were as many as 200,000 cesspools in London at that time. Sludges were periodically removed from cesspools and used as agricultural fertilizer, but sometimes they overflowed into city streets, roadside ditches, and nearby streams and rivers that flowed into the Thames. As a result, the Thames's tributaries became London's street sewers. Every time it rained, they received refuse, animal manure, and human sewage runoff from streets and alleys (pollution from **non-point sources**). Some of the tributaries were later covered over and became known as

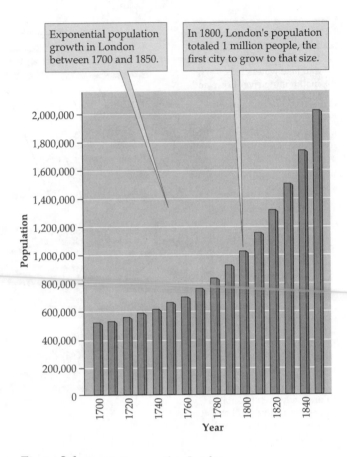

Figure 3.6   Population growth in London.

the "lost rivers of London." Pollution in the river, particularly as it passed through London, gave off a terrible stench that was often intolerable, especially during the summer months.

In 1810, London's citizenry was introduced to an invention called the water closet—what Americans call the toilet. One of its designers was Thomas Crapper. (Yes, Thomas Crapper!) Crapper's invention flushed sewage directly to nearby cesspools and made life more comfortable for many Londoners, who no longer had to venture forth to cold, damp, outhouses (Figure 3.7). But as more water closets were installed in London dwellings, the city cesspools frequently overflowed. To resolve that dilemma, the British Parliament passed a law in 1847 requiring that water closets bypass cesspools and discharge directly into street sewers. This meant that raw sewage was being more efficiently channeled to the Thames.

Some of the pollution problems were legendary. For example, London's Chelsea Fever Hospital discharged biomedical wastes directly into the Thames very close to where the Grand Junction Water Works Company withdrew drinking water and supplied it to London customers. (The hospital was a **point source** of

Figure 3.7 Indoor sanitary toilet facilities sold by Thomas Crapper & Co. in the early 1800s.

pollution.) Such offensive practices prompted the British House of Commons to state publicly:

> . . . the water taken from the river Thames at Chelsea, for the use of the inhabitants of the western portion of the Metropolis, being discharged with the contents of the great common sewers, the drainings from dunghills, and laystalls, the refuse of hospitals, slaughter houses, colour, lead and soap works, drug mills and manufactories, and with all sorts of decomposed animal and vegetable substances, rendering the said water offensive and destructive to health, ought no longer to be taken up by any of the water companies.

During the summer of 1858, the Thames smelled so bad that on several occasions members of the British Parliament, which stood directly alongside the river, decided to go home. It was called "The Year of the Great Stink."

Effects of Pollution

Decline of Fish and Wildlife

Before England's Industrial Revolution had seriously impacted the Thames estuary, the river was known far and wide for the diversity and abundance of its fish. Import and export enterprises had grown to depend on this productivity. But pollution led to a rapid decline in the number and kinds of fish in these waters. Few data exist documenting this decrease, but one family living on the Thames kept a careful record of the dwindling numbers of salmon caught between 1798 and 1821 (Figure 3.8). It is reported that the last Thames salmon of this era was caught just west of London Bridge in 1833. After that, salmon disappeared from the estuary for more than a hundred years.

Salmon were among the first fish species to retreat from the Thames as a result of pollution, but before long, all of the river's fish (except eels) had vanished. They were unable to survive in these waters because pollution had led to very low levels of dissolved oxygen, which they needed for survival. At the same time, there were noticeable declines in the numbers of waterfowl and sea birds normally found near the estuary. Their food supply, an abundance of fish in the Thames, was gradually being lost.

The diversity of fish and wildlife populations provides important insights into the environmental status of an ecosystem. In the case of the Thames estuary, marked declines in the numbers and diversity of the fish, waterfowl, and sea bird populations paralleled deteriorating water quality and signaled growing ecological problems.

London's Cholera Epidemic

In 1849, London's newspaper, *The Times*, printed the following letter, signed by more than fifty of the city's poor:

> We live in muck and filthe. We aint got no priviz, no dust bins, no drains, no water splies, and no drain or suer in the hole place. . . . We all of us suffer, and numbers are ill, and if the Colera comes Lord help us.

Many of the signers of this letter had survived a cholera epidemic that afflicted the city just three years earlier. In 1853 and 1854, cholera broke out again in London, and tens of thousands of citizens died.

Cholera, we know today, is a water-borne disease transmitted mainly through drinking water

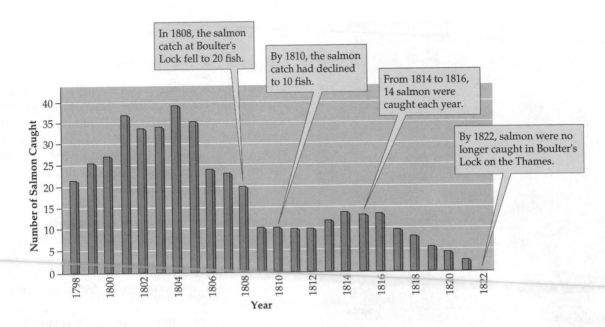

Figure 3.8 Number of salmon caught by the Lovegrove family in Boulter's Lock at Taplow on the Thames—a five-year running average spanning 1798 to 1821. By the 1830s, salmon were rare or totally absent in the Thames.
(*SOURCE: Data from M. J. Andrews, Thames Water Authority, Middlesex, England.*)

contaminated by a specific bacterium, *Vibrio cholerae*. It occurs when drinking water sources are polluted by human wastes. Its symptoms include violent diarrhea, loss of fluids, vomiting, muscle cramps, and eventual collapse. It is recorded that cholera runs so rapid a course that a man in good health at daybreak may be dead and buried by nightfall. Not uncommonly, its victims suffer serious dehydration and lose up to 21 quarts ($\approx$ 20 liters) of fluids.

Cholera was first reported in India in the early 1800s and then spread to the Near East, Russia, Europe, and North America. Millions of people died in one epidemic after another. In Paris, deaths occurred so quickly that bodies had to be buried in bags; there weren't enough coffins. In the 1850s, cholera killed a million Russians, including the great composer Peter Tschaikovsky. Some historians regard cholera as the greatest scourge of the past two centuries. The disease is rare today, but in the early nineteenth century, no disease was more feared.

Throughout most of the 1800s, it was not known that diseases such as cholera, typhoid, hepatitis, and dysentery were spread through drinking water. Nor was there any knowledge that microorganisms were responsible for certain illnesses. Many thought illnesses like cholera were contracted by bad smells rising from outhouses and dungheaps. This is known as the *miasmic theory* of disease. ("Miasma" is the German word for pollution and refers to foul odors from decomposing organic matter.) The true cause of cholera was not made clear until 1883 when the

German bacteriologist Robert Koch showed that bacteria of a particular species were responsible. However, well before that, the cause was partially identified, and the means whereby it is spread made clear due to the work of a London physician named John Snow.

John Snow John Snow, an anesthetist by training, was quite well known in London for having administered chloroform to Queen Victoria during childbirth, one of the first times medical anesthesia was performed. Dr. Snow was also intrigued by cholera and studied its cause for years.

In 1853, a particularly bad epidemic hit London districts south of the Thames. At that time, several water companies made regular deliveries of water drawn mostly from the Thames to various wells around the city. These wells were actually underground holding tanks made of bricks laid in lime mortar. They were located along various streets in London, and people would pump water from them for drinking and cooking. During seven weeks of the 1853 epidemic, Dr. Snow collected data on cholera fatalities. He found that death rates were 8.5 times higher in houses where people had used water supplied by the Southwark and Vauxhall Company than where they had used water supplied by the Lambeth Company. Table 3.2 summarizes Dr. Snow's data.

John Snow was convinced that some kind of "cholera poison" was being transmitted in drinking water. He knew that in earlier years both water

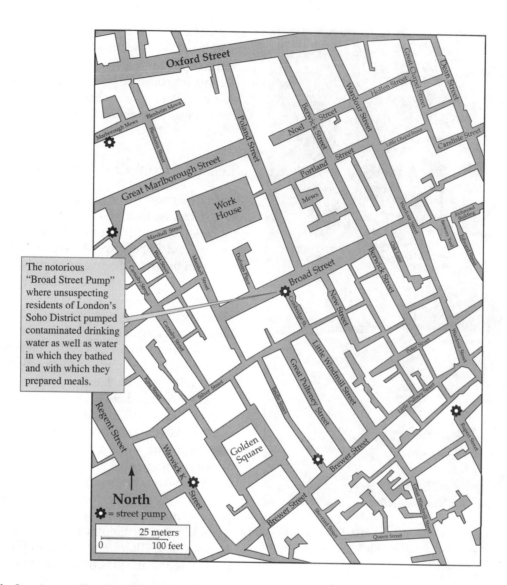

Figure 3.9 John Snow's map of London's Soho District during the cholera epidemic of 1854. Snow used this map to track how many cholera deaths had occurred in each house and from which street pumps water was obtained.

TABLE 3.2   Deaths from Cholera by Source of Water in London's South Districts during the 1853 Epidemic

Source of Water Supply	Number of Homes	Deaths from Cholera	Deaths per 10,000 Homes
Southwark and Vauxhall Company	40,046	1,263	315
Lambeth Company	26,107	98	37

SOURCE: Data from *Snow on Cholera*, The Commonwealth Fund, 1936.

companies had obtained water directly from grossly polluted sites on the Thames, but in 1852 the Lambeth Company had changed its water source from the tidal estuary to freshwaters above Teddington Weir.

In 1854, a serious cholera outbreak struck London's Soho district. There were 500 fatalities in just 10 days. Dr. Snow, who lived near Soho, investigated the outbreak. He drew a map, noting the locations of street

Figure 3.10 A replica of the Broad Street Pump stands on Broadwick Street in London's Soho District, the street once named Broad Street.
(Photo by Michael Cline.)

Many years later, experts concluded that contamination of the Broad Street well was not due to polluted water supplied by a water company, but to raw sewage seeping into the well from a nearby cesspool. However, that discovery didn't change Snow's hypothesis that cholera was transmitted through contaminated water.

Lessons from London's Experience

Today we know that many diseases are caused by specific microorganisms (germs). But it was not until John Snow's studies in London, Louis Pasteur's work in France, and Robert Koch's research in Germany that the germ theory of disease was firmly established. Pasteur, who is considered the founder of microbiology, showed that diseases were due to infectious agents, microorganisms such as bacteria. Koch demonstrated that tuberculosis, anthrax, cholera, and several other diseases were caused by specific bacterial species that could be identified under a microscope.

Following the landmark discoveries of Snow, Pasteur, Koch, and others, it was realized that polluted waters, like those in the Thames estuary, were not potable. After London's cholera epidemics, most of the city's drinking water was obtained from freshwaters above Teddington Weir. Today, 60 percent of London's drinking water comes from the nontidal Thames above Teddington Weir, 25 percent from the River Lee, and 15 percent from aquifers beneath the city. The water is required by law of the British Parliament to be filtered through **sand filters**, a way of purifying water discovered by John Gibb in Scotland (Insight 3.3).

Sewage Interceptors

By the time John Snow's cholera investigations were published, officials in London recognized that something had to be done about pollution in the Thames. In 1856, London's Metropolitan Board of Works directed its chief engineer, Joseph Bazalgette, to develop plans to prevent sewage wastes from entering the estuary near the city. Bazalgette's plan was to intercept wastewaters and divert them by pipeline to points downriver from London. This, he believed, would solve the river's problems.

Bazalgette began constructing sewer interceptors, three running parallel to the Thames on its north side, collecting wastes from north-side sewers, and two running parallel to the river on its south side, collecting wastes from south-side sewers. When completed in 1874, the northern interceptors carried wastes to the city of Beckton, about 18 kilometers (≈ 11 miles) east of London Bridge. The southern interceptors extended to

pumps where people obtained water, what households were afflicted by cholera, and how many fatalities occurred in each house (Figure 3.9). Almost all who died in Soho had pumped their water from the Broad Street Pump.

At the height of the Soho epidemic, Dr. Snow explained to local authorities that there was a cholera poison in Thames water being supplied to the Broad Street well (Figure 3.10). He convinced them to shut down the well by removing its handle. The epidemic ended in a few days. John Snow published his studies in 1855. His data and observations convinced the scientific community that cholera was a water-borne disease linked to contaminated water. This was the first time water had been shown to transmit disease. John Snow's investigations are regarded as the beginning of **epidemiology,** the study of the origin and transmission of human disease.

INSIGHT 3.3

Sand Filtration

In the late 1700s and early 1800s, drinking water in Paisley, Scotland, was obtained from the nearby River Cart. But the river was usually polluted with unsightly mud, scum, and floating solids. In 1804, John Gibb, a civil engineer, designed and built a filtering system to remove visible pollution from river water. His filter consisted of concentric rings of stacked rock, gravel, and sand through which river water was allowed to seep (Figure 3.A). Using this system, Gibb filtered, bottled, and transported water around Paisley by horse-drawn wagon, selling it for one-cent a gallon. This was the first water treatment filter to supply an entire city with drinking water.

In 1829, James Simpson, Inspector General of London's Chelsea Water Works located on the north bank of the Thames, designed and built a vertical sand filter to purify drinking water. His design became the standard model for almost all drinking water filtration plants worldwide (Figure 3.B).

It is easy to see how sand can filter visible contaminants from water. But can it remove disease-causing bacteria and other microscopic pathogens? That question was not even addressed until Louis Pasteur and Robert Koch showed that microorganisms can cause disease. In 1880, Koch developed a laboratory procedure for culturing bacterial colonies on a nutrient gel. This technique enabled scientists to identify and count numbers of bacteria present in water. Using this method, it was found that sand filters can remove as much as 98 percent of bacteria present in polluted waters!

Sand filters remove bacteria because individual sand granules become coated with a sticky slime of algae and other microorganisms when water trickles through them. The slime traps bacteria. Sand filtration, coupled with chlorination, has virtually eliminated epidemics like cholera. Today, cities and towns across the world have water filtration plants that purify drinking water using sand filters based on Gibb's and Simpson's designs.

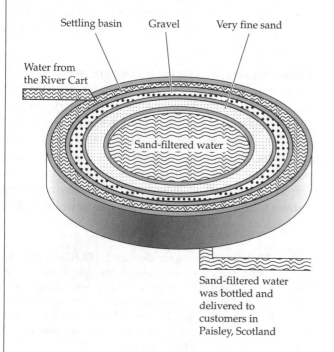

Settling basin   Gravel   Very fine sand

Water from the River Cart

Sand-filtered water

Sand-filtered water was bottled and delivered to customers in Paisley, Scotland

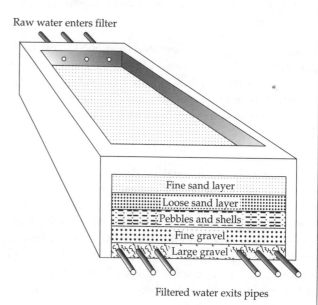

Raw water enters filter

Fine sand layer
Loose sand layer
Pebbles and shells
Fine gravel
Large gravel

Filtered water exits pipes

Figure 3.A John Gibb's sand filter, the first known filter to supply an entire city with drinking water. Gibb's filter was built in Paisley, Scotland, in 1804. This drawing is based on a description of the Paisley filter in *The Quest for Pure Water* by M. N. Baker, published by The American Water Works Association in New York in 1949.

Figure 3.B Cross-sectional view of London's first sand filter designed to purify drinking water and built by James Simpson at the Chelsea Water Works in 1829. This drawing is based on a description of the Simpson filter in *The Quest for Pure Water* by M. N. Baker, published by The American Water Works Association in New York in 1949.

Crossness, about 22 kilometers (≈ 14 miles) east of London Bridge. Sewage wastes were collected in large reservoirs at both locations and then periodically discharged to the Thames during recurring ebb (receding) tides to be carried out to sea (see Figure 3.4).

The Thames estuary near London became visibly cleaner and better smelling. A Royal Commission appointed in 1882 confirmed this, but reported that commercial fishing east of London, where sewage was being discharged to the river, had been virtually wiped out. The problem was that wastes discharged into the river at ebb tide were carried back upriver at flood tide. In addition, river conditions deteriorated badly following major rainstorms because the street sewers had limited capacity. Heavy rains simply bypassed street sewers, washing pollution directly into the Thames. While problems like these were obvious, it became clear that scientific studies were needed to better understand all the problems affecting the Thames.

Scientific Studies of the Thames Estuary

William Dibdin, a chemist working for the Royal Commission, began his research on the Thames estuary in 1882. His work focused mainly on *dissolved oxygen* and *biological oxygen demand*, two critically important water quality parameters.

Dissolved Oxygen   Dissolved oxygen (DO) is oxygen gas dissolved in water. Like other gases, oxygen dissolves in freshwaters and marine waters through hydrogen bonding. Hydrogen bonding links oxygen to water molecules, thereby holding it in solution:

$$O_2 \cdots\cdots H_2O$$

The maximum level of dissolved oxygen, referred to as *DO saturation*, depends mainly on water temperature and salinity. The higher the temperature, the less oxygen dissolves in water because heat increases molecular motion, breaking the hydrogen bonds that link oxygen to water. Greater salinity also lessens DO levels because salts such as sodium chloride compete for hydrogen-bonding sites in water.

DO saturation is the maximum concentration of oxygen gas that typically can dissolve in an aquatic system. DO saturation depends mainly on water temperature and salinity.

DO saturation commonly lies between 8 milligrams and 14 milligrams DO/L in freshwaters and between 7 mg and 11 mg DO/L in marine waters. DO levels this high are found only in well-oxygenated, clean, unpolluted aquatic environments. DO is important because most aquatic plants and animals require oxygen to carry out aerobic respiration, a biological activity in which oxygen is consumed and carbon dioxide is released.

Relatively high DO levels are beneficial to aquatic plants, plankton, finfish, shellfish, mammals, and other organisms that depend on aerobic respiration. Low DO levels are less beneficial, and very low levels may fail to support fish and other aquatic life. In polluted waters, DO levels are usually much lower than DO saturation levels. While some finfish and shellfish can tolerate DO as low as 3 or 4 mg/L, highly prized commercial and sport fish like salmon and rainbow trout require 6 to 8 mg DO/L. Most fish, however, require at least 5 mg DO/L to survive and reproduce.

William Dibdin found that DO levels in the Thames sometimes fell below 2 mg/L. During summer months, the problem was even worse: DO fell to zero—that is, there was no dissolved oxygen at all. Natural waters with no dissolved oxygen are described as **anoxic,** meaning that they are unable to support aquatic plant and animal life normally found in unpolluted environments. These observations explain why most species of fish were no longer living in the tidal Thames. But they don't account for why DO levels fell to such low levels.

Biological Oxygen Demand   Organic matter, including human and animal wastes, is **biodegradable**. That is, it is biologically decomposed or broken down into simpler substances by decomposers such as bacteria and fungi. Adding organic matter to aquatic environments stimulates the growth of decomposer populations. Since decomposers require oxygen for respiration, growth in their numbers creates increasing demands for dissolved oxygen, a process called *biological oxygen demand*.

Biological oxygen demand (BOD) is the demand for dissolved oxygen brought about by increasing numbers of bacteria and other decomposer organisms as they assimilate organic matter in aquatic environments. Technically, BOD is defined as the number of milligrams of oxygen needed to oxidize the organic matter present in one liter of a water sample.

Aquatic systems that possess high levels of dissolved oxygen support large numbers of decomposers and are capable of biodegrading (assimilating) most organic wastes discharged to them. This ability is referred to as natural **assimilative capacity.** Unpolluted aquatic environments are therefore able to purify themselves—up to a point—without observable damage. DO levels are replenished naturally, but often at a slow rate. The physical process of water rushing over

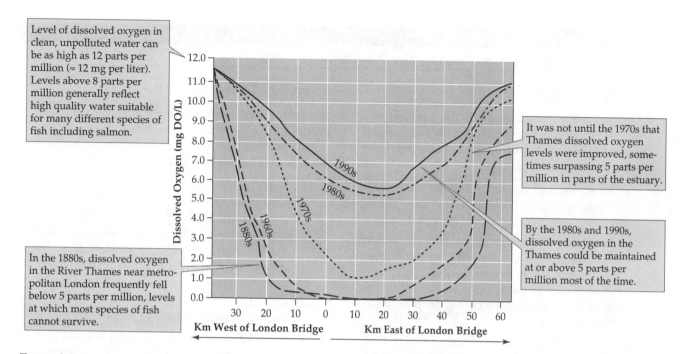

Level of dissolved oxygen in clean, unpolluted water can be as high as 12 parts per million (≈ 12 mg per liter). Levels above 8 parts per million generally reflect high quality water suitable for many different species of fish including salmon.

It was not until the 1970s that Thames dissolved oxygen levels were improved, sometimes surpassing 5 parts per million in parts of the estuary.

By the 1980s and 1990s, dissolved oxygen in the Thames could be maintained at or above 5 parts per million most of the time.

In the 1880s, dissolved oxygen in the River Thames near metropolitan London frequently fell below 5 parts per million, levels at which most species of fish cannot survive.

Figure 3.11   A series of oxygen sag curves in the Thames estuary near London Bridge. Data from the 1880s, 1960s, 1970s, 1980s, and 1990s are shown.

rocks (riffle areas) or being wind-driven by wave action mixes water with air. This causes atmospheric oxygen to dissolve in water. DO is also restored in aquatic environments through photosynthesis as aquatic plants such as algae release oxygen to natural waters. However, if the demand for dissolved oxygen is greater than the rate at which DO can be restored, DO levels will fall.

The impact of BOD in a river is often seen as a sag in DO levels downriver from the point where wastes are added, followed by recovery. Figure 3.11 shows a series of typical oxygen sag curves in the tidal Thames based on data from the 1880s to the 1990s. It can be seen that the problem of extremely low DO levels William Dibdin observed in the Thames was not overcome until the 1980s.

Sewage Treatment Plants

William Dibdin's research demonstrated the importance of dissolved oxygen in maintaining healthy fish populations. His work convinced the Royal Commission that untreated sewage, low dissolved oxygen levels, and sludge banks building up downriver east of London were causing most of the estuary's problems. Dibdin's studies showed that if sewage wastes were treated with the chemical lime (calcium hydroxide), a solid sludge separated out. Removing this sludge produced wastes with a lower BOD, wastes that created less demand for dissolved oxygen. Dibdin's work

marks the beginning of **primary** and **secondary wastewater treatment** (Insight 3.4).

Primary wastewater treatment was introduced at Beckton in 1889 and at Crossness in 1891. Sludges produced at these plants were transported by barges and dumped into the North Sea. Water quality improved greatly in the Thames estuary, and many species of migratory fish and waterfowl not seen for decades returned. However, these encouraging results were short-lived. By 1920, London had grown to 8 million people. High levels of pollution reappeared in the Thames simply because the city's sewers and wastewater treatment system failed to cope with the increased volume of sewage. As a result, 50 kilometers (≈ 31 miles) of the estuary again became putrid and devoid of fish. The problem worsened during World War II when the bombing of London inflicted major damage to the city's wastewater facilities.

The Thames Today

Recovery of Fish and Wildlife

After World War II, London's sewers were rebuilt and its sewage treatment plants modernized. Additional treatment plants were built, and in time higher DO levels prevailed in the Thames estuary. The improvement can be seen in oxygen sag curves plotted after the 1960s (see Figure 3.11). After the Crossness treatment plant was rebuilt and facilities at Beckton modernized,

INSIGHT 3.4

Primary and Secondary Wastewater Treatment

Primary Treatment

Primary wastewater treatment is the simplest and least costly approach to dealing with sewage and other waste matter (Figure 3.C). The principal objectives of primary treatment are:

- To screen out debris and grind solids that might interfere with plant operations. The resulting grit is collected and removed.
- To allow suspended solids, which are usually heavier than water, to settle to the bottom and form a primary sludge, which is later incinerated or transported to a landfill. The goal is typically 85 percent reduction in settleable solids.
- To disinfect primary effluents prior to their discharge to receiving waters. Chlorination is the usual disinfection agent.

    Reducing BOD levels is not a principal objective of primary treatment, but some BOD reduction occurs when suspended solids settle out.

Secondary Treatment

Secondary wastewater treatment is often required to produce effluents that can be safely discharged to surface waters. The principal objectives of secondary wastewater treatment are:

- To reduce BOD levels. The goal is typically 85 percent reduction in BOD. One technique, the activated sludge process, uses aeration to supply oxygen to a microorganism-rich sludge that is added to wastewaters. Increased numbers of microorganisms metabolize and consume organic matter, thereby lowering BOD levels.
- To further reduce levels of suspended solids.
- To disinfect secondary effluents prior to their discharge to receiving waters. Chlorination is the usual disinfection agent.

Secondary wastewater treatment was mandated in the United States and funded in part through the Federal Water Pollution Control Act of 1972 (the Clean Water Act) and its amendments.

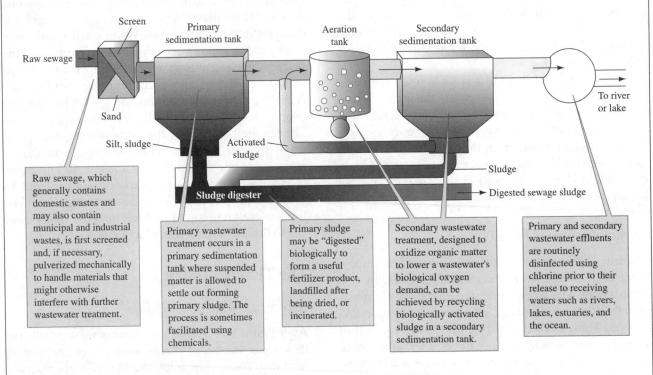

Figure 3.C The principal components of primary and secondary wastewater treatment.

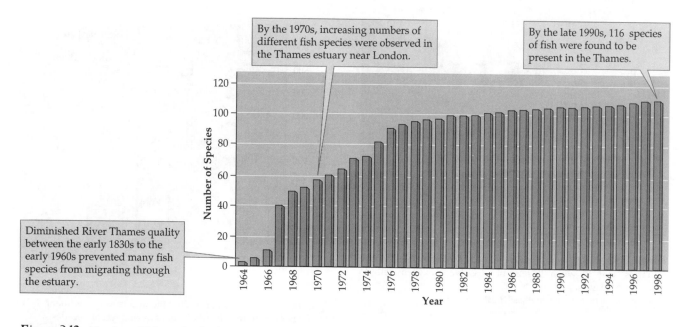

By the 1970s, increasing numbers of different fish species were observed in the Thames estuary near London.

By the late 1990s, 116 species of fish were found to be present in the Thames.

Diminished River Thames quality between the early 1830s to the early 1960s prevented many fish species from migrating through the estuary.

Figure 3.12 Number of fish species that have returned to the tidal Thames since 1964.

the estuary never suffered serious anoxic conditions again. Greater numbers of fish representing many different species slowly returned to the Thames (Figure 3.12). In 1974, after being absent for over 140 years, salmon returned to the river. A year later, rainbow trout were discovered. Salmon and trout are good **biomonitors,** indicators of water quality. They are not very tolerant of low levels of dissolved oxygen since they generally require DO levels of 6 mg/liter or higher, so their presence in natural waters indicates high water quality. By 1998, 116 species of fish had returned to the Thames estuary.

The recovery of the Thames is also reflected in the increasing numbers of nesting birds and waterfowl now commonly seen there. These include mute swans, great crested grebes, cormorants, kingfishers, oystercatchers, blackheaded gulls, and herons. Evidence of increasing biodiversity in the Thames region reflects continuing restoration of the ecosystem.

The Thames Bubbler

Despite the many advancements in London's sewage treatment facilities, problems still occur in the Thames estuary. Heavy rainstorms overload the hydraulic capacity of city sewers, causing them to periodically overflow and wash untreated wastes into the river. The problem is particularly acute during summer months because bacterial populations cause biological oxygen demand to grow more rapidly in warm waters. Therefore, storms put the tidal Thames fishery in jeopardy, increasing the likelihood of massive fish kills due to low DO levels.

One way to solve the problem would be to totally rebuild London's sewer system and expand its capacity to handle storms. But because the cost of doing this has been estimated at £1 billion (≈ U.S. $1.5 billion), the following alternative river management plans were adopted in 1980:

- Eight water-quality-monitoring stations were installed in the estuary to measure DO and other parameters. The data are transmitted to central agencies near London where decisions are made to deal with oxygen sag problems.

- An experimental "bubble barge" was built with on-board equipment to produce oxygen-enriched air. Whenever a serious oxygen sag forms on the Thames, the barge is dispatched to that location, and purified oxygen is bubbled into the water.

Although its capacity to pump oxygen into the Thames was limited, the experimental bubble barge was able to deal with oxygen sag events. But in 1986, while the barge was in dry dock, 1.5 million fish died following a major storm. This led to the construction of a larger, more capable vessel—the *Thames Bubbler*—which became operational in 1989. Its cost was £3.5 million (≈ U.S. $5.2 million). On board the *Bubbler*, air is pumped into three pressurized tanks containing zeolite, a mineral that traps nitrogen from compressed air, thereby providing a way to make oxygen-enriched air. Using this technique, the *Bubbler* can produce 30 metric tons (≈ U.S. 33 tons) of oxygen-enriched air daily, containing about 93 percent oxygen. When data from monitoring stations signal that an oxygen sag is

Figure 3.13 The Thames Bubbler is used to inject oxygen into the estuary's waters when levels of dissolved oxygen fall to unacceptable levels.

(Photo courtesy of the Environment Agency, London, England.)

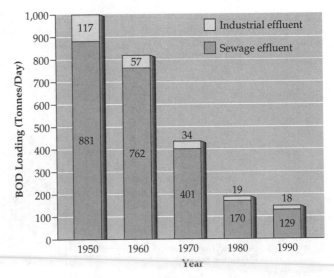

Figure 3.14 Declining industrial and sewage effluent BOD loading added to the Thames Estuary during recent decades.

forming in the river, the *Bubbler* is dispatched to where DO is at a minimum. There, river water is pumped into the vessel, mixed with oxygen-enriched air, and discharged back to the river through 160 underwater nozzles along the vessel's hull (Figure 3.13).

Since 1989, no major fish kills have occurred in the Thames estuary. The *Bubbler's* success has led to plans to build another bubbler in the near future. Nevertheless, some government officials and private citizens regard the *Bubbler* as a "river ambulance" sent periodically on missions of mercy to save fish before seriously low oxygen sags develop in the river. They argue that this is a "band-aid" approach and that the river's real problems are not being dealt with. The two principal problems are:

1. The limited capacity of London's sewers causes major, recurring storm-water runoff problems. This, in turn, leads to sharp declines in DO in parts of the estuary. To solve this problem, the city's sewer system would have to be totally rebuilt.

2. Joseph Bazalgette's initial plan to discharge London's sewage wastes 22 kilometers (≈ 14 miles) or so downriver from the city's center is flawed. He assumed that London's wastes would be transported out of the estuary with each low tide. Unfortunately, this is not happening.

Other officials and citizens point out that Great Britain has invested heavily in several new and improved wastewater treatment plants, especially in the Thames region. These facilities, which have greatly reduced total BOD loading in the estuary, have led to improved water quality (Figure 3.14).

The Future of the River

Scientists, engineers, ecologists, and private citizens regard the improvements in the Thames estuary as a success story. Many species of fish and wildlife that inhabited these waters and this region in earlier times have returned. Epidemics no longer plague London, and the city's drinking water, most of which is drawn from the freshwater Thames, meets or exceeds the highest standards set by industrialized nations. Levels of estuarine dissolved oxygen are high enough most of the time to sustain the river's fish populations—and when DO levels fall to unacceptable levels, the *Thames Bubbler* is used to deal with the problem.

Water pollution began in the Thames estuary hundreds of years ago, and insights into these problems were slow in coming. But today the estuary serves as receiving waters for municipal and industrial wastes, functions as an industrial shipping lane, and supports a viable commercial and sport fishery. Engineers and scientists have learned how to protect much of the river's ecology. It was on the Thames that the world's first wastewater collection and treatment facilities were designed and built. And it was in the Thames region that safe drinking water was first provided on a large scale to meet the needs of a major city. In 1989, British Prime Minister Margaret Thatcher called the Thames "the cleanest metropolitan estuary in the world."

The Thames estuary is a restored ecosystem. The work accomplished reflects what can be achieved when environmental problems are studied, understood, and addressed.

REGIONAL PERSPECTIVES

Canada's Mackenzie River: Contamination from Tar Sands

The Mackenzie River flows from Great Slave Lake in Canada's Northwest Territories to the Beaufort Sea, a distance of approximately 1,700 kilometers ($\approx$ 1,060 miles). It is among the top ten rivers in the world in terms of discharge volume to the sea. Alexander Mackenzie discovered the river in 1789 as he pursued trade explorations on behalf of the Montreal Fur Company. He called it the "River of Disappointment" when he found that it didn't lead to the Pacific Ocean, as he had hoped. This would have provided a northwest passage to the Indies. The river was later named after Mackenzie himself.

As he explored the river, Mackenzie came across waters that were visibly contaminated with dark, floating, oily slicks. He also found tarlike, aromatic-smelling material oozing from cliffs along the river's banks. We know today that Mackenzie had discovered Canada's immense deposits of tar sands, one of the richest *hydrocarbon* reserves in the world.

Hydrocarbons are organic chemicals composed of the elements hydrogen and carbon. They are derived principally from fossilized plant matter that has been transformed over geologic time into methane (natural gas), crude oils like petroleum, and waxy solids like paraffin. Canadian tar-sand hydrocarbons were formed from marine algae.

Traces of tar-sand oils contaminate parts of the Mackenzie River even today. In some places, the water is not potable, failing to meet drinking-water quality standards set by the World Health Organization. Should this natural seepage of hydrocarbons into the Mackenzie River be considered "pollution"? It depends on your definition of pollution. While Industrial-Revolution-driven contamination of the River Thames is clearly river pollution, it is probably not useful or reasonable to describe the totally natural Mackenzie River environment encountered by Alexander Mackenzie as polluted.

Milwaukee, Wisconsin: *Cryptosporidium* in Drinking Water

Potable water supplies in most industrial societies today are regarded as safe because they are filtered, disinfected, and tested. Sand filtration is commonly employed, together with chlorination, ozonation, or some other chemical agent, to kill bacteria. Testing is usually limited to analyses for water turbidity, *coliform* bacteria, and nitrates.

The 1993 outbreak of cryptosporidiosis in Milwaukee, Wisconsin, caused close to 400,000 cases of severe gastrointestinal illness and 100 deaths. The epidemic was unexpected because the city's drinking water is sand-filtered and chlorinated; water is drawn from Lake Michigan and processed at two municipal waterworks, one to the south of the city and one to the north. So why did this outbreak occur?

Very simple analyses for water **turbidity** turned out to be the clue. According to EPA water quality standards, turbidity, a measure of suspended solids in water, should not exceed 1.0 NTU (nephelometric turbidity units). Higher levels than this can cause two problems:

- Increased difficulty of effective water filtration
- Interference with effective chlorine disinfection

Milwaukee's two waterworks, like all modern water filtration plants, were usually able to maintain turbidity levels below 1.0 NTU—typically, 0.45 NTU or lower. But in March 1993, the city's southern treatment plant supplied drinking water that had turbidity levels as high as 1.7 NTU. The problem was caused by improperly installed equipment designed to add a chemical coagulant to incoming water prior to filtration. The coagulant helps remove suspended matter during water treatment.

The increased turbidity of Milwaukee's water was due to unusually high levels of runoff from pasturelands and slaughterhouses close to the city's southern waterworks. It is believed that this pollution was the source of microbiological protozoan parasites responsible for the cryptosporidiosis outbreak. The protozoa were present as cryptosporidium oocysts, an extremely small growth stage capable of penetrating sand filters and known to be resistant to chlorination.

The need to control water turbidity in preparing safe drinking water was underscored by the Milwaukee epidemic.

Des Moines, Iowa: Nitrates in Drinking Water

It has long been known that nitrates in drinking water can pose a public health hazard. The nitrate problem is usually found only in rural agricultural areas where chemical fertilizers are used to enhance crop production. When excess nitrate is applied to certain soils, leaching occurs, whereby nitrates infiltrate soil horizons and contaminate groundwaters. Rural water supplies often depend on relatively shallow wells that in some cases deliver drinking water containing unhealthy levels of nitrate.

The nitrate problem generally affects only infants under six months of age. In immature gastric systems, nitrate (NO_3^-) in drinking water or baby formula is biochemically reduced to nitrite (NO_2^-). Nitrite enters the bloodstream and oxidizes hemoglobin to methemoglobin, a form of hemoglobin that is unable to transport oxygen. The condition is referred to as the "blue baby syndrome." The nitrate problem is managed in the United States by the Environmental Protection Agency (EPA), which has established a nitrate standard of 10 mg of nitrate-nitrogen per liter of drinking water. As long as the standard is not exceeded, no public health problem is likely.

Nitrate levels in surface waters rarely exceed the EPA standard. But in Iowa, a state heavily engaged in agriculture, surface runoff of chemical fertilizers containing nitrate can seriously contaminate rivers. In the 1980s, rising levels of nitrate were found in the Des Moines and Raccoon Rivers, sources of Des Moines's drinking water. Removing nitrates from water isn't easy. All the standard methods of water treatment fail. But one technique is almost ideal—ion exchange. In the Iowa ion-exchange treatment system, small resin beads with internal, spongelike structures are first treated with a sodium chloride (salt) solution. Chloride ions attach to the beads. Then, when river water containing nitrate is passed through the ion-exchange resin, nitrate ions are bound to the resin, releasing chloride ions in their place. In this way, nitrate levels are reduced to safe levels.

KEY TERMS

anoxic

aquifers

assimilative capacity

biodegradable

biological oxygen demand (BOD)

biomonitors

condensation nuclei

dissolved oxygen (DO)

DO saturation

epidemiology

estuary

evaporation

evapo-transpiration

freshwater

hydrocarbons

hydrologic cycle

non-point sources

point source

potable water

primary wastewater treatment

runoff

sand filters

secondary wastewater treatment

transpiration

turbidity

watershed

weir

DISCUSSION QUESTIONS

1. Since aquatic systems have a natural assimilative capacity, a river receiving wastewater discharges may display no observable damage. DO levels may decline periodically, but not to dangerously low levels. Suppose that a small town on the Hudson River in New York State claims that wastewater from its primary treatment plant is causing no significant harm to river ecology. Should the town be exempt from having to build an expensive secondary sewage treatment plant?

2. Most fish species survive even when dissolved oxygen levels fall as low as 5.0 mg DO/L. Therefore, why should a state or federal agency require that certain rivers and lakes maintain DO levels of 7.0 mg/L or higher?

3. Current engineering technology can remove or kill pathogenic microorganisms in drinking water by using filtration techniques and either chlorine or ozone as a disinfectant. Therefore, why should we be concerned if untreated sewage and industrial wastes enter rivers and lakes?

4. Which of the following do you think is a better way to assess environmental quality in an aquatic system?

    A. Monitor the ecological well-being of fish or other aquatic organisms in that system.

    B. Carry out chemical, biological, and physical water quality tests and compare results with established standards.

5. Soil erosion is a normal, natural process that carries nutrients and minerals to rivers, lakes, and estuaries. This is beneficial to many organisms. But storms can trigger large-scale soil erosion

that has devastating impacts on the land and on aquatic environments as well. Given this perspective, how should we define pollution in aquatic environments?

INDEPENDENT PROJECT

Just about every town and city has a wastewater treatment plant to handle public and sometimes industrial wastes before discharging treated effluents to a river, lake, estuary, or ocean. Plan a visit to a nearby wastewater plant and collect information concerning:

- Types of different wastes handled on a daily basis.
- Typical treatment given to these wastes.
- If secondary treatment is involved, how is it done?
- Is nutrient reduction (phosphorus, nitrogen, or both) a part of the treatment process?
- How are various sludges (primary or secondary) disposed of?
- Are sludges composted for use on farms?
- Are the plant's effluents chlorinated?
- Are any industrial wastes handled by this plant?
- What happens during major storms? Is raw sewage discharged untreated?
- Are there any plans to upgrade or improve this plant?

Write a short report based on your visit, detailing the information you have collected. What are your impressions and insights based on your visit?

INTERNET WEBSITES

Visit our website at http://www.mhhe.com/environmentalscience/ for specific resources and Internet links on the following topics:

A historical account of the River Thames with photographs

The tidal Thames

Cleaning up London's rivers

Live video camera—London real-time

Britain's Environment Agency

SUGGESTED READINGS

Andrews, M.J. 1984. Thames estuary: Pollution and recovery. Pp. 195–227 in *Effects of Pollutants at the Ecosystem Level*, edited by P.J. Sheenan, D.R. Miller, G.C. Butler, and Ph. Bourdeau. London: John Wiley & Sons, Ltd.

Astor, Gerald. 1983. *The disease detectives*. New American Library of Canada Limited. New York.

Baker, M. N. 1949. *The quest for pure water: The history of water purification from the earliest records to the twentieth century*. New York: The American Water Works Association, Inc.

Barton, Nicholas. 1992. *The lost rivers of London*. London: Historical Publications LTD.

Binnie, G. M. 1981. *Early victorian water engineers*. London: Thomas Telford Limited.

Caudwell, S. 1990. Along the Thames to the Great Port of London. *The New York Times Magazine* 140: 866.

Fish found in the tidal Thames. 1998. Reading, England: The Environment Agency.

Jaret, Peter. 1991. The disease detectives. *National Geographic* 179(1): 116–40.

Lilienfeld, Abraham M., and David E. Lilienfeld. 1980. *Foundations of epidemiology*. 2d ed. Oxford: Oxford University Press.

Milne, R. 1991. London's water off to a clean start. *New Scientist* August 3: 131. 1991.

Partnership in planning: Riverbank design guidance for the tidal Thames. 1998. Reading, England: The Environment Agency.

Pelling, Margaret. 1978. *Cholera, fever and English medicine 1825–1865*. Oxford: Oxford University Press.

Pure and Wholesome. 1982. A collection of papers on water and wastewater treatment at the turn of the century. American Society of Civil Engineers.

Snow on Cholera. 1936. A reprint of two papers by John Snow, M.D. New York: The Commonwealth Fund.

The Thames tideway and estuary fact book: Teddington to Shoebury. 1998. Reading, England: The Environment Agency.

Van der Leeden F., F. L. Troise, and D. K. Todd. 1990. *The Water Encyclopedia*. 2d ed. Chelsea: Lewis Publishers.

CHAPTER 4

Estuaries, Seagrass Meadows, and Saltwater Ecosystems

Digital Stock/Undersea Life

... a faire Bay compassed but for the mouth with fruitful and delightsome land. Within is a country that may have the prerogative over the most pleasant places of Europe, Asia, Africa or America, for large and pleasant navigable rivers. Heaven and earth never agreed better to frame a place for man's habitation.

Captain John Smith, an organizer of the Virginia Company of London, first governor of the Jamestown Colony, and early explorer of the Chesapeake Bay in 1607 and 1608

Chesapeake Bay, one of the world's most biodiverse and productive eocsystems, has served as a rich commercial and recreational fishery, a wildlife sanctuary, and a historically significant part of American history. But during the twentieth century, two major declines in bay ecology took place, ultimately leading to economic and aesthetic losses that included deteriorating water quality, collapse of the bay's oyster industry, and reduced harvests of several species of finfish and shellfish. In addition, the numbers of native and migratory waterfowl fell in marshes and wetlands surrounding the bay, and the bald eagle, an American symbol, all but disappeared. A way of life, a cultural heritage, and a traditional means to earn a living were at risk.

Recognizing the bay as a national treasure, major commitments have now been undertaken by state governments in partnership with the federal government to restore this ecosystem. Urban residents, farmers, college and university scientists, and water resource managers are at work to better the bay. Thus far, the restoration of Chesapeake Bay is a qualified success, but the question remains: Will enough people continue allocating the time and resources necessary to restore this complex estuarine ecosystem?

We will begin this chapter by describing the bay's ecosystem and exploring how watersheds and estuaries function in general.

BACKGROUND

Overview of Chesapeake Bay

Chesapeake Bay is North America's largest estuary. It stretches along the U.S. eastern seaboard from Harve de Grace at the mouth of the great Susquehanna River in northeastern Maryland to Norfolk, Virginia, 310 kilometers (≈ 195 miles) to the south (Figure 4.1).

The bay acquired its present configuration about 3,000 years ago, but the process started during the Pleistocene Epoch, beginning about two million years ago, when periods of planetary cooling and warming caused continental glaciers to advance southward and then retreat northward. Warming in the latter part of the Pleistocene caused sea levels to rise worldwide as polar ice, ice fields, and glaciers melted. In North America, rising sea levels flooded broad river basins carved out earlier by the Susquehanna, Potomac, and James Rivers. As they filled in, Chesapeake Bay took shape.

Since the earliest days, Chesapeake Bay was known for its spectacular abundance of finfish, shellfish, and waterfowl. Settlers were drawn by the promise and potential of these waters and by the beauty and tranquility of the bay itself. European settlement in the Chesapeake region can be traced to 1572 when English aristocracy settled there, establishing grand country estates and tobacco plantations. Poor people came too, but they possessed few resources and owned no land. They camped on bay shorelines, marshes, and islands, pursuing a difficult subsistence way of life. Their daily needs were met by fishing, hunting, trapping, and crabbing. These watermen of the Chesapeake could be seen every day plying bay waters in small boats. They fished for striped bass and caught crabs in pots made of chicken wire. Other watermen fished for soft- and hard-shelled clams and oysters.

In the early 1800s, two developments changed the watermen's way of life: the coming of the railroad and the introduction of ice-cooled railway boxcars to transport food. The Baltimore and Ohio (B&O), one of the first American railways, began operations in 1830. The B&O first ran trains between stations along the bay and then expanded service to Baltimore and Philadelphia. Demand for oysters and crabs grew, and before long railroads were carrying Chesapeake seafoods to markets and restaurants in New York, San Francisco, and other major cities. By the early 1900s, more than 9,000 watermen were working the Chesapeake. The bay had become a gold mine.

Today, more than 15 million people live in the region, mostly in cities such as Richmond and Norfolk, Virginia; Baltimore, Maryland; Washington, D.C.; and Fayetteville, Pennsylvania. Almost half of the land that was originally forested has been converted into urban, suburban, commercial, industrial, and agricultural developments. Recreational uses of the bay region include hunting and fishing, power and sail boating, crabbing, swimming, and camping. More than 200,000 pleasure craft are registered in Maryland and Virginia, and close to one million people fish bay waters and nearby rivers each year.

The bay is a world-class commercial waterway. Notable industrial, manufacturing, and shipping centers are near it, and about 100 million tons of cargo are shipped annually on the Chesapeake. Major port facilities are located at Baltimore, Maryland, and at Hampton Roads, Virginia. Coal mined in nearby Appalachia is exported overseas from these ports and accounts for a significant share of U.S. international trade. Newport News, a part of the Hampton Roads complex, is a key naval construction site where ships, submarines, and aircraft carriers are built.

Bay Biodiversity

As the Chesapeake's once-bountiful seafood harvests testify, estuaries are biologically productive ecosystems. They provide habitat, food, and protection for a wide variety of plants and animals. They also serve as a nursery for innumerable species of finfish and shellfish.

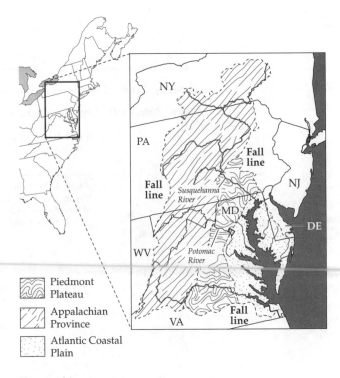

Figure 4.1   The Chesapeake Bay Watershed
(*Source: The Chesapeake Bay Program, U.S. Environmental Protection Agency, Washington, D.C.*)

Finfish   As many as 250 species of finfish are found in the Chesapeake. They include year-round inhabitants, such as small killifish and bay anchovy (the most abundant fish in the bay), and migratory species, such as striped bass, herring, shad, and sturgeon. Sturgeon eggs, a source of the delicacy caviar, were once so plentiful that it was not unusual to harvest over a million pounds annually. But sturgeon are rarely found in the bay today. Striped bass, another important species, are *anadromous fish* that migrate to the Chesapeake in May from as far north as Labrador and as far south as North Carolina. They spend several years in the estuary as immature fish, migrate to ocean coastal waters to mature, and then return to bay freshwater tributaries each spring to spawn (Insight 4.1). Chesapeake Bay eels, on the other hand, are *catadromous fish*. They reproduce in Atlantic ocean waters but die immediately after spawning.

Anadromous fish spend most of their life cycle in marine and estuarine waters but migrate to freshwaters to spawn and reproduce. The Chesapeake's striped bass are an anadromous fish species. Pacific and Atlantic salmon are also anadromous.

Catadromous fish spend the greater part of their life cycle in estuarine and freshwaters but migrate to ocean waters to spawn and reproduce. Chesapeake eels are catadromous.

INSIGHT 4.1
Striped Bass

The abundance of Chesapeake Bay striped bass, or "rock fish" as they are also called, was noted by Captain John Smith in the early 1600s. They are common to Atlantic Ocean coastal waters, live as long as 30 years, and can weigh as much as 34 kilograms (≈ 75 pounds).

After hatching in the bay, juvenile striped bass spend several years eating larvae, insects, and worms. They mature after migrating to coastal waters off southern New England and the Gulf of Maine to feed on anchovy, menhaden, herring, and other small fish. Being anadromous, they return to Chesapeake waters each spring to reproduce. They are one of the most popular sport and commercial finfish in the Chesapeake.

Chesapeake menhaden, members of the herring family, are of considerable value, and robust catches are gathered every year. These fish are used as a protein source in animal feed, as fishing bait by sport and commercial fishers, and as an ingredient in commercial fertilizers. Young menhaden spend the early stages of their life in Atlantic coastal waters near the mouth of the bay.

Shellfish For more than a hundred years, Chesapeake Bay yielded impressive harvests of shellfish, including blue crab, soft- and hard-shelled clams, and oysters. Atlantic blue crab are in high demand all around the world. In a good year, 100 million pounds, worth more than $200 million, are harvested, more than half the total U.S. catch. Economically, they are the most important of the bay's living resources. They utilize many of the bay's diverse habitats, from deep channels to shallow edges and from salty southern waters to freshwaters in the north. Even with today's high fishing pressure, the blue crab continues to thrive in the Chesapeake Bay (Insight 4.2).

Perhaps the best example of the Chesapeake's productivity is found in the historic catches of Eastern oysters once harvested from these waters. Oysters were extremely important both economically and culturally. In the 1880s, annual catches of 50 million kilograms (≈ 110 million pounds) or more were common. But a dramatic decline has taken place since then, and only 1 percent of those legendary catches are obtained today. We will examine the reasons for the decline later in this chapter.

Birds Besides finfish and shellfish, the bay and its coastal *wetlands* provide habitat for many bird species migrating along the Atlantic Flyway.

Wetlands are bogs, swamps, river deltas, and marshes, whose soils are periodically flooded or water-saturated most of each growing season (Insight 4.3).

Major nesting areas for bald eagles and osprey were common in the bay area. Today, on the average, 300,000 Canada geese, 28,000 tundra swans, and 650,000 ducks, including canvasbacks, pintails, scoters, widgeon, redheads, and eiders, find winter food and shelter near Chesapeake Bay each year.

How Watersheds Function

The water quality in surface waters can reflect environmental disturbance in watersheds. Generally, water of high quality occurs in runoff from undisturbed watersheds and can serve as a source of drinking water and as habitat for fish and wildlife. But runoff from disturbed watersheds often fails to meet these purposes. Most watersheds are known for diverse species of vegetation—trees, shrubs, vines, herbs, and grasses. Vegetation influences watershed ecology and affects the quality and quantity of watershed runoff in at least three ways:

1. Woodland soils are protected from the impacts of rainfall by grass, shrubs, overarching tree branches, and leaves. Soil erosion, which can lead to high turbidity in watershed runoff, is minimized in vegetated areas.

2. Trees, shrubs, and vegetation in general create spongelike, permeable soils and soil litter capable of storing water. This controls watershed runoff by limiting peak flows when it rains and by maintaining higher base flows in streams during dry weather.

3. Trees, shrubs, and grasses continuously consume soil nutrients, thereby limiting nutrient release in watershed runoff.

Runoff from undisturbed watersheds carries only limited quantities of essential nutrients to receiving waters to support aquatic plant life, including algae in

INSIGHT 4.2
The Atlantic Blue Crab

The Atlantic blue crab (*Callinectes sapidus*, which means tasty, beautiful swimmers) is named for the pale blue coloring in its walking and swimming legs (Figure 4.A). Crabs are **crustaceans,** joint-footed, invertebrate animals that grow an external covering, or *exoskeleton*. Exoskeletons are made of chitin, a calcium-containing, fingernail-like material that is at first soft but hardens within a few days. Worldwide, there are over 30,000 species of crustaceans, including lobsters, shrimp, and crayfish.

Blue crabs begin their life as larvae that hatch during summer months from eggs that are produced by and remain attached to the females. Water currents transport the larvae from the bay to ocean coastal waters where further development occurs. Juvenile crabs return to the bay in the fall and later scatter throughout the estuary and its tributaries. Blue crabs grow to about 20 centimeters ($\approx$ 8 inches) across through several stages over a two-year period, casting off one exoskeleton after another (molting).

Bay blue crabs feed on fish fry, minnows, worms, and epiphytic fauna, all of which are found

Figure 4.A   The Atlantic blue crab.
(E.R. Degginger/Photo Researchers, Inc.)

in eelgrass beds that serve as key nursery, molting, and foraging habitats. Crabs harvested before their exoskeleton hardens are called soft-shelled crabs. More than half of all U.S. soft-shelled crabs come from the Chesapeake Bay.

INSIGHT 4.3
Wetlands

Wetlands are terrestrial areas where soils are flooded and water-saturated at least part of each year. They are noted for species of **hydrophytic** (water-loving) **plants** and specialized soil types known as hydric soils.

Freshwater wetlands support plant communities such as wild rice, arrow arum, pickerel weed, pond lily, and cattail. Forested freshwater wetlands are often called swamps.

Tidal wetlands are coastal areas inundated by recurring tides and varying in water salinity. They are of two kinds: low marshland wet zones flooded daily by normal high tides and high marshland wet zones flooded infrequently by very high tides. Tidal wetland vegetation includes cordgrass, needlerush, saltgrass, and marsh elder.

lakes and seagrasses in estuaries. The phosphorus, nitrogen, potassium, sulfur, iron, and other plant nutrients needed to support plant growth are derived from watersheds via weathering of minerals, decaying plant matter, animal wastes, and eroding soils.

Land use changes resulting from agricultural developments, urbanization, timber harvesting, industrial growth, and the building of highway infrastructures all contribute to watershed disturbance. The most common outcome of watershed development is accelerated rates of soil erosion and nutrient runoff to streams, rivers, and lakes. As a result, what was once *natural eutrophication* of aquatic systems becomes *cultural eutrophication*.

TABLE 4.1 Typical Rates of Soil Erosion in Temperate Regions of the World

Soil Erosion*	Kg/Hectare/Year
Undisturbed forest	84
Undisturbed prairie	840
Cultivated farmland	16,800
Clear-cut forest	42,000

*Soil formation: 1,000 Kg/Hectare/Year
SOURCE: Data from The Chesapeake Bay Program, U.S. Environmental Protection Agency, Washington, D.C.

Natural eutrophication is slow aging of surface waters, especially lakes. It results from the gradual addition of soil and nutrient runoff from undisturbed watersheds. No rapid or dramatic changes in water quality, fish populations, or aquatic wildlife occur.

Cultural eutrophication is rapid aging of surface waters usually resulting from human, agricultural, or cultural activities. Cultural eutrophication often leads to increased water turbidity (loss in water clarity), excessive growth of algae and weeds, and changes in fish and wildlife.

It is clear that soil erosion plays an important part in controlling the quality of watershed runoff, which in turn can affect rates of eutrophication in surface waters like Chesapeake Bay. Table 4.1 provides data on rates of soil erosion in temperate climates based on land use.

How Estuaries Function

Estuaries are complex physical, chemical, and biological ecosystems. They are relatively warm, shallow, aquatic environments where nutrients from watershed runoff mix with tidal marine waters. Estuaries support diverse populations of plankton, seagrasses, finfish, shellfish, and waterfowl. Chesapeake Bay is influenced by two major sources of water:

- Freshwater enters the bay through a number of major rivers connected to literally hundreds of tributaries that drain a very large watershed. The watershed spans areas of Maryland, Virginia, West Virginia, Pennsylvania, Delaware, New York, and the District of Columbia. The rivers flowing to the bay include the Susquehanna,

Potomac, James, York, Patuxent, Rappahannock, and Choptank. The Susquehanna alone accounts for about half of all the freshwater entering the Chesapeake (see Figure 4.1).

- Saltwater enters the bay from the Atlantic Ocean. Daily tides pulse up and down the estuary and can be seen as far north as 160 kilometers ($\approx$ 100 miles) up-estuary from the Atlantic. Because of their higher density, marine waters tend to wedge beneath freshwaters flowing down-estuary.

Figure 4.2 is a cross section of a typical estuary. Here, freshwater runoff is shown flowing down-estuary above a saltwater layer. The two layers mix, depending on winds and tides. Their mixing is part of the estuary's complexity, which enables it to support biodiverse habitats, food webs, and nursery areas unparalleled in most other ecosystems. To better understand how estuaries work, we will consider the major factors that contribute to their ecology: light, turbidity, nutrients, salinity, and the presence of submerged seagrass meadows.

Light

In almost all ecosystems, the energy required to support living systems comes from sunlight. Particular wavelengths of light energy are captured by green plant pigments like chlorophyll *a*, enabling photosynthesis to take place (Insight 4.4). Through photosynthesis, green plants convert inorganic nutrients into organic matter (plant biomass), a process called primary production.

Primary production in Chesapeake Bay is accomplished mainly by algae and submerged seagrasses. Sunlight is easily available to algae since they occupy surface waters. But submerged seagrasses depend on the ability of light to penetrate water. Normally, this is not a problem and light can easily reach the shallow seagrass beds in bay and river coastal areas. However, the ability of light to penetrate estuaries can become severely limited by water *turbidity* caused mainly by watershed soil runoff.

Turbidity

Studies of Chesapeake Bay include measurements of water **turbidity,** which limits sunlight penetration to deeper waters. Turbid water is cloudy, murky, or muddy due to the presence of soils eroded from disturbed watersheds or by algal blooms. When sunlight penetrates water, most infrared and ultraviolet wavelengths are absorbed in the upper 1 meter ($\approx$ 3.3 feet) of the water column. Red, orange, and yellow light are

INSIGHT 4.4

Visible Light and the Electromagnetic Spectrum

Visible light is part of a larger **electromagnetic spectrum** spanning many wavelengths. Longer wavelengths include infrared radiation and radio waves; mid-range wavelengths make up visible light; and shorter wavelengths include ultraviolet light and X rays (Figure 4.B).

Infrared radiation (heat rays) consists of wavelengths above 700 nm (nanometers). (One nanometer is 10^{-9} meter; there are one million nanometers in a millimeter.) We can feel their warmth in sunlight on a sunny day. Visible light extends from about 400 nm to 700 nm. Radiation in this range appears to us as different colors of light: Violet is about 410 nm; indigo, 440 nm; blue, 475 nm; green, 510 nm; yellow, 550 nm; orange, 625 nm; and red, 700 nm.

Chlorophyll and other photosynthetic plant pigments absorb solar energy at particular wavelengths. Chlorophyll *a* absorbs at 430 nm and 662 nm; chlorophyll *b* absorbs at 453 nm and 642 nm. Because the two chlorophylls absorb mainly blue and red light, they appear green. Carotenoid pigments, found for example in carrots, absorb indigo light and appear orange. The colors seen in fall foliage (orange, red, and brown) are due to carotenoid pigments masked most of the year by chlorophylls.

Radiation between 200 nm and 400 nm is ultraviolet (UV) radiation. Because they have shorter wavelengths than visible light, UV rays possess more energy and are potentially harmful to living systems. (See Chapter 13 for more information about UV.)

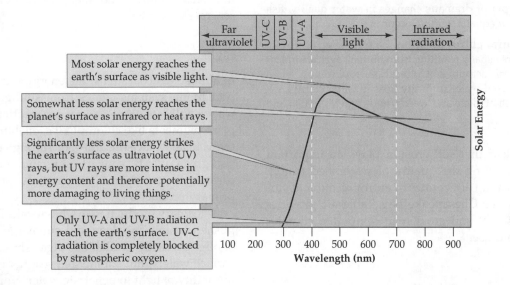

Most solar energy reaches the earth's surface as visible light.

Somewhat less solar energy reaches the planet's surface as infrared or heat rays.

Significantly less solar energy strikes the earth's surface as ultraviolet (UV) rays, but UV rays are more intense in energy content and therefore potentially more damaging to living things.

Only UV-A and UV-B radiation reach the earth's surface. UV-C radiation is completely blocked by stratospheric oxygen.

Figure 4.B Solar electromagnetic energy reaching the earth's surface. Wavelengths increase from far ultraviolet radiation on the left to infrared radiation on the right. Solar energy is available to green plants at certain visible spectrum wavelengths between approximately 400 and 700 nanometers.

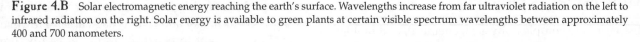

partly absorbed, allowing mostly green and blue light to reach deeper waters. The survival of an estuary's submerged seagrasses depends on the availability of green-blue light.

Chesapeake Bay's turbidity is due mainly to plankton (mostly algae) and suspended soils, bottom sediments that are resuspended in the estuary during storms. Prior to the 1960s, turbidity in Chesapeake Bay was moderate enough to allow sunlight to reach seagrass habitats as deep as 3 meters ($\approx$ 9.9 feet). As a result, submerged seagrass meadows occupied more

than 200,000 hectares ($\approx$ one-half million acres) of bay and river coastal areas.

Turbidity in the Chesapeake used to be controlled by immense numbers of **filter feeders,** including oysters, clams, and sponges. In their quest for food, they literally filtered plankton and suspended matter from the water. Oysters in particular played a major role in regulating bay turbidity. It has been calculated that a hundred years ago there were enough oysters to filter the bay's entire water volume in only three days. Today's decimated oyster population would require

INSIGHT 4.5

Liebig's Law of the Minimum

In 1840, the German chemist and educator Justus von Liebig found that many nutrients are required for plant growth. However, one nutrient is usually in short supply—*the limiting nutrient*. Liebig showed that plant growth is regulated by the availability of the limiting nutrient.

In most freshwaters, phosphorus, occurring mainly as phosphates, is the limiting nutrient. Its relative unavailability limits algal production and prevents algal blooms (exponential growth in algal populations). But if the supply of phosphorus increases, algal blooms can occur, along with a shift in algal species that favors more offensive species like blue-green algae. If this happens, water quality declines. A typical phosphate level below which algal blooms are not likely to occur is 0.050 mg phosphorus per liter (0.050 ppm).

In ocean waters, nitrogen, occurring mainly as ammonia and nitrate, is often the limiting nutrient. However, estuaries are more complex, and both phosphorus and nitrogen can function as limiting nutrients in different parts of an estuary at different times of the year.

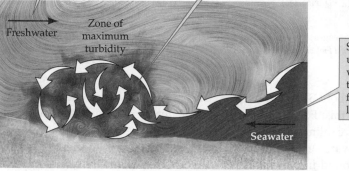

Freshwaters entering Chesapeake Bay through its several tributaries flow generally southward, down-estuary and tend to flow above seawater which is more dense.

Freshwaters and seawater continuously mix as a result of tides, winds, and periodically large inflows of freshwater following watershed precipitation. Mixing lifts bottom sediments into estuarine waters causing increased water turbidity.

Freshwater → Zone of maximum turbidity

Seawater, which moves up-estuary twice daily with each high tide tends to flow below freshwaters which are less dense.

Seawater

Figure 4.2   Cross section of an estuary showing less dense freshwater (on the left) flowing above more dense seawater (on the right) causing water turbulence and turbidity.
(Source: The Chesapeake Bay Program, U.S. Environmental Protection Agency, Washington, D.C.)

about a year to do the same job if it could. What oysters once accomplished in sustaining the quality of Chesapeake Bay waters is an example of **negative feedback,** any process that helps maintain a system's equilibrium. By continuously filtering the estuary's water, oysters preserved water clarity, allowing sunlight to reach the submerged seagrasses

Nutrients

The availability of nutrients to primary producers such as algae underlies an important part of estuarine productivity. If a particular nutrient is in short supply, it may limit algal growth, thereby affecting the entire aquatic food web. This is known as the **limiting nutrient principle.** Phosphorus is generally the limiting nutrient in freshwaters, while nitrogen is often the limiting nutrient in marine waters. In estuaries, however, both nutrients can affect primary production. The behavior of limiting nutrients follows Justus von Liebig's Law of the Minimum (Insight 4.5).

Salinity

Salinity is a measure of the saltiness of oceans and estuaries. This saltiness is due mainly to dissolved minerals such as sodium chloride, potassium chloride, sodium bicarbonate, and potassium sulfate. Salinity is reported in terms of parts per thousand or grams of dissolved solids per kilogram of water (g/kg). Most ocean

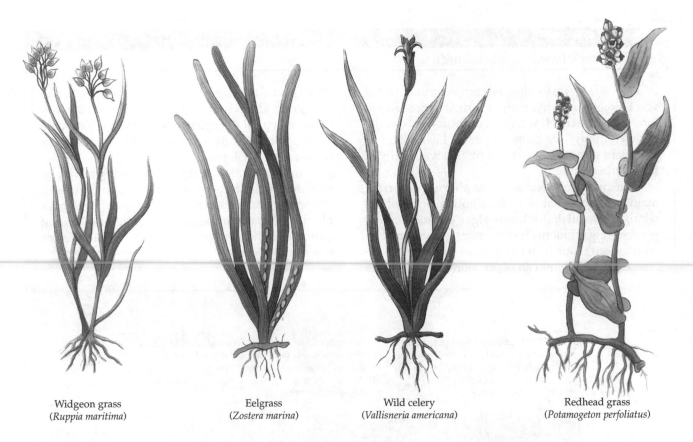

Widgeon grass
(*Ruppia maritima*)

Eelgrass
(*Zostera marina*)

Wild celery
(*Vallisneria americana*)

Redhead grass
(*Potamogeton perfoliatus*)

Figure 4.3   Four common Chesapeake Bay seagrasses: widgeon grass, eelgrass, wild celery, and redhead grass.
(Source: The Chesapeake Bay Program, U.S. Environmental Protection Agency, Washington, D.C.)

waters have an average salinity of about 35 g/kg, but some exhibit salinities greater than this. For example, the Red Sea and the Persian Gulf have salinities as high as 40 g/kg due to high rates of water evaporation. Lower than average salinities are found in polar seas where melting ice dilutes the sea's salt concentration. In estuaries, where ocean and freshwaters mix, salinities typically fall within the range of 10–25 g/kg.

Estuaries generally display salinity gradients or layers, with increased salinity occurring at lower depths, creating a vertical spectrum of estuarine habitats. They also exhibit horizontal salinity profiles. While low salinities prevail up-estuary resulting from freshwater tributaries, higher salinities are found down-estuary where ocean waters enter. Life cycles of Chesapeake Bay finfish and shellfish and the habitats where various life cycle stages occur are usually related to salinity gradients. For example, Chesapeake hard clams are only found in waters of 12 g/kg salinity or greater.

Submerged Seagrass Meadows

As first-trophic-level photosynthetic primary producers, submerged seagrasses play a vital role in Chesapeake Bay and other estuaries. They convert nutrients, carbon dioxide, and water into plant biomass and add oxygen to the aquatic environment. Approximately 50 seagrass species are known worldwide, and at least 12 of them are commonly found in the Chesapeake region. They represent a diverse fusion of many different species of submerged aquatic vegetation that includes widgeon grass, eelgrass, wild celery, and redhead grass (Figure 4.3). Their distribution beneath the bay's water depends on light availability, the behavior of individual species, and prevailing salinity gradients. Eelgrass tolerates high-salinity waters and is abundant in the central and lower bay. Water milfoil is not as salinity tolerant and is limited to the central bay and some rivers. Bay seagrasses and their distributions are shown in Table 4.2.

Submerged seagrasses perform a number of important ecological functions:

1. Through photosynthesis, seagrasses produce food for other organisms and add dissolved oxygen to bay waters.

2. Seagrasses create benthic (bay-bottom) nursery habitats where small organisms such as aquatic

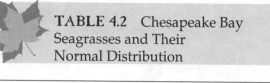

TABLE 4.2 Chesapeake Bay Seagrasses and Their Normal Distribution

Common Name	Normal Distribution
Eelgrass	Central and lower bay
Water milfoil	Central bay and rivers
Sago pondweed	Entire bay and some rivers
Redhead grass	Northern and central bay
Horned pondweed	Entire bay and some rivers
Common elodea	Northern and central bay
Coontail	Northern and central bay
Southern naiad	Northern bay and rivers
Wild celery	Central and northern bay
Water stargrass	Northern bay and rivers
Widgeon grass	Central and lower bay
Hydrilla	Upper bay

SOURCE: Data from The Chesapeake Bay Program, U.S. Environmental Protection Agency, Washington, D.C.

invertebrates, finfish fry, and shellfish larvae can find cover and protection from predators.

3. Because they are rooted aquatic plants, seagrasses abate water currents and wave action, thereby stabilizing bottom sediments and protecting shorelines from erosion.

4. Decaying seagrass leaves are a food source for zooplankton such as copepods, which in turn are eaten by small fish, clams, crabs, and oysters.

5. Seagrass leaves serve as surface habitats where small aquatic organisms called **epiphytes** colonize and grow. The epiphytes, which include diatoms, zooplankton, crustaceans, and other microscopic life, are a food source for snails and shrimp that then become food for finfish and shellfish.

CASE STUDY: THE CHESAPEAKE BAY

Until the late 1960s, marine scientists and commercial fishers considered Chesapeake Bay among the most biologically productive estuaries in the world. Year after year, these waters yielded huge catches of oysters, striped bass, clams, hard- and soft-shelled crabs, and sturgeon—seafood exceeding $1 billion a year in value. Wetlands and marshlands surrounding the estuary served as nesting, feeding, staging, and wintering grounds for a wide variety of birds, including ducks

such as widgeon and redheads, bald eagles, great blue heron, white swans, and Canada geese. But by 1970, a downturn in the bay's living resources had become evident, and it paralleled losses in the estuary's submerged seagrass meadows.

Evidence of Decline

Seagrasses

The lush, expansive seagrass meadows that once flourished beneath the shallow waters of Chesapeake Bay served as habitat for prodigious numbers of attached algae and plankton, the biological building blocks of estuarine productivity, and as sources of food and nursery habitats for finfish, shellfish, and waterfowl (Figure 4.4). But Chesapeake Bay was devastated by two major seagrass dieoffs, the first in the 1930s and the second starting in the late 1960s.

The First Seagrass Decline   During the 1930s, a massive decline in submerged seagrasses took place in both European and North American estuaries, including Chesapeake Bay. It was called the "wasting disease." Large areas of Chesapeake Bay seagrasses died out, including eelgrass, one of the most important. The population crash was both dramatic and drastic. Recovery from the wasting disease was slow, and not until the 1960s did significant regrowth occur.

There are many theories to explain the cause of the wasting disease. Some scientists blame a decline in precipitation during the 1930s. Others think herbicides like 2,4-D, introduced into agriculture in the early 1930s, were the cause. The scientific literature indicates that 2,4-D is toxic to eelgrass but only at high concentrations. Ecologists studying the Chesapeake found that 2,4-D levels likely to have been present in bay waters during the 1930s were not high enough to kill seagrasses. More recent studies suggest that the wasting disease was caused by the microbial slime mold *Labyrinthula*. But no one knows how the slime mold might have invaded estuaries on both sides of the Atlantic.

The Second Seagrass Decline   By 1965, considerable regrowth of submerged seagrasses had occurred. The recovery was significant, but it didn't come close to the 200,000 hectares of seagrasses believed to have occupied bay waters before 1930. Then, in the late 1960s, a second major decline in Chesapeake seagrass meadows began. This dieoff, unlike the wasting disease, occurred *only* in the Chesapeake and its tributaries. In the Potomac River, submerged vegetation, including common elodea, coontail, and wild celery, disappeared

Phytoplankton

Zooplankton

Bluefish

Striped bass

Striped killifish

Bay anchovy

Figure 4.4 Submerged seagrasses form the basis of the estuarine food web, supporting phytoplankton, zooplankton, small fish, and large fish.

completely—a 100 percent die-off. In the Choptank River, the largest tributary on the bay's eastern shore, there was a 98 percent decline in widgeon grass, red-head grass, sago pondweed, water milfoil, and horned pondweed.

The Chesapeake's second dieback reached its greatest extent in 1984, leaving scattered fringes of submerged vegetation in a few bay and tributary areas. Aerial surveys showed that only 15,800 hectares (≈ 39,000 acres) of seagrass beds, representing about 8 percent of the original 200,000 hectares, remained. An overall 92 percent decline in seagrasses had occurred.

Dissolved Oxygen

To make matters worse, levels of dissolved oxygen (DO) in parts of the Patuxent, Potomac, and Rappahannock rivers were lower than normal from May to September. Low DO levels were also observed in the bay itself, especially in its deeper channels, and, at

TABLE 4.3 Processes Affecting DO Levels in Aquatic Environments	
Add DO	**Reduce DO**
Seagrass photosynthesis	Plant and animal respiration
Algal photosynthesis	Increased BOD levels
Wind and wave action	Elevated water temperatures
Runoff of dissolved-oxygen-rich waters	Stratification limiting deep-water DO

SOURCE: Data from The Chesapeake Bay Program, U.S. Environmental Protection Agency, Washington, D.C.

times, no oxygen at all was present in these channels. In the summer of 1993, bay researchers found that DO levels were at or near zero in 15 percent of the bay's deeper waters, a totally unacceptable condition that indicated the estuary and its living resources were under severe environmental stress.

The decline in DO was significant because aquatic plants and animals require oxygen gas physically dissolved in natural waters such as rivers, lakes, streams, estuaries, and oceans. In these aquatic environments, DO is normally present at levels between 5 and 12 parts per million (ppm). (In freshwaters, ppm is the same as milligrams per liter, mg/L; in marine waters, ppm is approximately the same as mg/L.) By way of comparison, the level of oxygen in the earth's atmosphere is about 210,000 ppm. While 5 ppm of DO doesn't sound like much, it is enough to meet the needs of nearly all aquatic organisms. Most finfish and shellfish complete their life cycle successfully if DO levels of at least 5 ppm prevail. However, if DO falls below 5 ppm, many aquatic species cannot survive. Oysters are an exception because they can survive in waters where DO is as low as 3 ppm.

Natural waters generally maintain adequate DO levels, but when oxygen-consuming wastewaters are discharged into them, DO levels can decline significantly. Many wastewaters create a demand for dissolved oxygen through a process known as *biological oxygen demand* (BOD).

Biological oxygen demand is a microbiological process that reduces dissolved oxygen levels in aquatic systems. It occurs when organic matter in wastewaters, such as municipal and industrial discharges, is added to aquatic environments triggering the growth of bacterial populations. Bacterial respiration by increasing numbers of bacteria creates a rising demand for dissolved oxygen. (For more information about BOD, see Chapter 3.)

The level of dissolved oxygen in an aquatic system results from an equilibrium between processes that add oxygen and processes that diminish oxygen (Table 4.3). If DO falls to 2 ppm or less, the result is **hypoxia,** an unacceptably low DO level potentially harmful to finfish and shellfish. If DO falls to zero, the result is **anoxia,** the complete loss of dissolved oxygen, which results in fish kills and undesirable chemical changes in aquatic environments. Hypoxia and anoxia can occur in natural waters if photosynthesis is restricted by water turbidity, if high BOD levels prevail, or if *reaeration* is blocked.

Reaeration consists of physical processes that add oxygen to aquatic systems, especially following a decline in DO. Natural reaeration takes place in stream riffle areas, waterfalls, and turbulent waters.

Animal Life

Further evidence of estuarine decline was reflected in bay finfish and shellfish populations which had declined to the point where their future was in doubt. While the loss of the seagrasses contributed to this dilemma, other problems came to light.

Eastern Oyster The most dramatic proof of deteriorating water quality is the virtual demise of Chesapeake Bay oysters, the Eastern oyster. Annual oyster catches of 50 million kilograms (≈ 110 million pounds) or more were common in the 1890s, but in the 1990s less than 0.5 million kilograms (≈ 1.1 million pounds) were harvested each year, a 99 percent decline! This is a classic example of a **population crash,** a precipitous fall in the numbers of a particular species. One of the reasons oyster populations had been so successful in the Chesapeake was their ability to tolerate variations in water temperature, dissolved oxygen, turbidity, and salinity—but there were limits to the oyster's tolerance of changes in the bay. We now know that the crash of the bay's oyster population was due to several factors: overharvesting, diminished water quality, and two parasitic diseases, Dermo and MSX (Insight 4.6).

The population crash of the Eastern oyster has put increased fishing pressure on Atlantic blue crabs. Now the bay's most important seafood product, crabs are being harvested more intensively than ever before. It is estimated that each year Chesapeake crabbers catch about 75 percent of the bay's entire adult blue crab population. Baywide surveys reveal lower and lower crab harvests, especially in light of the number of crabbers. Owing to declining crab harvests in Mary-

INSIGHT 4.6
Dermo and MSX

Two microscopic parasites, Dermo and MSX, attack Eastern oysters, the Chesapeake Bay oyster species. Dermo, a fungal disorder, known to have existed in the Gulf of Mexico since the 1950s, destroys oyster larvae. It was first identified in the Chesapeake in 1954 but didn't become a major problem until the late 1980s when it was found entrenched in bay-bottom sediments. Dermo releases enzymes called proteases that break down oyster tissues, causing oysters to die within two years. MSX, or multinucleate spore X, appeared in bay waters in 1957, destroying oyster beds in Virginia and Maryland.

The rates of Dermo and MSX infection in bay oysters are high, and no one knows how the parasites entered these waters. Research funded by the U.S. Congress (Oyster Disease Research Program) is aimed at understanding why Eastern oysters are susceptible to these attacks. The recent ability to culture Dermo in the laboratory may lead to strategies to control it. Also, genetic breeding is underway to develop a new strain of Eastern oyster more resistant to MSX.

TABLE 4.4   Harvests of Migratory Fish in Chesapeake Bay

Migratory Species of Bay Finfish	Average Annual Catch (kilograms) 1963–1977	Average Annual Catch (kilograms) 1978–1992	Average Decline (Percent)
American eel	427,200	370,800	13
River herring	9,524,000	553,000	94
Shad	1,334,000	195,800	85
Striped bass	2,298,000	508,700	78
White perch	675,100	343,500	49
Yellow perch	50,300	22,500	55

SOURCE: Data from The Chesapeake Bay Program, U.S. Environmental Protection Agency, Washington, D.C.

land and Virginia bay waters, state agencies have imposed a 3.5-inch size limit to help conserve blue crabs.

Striped Bass   Consider what happened to the Chesapeake's most highly prized finfish, the striped bass. The Chesapeake provides spawning grounds and juvenile habitat for perhaps 90 percent of all U.S. East Coast striped bass. They were once the mainstay of bay watermen, who netted almost 2.3 million kilograms ($\approx$ 5 million pounds) of striped bass each year between 1963 and 1977. But the annual catch dropped to about 0.5 million kilograms ($\approx$ 1.1 million pounds) between 1978 and 1992 , a 78 percent decline. Table 4.4 compares declines among six of the bay's most valuable finfish between these two 15-year periods.

The likelihood that overfishing in the Chesapeake was causing the striped bass decline was suspected by the Atlantic States Marine Fisheries Commission (ASMFC) as early as 1980. The ASMFC

developed management plans to restrict striped bass harvests along most of the Atlantic coast, and in 1985 a fishing moratorium was adopted in Maryland, Virginia, and Delaware. This proved so successful that the fishery was reopened in 1990. In 1995, the ASMFC declared that the bay's striped bass population had been restored. Overfishing may also be responsible for the decline of other bay finfish. The success of the striped bass management program will undoubtedly serve as a model for recovery efforts aimed at other species.

Birds and Waterfowl   Thirty species of waterfowl, including great blue heron, wild geese, and swans, were once commonly seen in Chesapeake Bay wintering, breeding, and staging areas. Canada geese were the most abundant and still are today (Insight 4.7). They are attracted to open grain fields close to the water. Unfortunately, much of this avian diversity has been lost due to declines in seagrass and wetland areas where

INSIGHT 4.7

Canada Geese

Perhaps you have seen Canada geese flying in their legendary V formations and heard them honking loudly. It may have been a northern migration to where geese nest, feed, and raise their young. Canada geese lay four to ten eggs a year. Ganders forage for food often leading gosling chicks to favorable foraging sites where seeds, grasses, and insects can be found.

It is thought that shorter days and cooler temperatures toward the end of summer signal the geese to begin their migration south. They traverse ancient flyways connecting northern nesting sites with southern wintering grounds like those along the Chesapeake's eastern shore. For example, every winter flocks of geese forage and feed along Maryland's Choptank River.

Canada geese are the most abundant waterfowl species on the Chesapeake. They are easily recognized by their black head, white collar, and long black neck. They are large birds, up to one meter (≈ 3.3 feet) long, and are related to ducks and swans. Their nests are found on dry ground and in trees. They can easily fly up to 8,000 kilometers (≈ 5,000 miles) in one season.

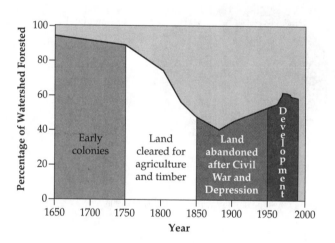

Figure 4.5 Deforestation in the Chesapeake's watershed between 1650 and 1998. In the 1600s about 95% of the basin was forested. In 1990, about 59% was forested. The forests that regrew earlier this century are now steadily declining. Current losses represent permanent conversions. Forest area may be as important as quality: proximity to water; species diversity; ecosystem resilience; habitat fragmentation; and economic viability.
(Source: From the U.S. Forest Services.)

food resources such as crustaceans, juvenile fish, and frogs are normally found.

Examples of bay migratory birds include the eagle and the canvasback duck, one of North America's most important game birds. In the 1950s, up to a quarter-million canvasbacks found refuge in the bay area. Today, their population has fallen to about 50,000, only 20 percent of earlier numbers. Bald eagles, estimated to have once numbered 3,000 birds, had decreased to about 90 nesting pairs by 1970 due to habitat destruction and the toxic effects of the pesticide DDT, which had accumulated in eagles, peregrine falcons, and other birds feeding on DDT-contaminated fish.

After DDT was banned in the United States in 1972, a remarkable recovery in the numbers of bald eagles was seen. By 1997, scientists counted 400 eagle nesting sites in the bay area, and the number of eagles had increased to 600 birds. The overall increase in American bald eagle populations has led to their being uplisted from "endangered" to "threatened" status in the lower 48 states.

Trees

The Chesapeake's drainage basin, which spans almost 17 million hectares (≈ 64,000 square miles), was once mostly forested land. In the 1600s, the Chesapeake Bay region was largely forested and undisturbed. It is estimated that trees covered 95 percent of the bay's immense watershed. As long as these forests remained, soil erosion and nutrient runoff were minimized, and the bay's living resources thrived. But by 1750, population growth and agricultural development began to change this picture, and by 1880 forests covered only 40 percent of the bay watershed. Later there was some regrowth, but more recently there has again been a downward trend in tree cover (Figure 4.5).

Water Pollution

Development in the Chesapeake Bay watershed has created both point sources and non-point sources of pollution to the bay. In 1970, the Potomac River, an important bay tributary and one of the nation's historic rivers, was in a degraded and polluted condition. The cause was Washington, D.C.'s sewage treatment plant, which discharged inadequately treated wastewater to the Potomac. The result was excessive nutrient enrichment of the river's water. Unsightly, smelly blue-green

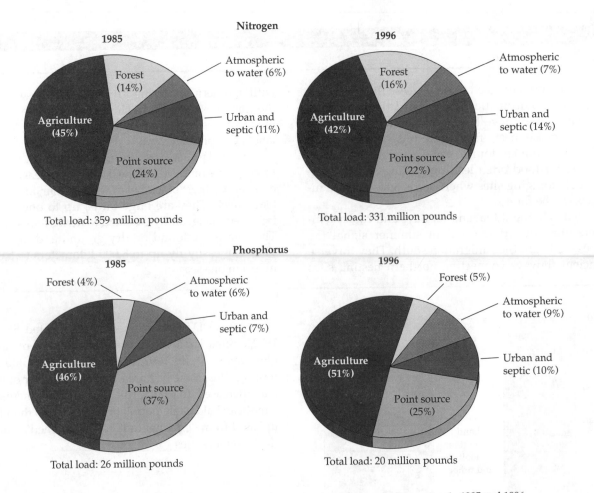

Figure 4.6 Point and non-point sources of nitrogen and phosphorus entering Chesapeake Bay waters in 1985 and 1996. *(Source: Data from The Chesapeake Bay Program, Annapolis, Maryland, 1998.)*

algae, a particularly undesirable kind of algae, bloomed frequently. While the Potomac's degradation was an embarrassment to people in the District of Columbia, it was but one example of worsening pollution in many of the rivers flowing to the bay. Indeed, more than 500 sewage treatment plants discharge nutrient-rich wastes to the estuary every day. In addition, hundreds of industrial plants discharge wastewaters to the bay and its tributaries.

A 1990 study of nutrient runoff from agricultural areas of Chesapeake Bay yielded important insights into the problem of non-point source runoff. The application of 318 million kilograms (≈ 350,000 tons) of commercial fertilizers to 3.2 million hectares (≈ 8 million acres) of pasture and croplands resulted in the runoff of 6.4 million kilograms (≈ 7,000 tons) of phosphorus-based nutrients and 66.3 million kilograms (≈ 73,000 tons) of nitrogen-based nutrients into the bay.

The distribution of nitrogen and phosphorus entering Chesapeake Bay from point and non-point sources is summarized for 1985 and 1996 in Figure 4.6. It is interesting to note how important agricultural

runoff is. In 1996, it accounted for 51 percent of all the phosphorus and 42 percent of all the nitrogen entering the bay. More importantly, the data reflect a decline of about 23 percent in phosphorus and 8 percent in nitrogen over this 11-year period.

Searching for Causes

Identifying the specific causes of the Chesapeake's deterioration was not easy. The pattern of seagrass loss provided some clues. Seagrass beds declined first in tributaries like the Potomac. Later, losses were seen in the estuary itself, first up-estuary and then mid-estuary. No major losses were seen down-estuary where the bay connects with the Atlantic. The pattern suggested that what was most damaging to seagrasses must enter through tributaries. The estuary behaves as a sink so that eroded soils, nutrients, and contaminants settle out and remain in the bay for long periods of time.

To investigate the reasons for the area's ecological decline, the Chesapeake Bay Program, a comprehensive research and education effort, was initiated by the U.S. Environmental Protection Agency (EPA) in 1975. Many professionals, officials, and citizens participated in this work, including scientists at Virginia's College of William and Mary, the University of Maryland, the U.S. Fish and Wildlife Service, the National Oceanic and Atmospheric Administration, and the EPA. Their highest priority was to find the cause of the bay's vanishing seagrass meadows, and eventually three theories emerged.

Theory 1—Heavy Metals

Heavy metals such as mercury, cadmium, and chromium are used in many different industries. Mercury is employed to make chlorine, a chemical used to bleach paper and disinfect drinking water and wastewater. Cadmium is a component of Nicad batteries. Chromium is important in metal plating and fabricating stainless steel alloys. One of the first theories regarding seagrass decline suggested that industrial wastes containing heavy metals might be toxic to seagrasses. But a detailed review of the scientific literature failed to uncover any data to support this idea, and experiments to test the effects of various heavy metals on seagrasses at concentrations similar to those in bay waters showed no toxic effects.

Theory 2—Pesticide Runoff

Pesticides include herbicides, insecticides, fungicides, and other types of pest-control agents. Herbicides (weed killers) interested bay researchers because they are designed to kill plants, so there was a possibility that they were killing seagrasses. Two herbicides in particular, atrazine and linuron, were widely used to control weeds on cornfields. In 1980, scientists at the EPA, Maryland's Horn Point Environmental Laboratory, and Virginia's Institute of Marine Science analyzed samples of runoff from different watershed sites, especially in agricultural areas. Low levels of atrazine, linuron, and other agricultural chemicals were found in farm ditches and nearby streams, and even lower levels were found in major tributaries and the bay itself. But these levels were too low to kill seagrasses.

Theory 3—Diminished Light Transmission

By 1990, a theory was being considered linking seagrass decline with cultural eutrophication. The theory suggested that the main factor responsible for the die-off of submerged vegetation is the inability of enough sunlight to penetrate the water and support seagrass photosynthesis. Scientists working on the Chesapeake had already observed rising levels of water turbidity. It seemed possible that as water turbidity increased, sunlight needed by deeper seagrasses might not be penetrating to required depths.

Research demonstrated that the bay's turbidity originated from watershed runoff carrying large quantities of suspended solids, mostly fine-grained silts and clays, into the estuary. Because of their small size, suspended solids like these settle out only very slowly and add significantly to turbidity. Another source of turbidity was recurring algal blooms triggered by agricultural fertilizers and animal manures in runoff from watershed farms. Rising water turbidity sharply limited the extent to which sunlight penetrated bay waters.

It is thought that in earlier years sunlight easily penetrated to 3-meter ($\approx$ 9.9-feet) depths where most of the bay's meadows grew, and there was less estuarine turbidity at that time. But as watershed development increased, water turbidity increased and the seagrasses retreated to more shallow waters. Today, the light needed to support seagrass meadows penetrates bay waters to depths averaging only about 1 meter ($\approx$ 3.3 feet). Thus, only patches of seagrass meadows have survived.

Another factor influencing the survival of seagrass meadows has been identified. Seagrass leaves serve as platforms on which epiphytes like diatoms and crustaceans colonize. The epiphytes grow as layers on seagrass leaves. When examined under an x-ray microscope, seagrass leaves today display excessively thick epiphytic growth. Nutrient enrichment of bay waters has apparently caused an overpopulation of leaf epiphytes. In some cases, as much as 94 percent of available light is prevented from reaching the leaves. This factor adds to the problem of seagrass survival.

It is now believed that soil erosion and nutrient enrichment in Chesapeake Bay are the main forces behind the current decline in the bay's ecology and living resources.

Searching for Solutions

Efforts to deal with Chesapeake Bay problems began in 1968 when the governors of Virginia and Maryland inaugurated Chesapeake Bay Week. A water quality conference was held, and public attention was focused on the bay—its cultural history, its economic importance, and its environmental decline. Actions to better understand the bay's ecology and its problems were slow in coming, but environmental awareness was increasing across America. This became evident in 1970 when the nation celebrated its first Earth Day. That year the state of Maryland passed a law to protect tidal wetlands, and two years later the U.S. government adopted the

INSIGHT 4.8
The Federal Water Pollution Control Act (FWPCA)

The FWPCA of 1972, which later became known as the Clean Water Act, aimed "to restore and maintain the chemical, physical, and biological integrity of the nation's waters." The discharge of pollutants into navigable waters was to be eliminated by 1985, making America's waterways once again "fishable and swimable." The act's two main strategies were:

1. To protect surface waters by establishing acceptable pollution limits for industries and other point-source polluters. The National Pollutant Dis-

charge Elimination System (NPDES), administered by the EPA, issued discharge permits to regulate pollution.

2. To provide federal funds to help municipalities construct and operate improved wastewater treatment facilities.

As a result, sewage treatment plants were built in towns that had never had them. In other cities, treatment facilities were improved, and more effective wastewater treatment plants were constructed.

Federal Water Pollution Control Act (FWPCA) of 1972 (Insight 4.8).

Chesapeake Bay Agreement of 1983

The official cleanup of Chesapeake Bay began in 1983 when the State of Maryland, the Commonwealths of Pennsylvania and Virginia, the District of Columbia, and the EPA signed the first Chesapeake Bay Agreement, a historic compact aimed at securing a better understanding of the ecology of the Chesapeake and its watershed. What was causing the continuing decline in the bay's living resources? What part did regional agriculture, industry, population growth, and possibly overfishing play? What practical, cost-effective strategies could be implemented to reverse this downturn in the Chesapeake? The parties to the agreement believe that the bay and its region have national significance and that citizens across America have a stake in its future. An excerpt from the agreement reads:

> We recognize that the findings of the Chesapeake Bay Program have shown a historical decline in the living resources of the Chesapeake Bay and that a cooperative approach is needed among the Environmental Protection Agency, the State of Maryland, the Commonwealths of Pennsylvania and Virginia, and the District of Columbia (the states) to fully address the extent, complexity and sources of pollutants entering the Bay. We further recognize that EPA and the states share the responsibility for management decisions and resources regarding the high priority issues of the Chesapeake Bay.

This agreement set the stage for cooperative actions based on the results of ongoing research. The highest

Figure 4.7   Blue Plains Water Treatment Plant near Washington, D.C. This plant is an advanced wastewater treatment facility on the Potomac River.
(Photo courtesy of the Department of Public Works, Washington, D.C.)

priority was to deal with the problem of phosphorus- and nitrogen-based nutrient loading in the estuary. An important step had already been taken in 1972 when funding through the FWPCA enabled over $1 billion to be spent on new facilities to convert Washington, D.C.'s Blue Plains Water Treatment Plant on the Potomac to advanced wastewater treatment methods capable of removing phosphorus. The plant processes District of Columbia wastes and is the principal point source of nutrients entering the Potomac (Figure 4.7). Since phosphorus is the limiting nutrient in Potomac waters, its control resulted in considerably better river water quality.

The 1983 agreement has led to improvements in other treatment plants that discharge wastewaters to the bay, improvements aimed at reducing BOD and phosphorus nutrients entering the Chesapeake. These measures have visibly benefited bay waters because most of the existing treatment plants provided only primary wastewater treatment (see Chapter 3), a process that removes solid matter from wastewaters but doesn't reduce BOD and nutrient levels significantly. In many cases, FWPCA provided funding to upgrade existing primary plants to secondary wastewater treatment. The main objective of secondary treatment is BOD reduction. In some areas of the bay, **advanced wastewater treatment** techniques were implemented. Such methods are designed to reduce nutrient levels in treated wastewaters, primarily phosphorus and sometimes nitrogen, to eliminate algal blooms.

It is estimated that phosphorus and nitrogen loading to the Chesapeake doubled between 1950 and 1980. Although there was little doubt about the need to limit nutrient loading, there was considerable uncertainty regarding the need to control both phosphorus and nitrogen (see Insight 4.5). Political and economic factors influenced what and where nutrient control strategies should be implemented. For example, because it is cheaper to carry out phosphorus reduction than nitrogen reduction, federal, state, and local agencies focused most of their attention on point-source phosphorus control.

The 1983 agreement also addressed the issue of non-point sources of pollution by setting forth a series of **best management agricultural practices** aimed at controlling soil erosion and the runoff of fertilizers and animal wastes into bay tributaries. These practices include:

- No-till and low-till soil cultivation techniques instead of traditional plowing. Furrows are disked in soils to receive seeds and fertilizers. Crop stubble from the previous year's harvest is left to hold soils in place. Less soil is exposed to wind and water erosion, and less chemical fertilizer is needed. However, farmers depend more on herbicides because no-till and low-till practices do not control weeds effectively.

- Protection and, if necessary, restoration of stream-side vegetation and hedgerows to minimize nutrient runoff and stream-bank erosion.

- Use of animal watering troughs and fences to limit farm animal access to streams and rivers, thus preventing direct input of animal wastes and resuspension of bottom sediments.

- Use of protected concrete slab areas and lined pits to store animal manure for later land application.

- Planting winter cover crops like ryegrass to protect soils from wind and water erosion between growing seasons.

- Minimizing chemical fertilizer use by taking into account the mineral and nutritional values of the previous year's crop residues.

- Planting trees, shrubs, and grasses in areas susceptible to erosion.

Chesapeake Bay Agreements of 1987 and 1992

By 1987, a sophisticated computer model of Chesapeake Bay had been developed to forecast future water quality scenarios, including dissolved oxygen levels, turbidity, and seagrass areas. The model promises to be useful in watershed decision-making. For example, the model projects that, without additional nutrient reduction, areas of the bay currently experiencing low oxygen levels will increase by 15 percent to 20 percent over the next decade. These losses could trigger additional problems.

The parties to the Chesapeake Bay Program met in 1987 and adopted a new Chesapeake Bay Agreement based on scenarios developed using the computer model. The 1987 agreement aims to reduce all "controllable" nutrients entering the bay by 40 percent by the year 2000. Controllable sources are defined as agricultural runoff as well as municipal and industrial wastewaters originating from the states that signed the agreement. As much as 50 percent of nitrogen-based and 70 percent of phosphorus-based nutrients that enter bay waters are considered controllable. Noncontrollable sources include runoff from states not included under the bay agreement and from atmospheric deposition (see Figure 4.6). The computer model projects the following outcomes for a 40 percent reduction in controllable nutrient levels:

1. Bay areas subject to periodic anoxia will be reduced by 25 percent.

2. Estuarine cultural eutrophication will be reversed.

3. Water turbidity will decline, and seagrasses will revegetate in deeper waters.

4. The overall ecology of the bay will improve.

Phosphorus Reduction   Reduction of controllable nutrients from point sources is measured against point-source discharges in 1985, the baseline year. Non-point source reductions are measured against a three-year average spanning the mid-1980s. By 1989, phosphates in laundry detergents had been banned in Maryland, Virginia, and Pennsylvania. This, together with improved phosphorus removal techniques at wastewater treatment plants, produced a 27 percent decline in total

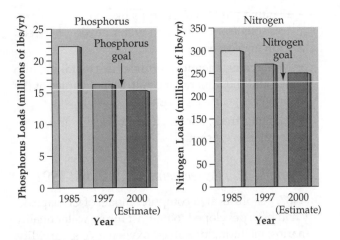

Figure 4.8 Total phosphorus and nitrogen nutrient loading to Chesapeake Bay from all sources, controllable and noncontrollable: 1985, 1997, and 2000 (estimates). Nitrogen loads declined 32 million lbs./yr. Maintaining reduced nutrient levels after 2000 will be a challenge due to expected population growth in the region.
(*Source: Data from the Chesapeake Bay Program Phase IV Watershed Model, Annapolis, Maryland, 1998.*)

phosphorus added to the bay from all sources. The computer model predicts that a 40 percent decline in phosphorus from controllable sources will lower total phosphorus by about 31 percent. Therefore, the phosphorus reduction goal is in sight and is expected to be reached by 2000 (Figure 4.8).

Nitrogen Reduction   Reduction in nitrogen is another matter. It is inherently more difficult and more expensive to remove nitrogen-based nutrients from municipal and industrial wastewaters. Nevertheless, there has been an overall decline of about 11 percent in total nitrogen added to the bay from all sources, controllable and noncontrollable (see Figure 4.8). The computer model predicts that a 40 percent decline in nitrogen from controllable sources alone will lower total nitrogen by about 23 percent. It appears that the goal for nitrogen reduction will not be achieved by 2000.

Hopefully, both goals will be achieved in the near future, but it will be difficult to maintain reduced nutrient levels, given the bay area's rising population, increasing agricultural activity, and urban/suburban growth.

Fishery Management   In addition to nutrient reduction goals, the 1987 agreement established fishery management plans for 19 critically important finfish and shellfish species, including striped bass, blue crab, oysters, summer flounder, American eel, yellow perch, largemouth bass, shad, and herring. In the case of blue crabs, a size limit was placed on harvested crabs—five inches for males and immature females. Special rules

were adopted to protect female crabs and ensure future spawning stocks. Recreational catch limits were set, and numbers of commercial harvesters limited. In the case of striped bass, both Maryland and Virginia restricted commercial and sport fishing for a number of years. A remarkable recovery followed, and the fishery was reopened. In fact, the recovery occurred so quickly that biologists decided the problem was overfishing, not estuarine decline.

The parties to the Chesapeake Bay agreements met again in 1992 and reaffirmed their 40 percent nutrient reduction goals. In addition, they adopted:

- Specific nutrient reduction goals for each of the bay's major tributaries including non-point atmospheric sources to reduce nitrogen loading.

- A plan to explore working relationships with the other three watershed states—New York, West Virginia, and Delaware—to develop additional strategies for nutrient reduction in the Chesapeake.

The Chesapeake Today: Indicators of Recovery

Those involved in the Chesapeake Bay Program agree that the best indicator of the bay's recovery will be the regrowth of its seagrasses. With this in mind, the executive council of the program set as a goal the restoration of 46,000 hectares (≈ 114,000 acres) of bay seagrasses by the year 2005. Progress toward meeting that goal had already been seen in the early 1990s. From a historic low of 15,800 hectares (≈ 39,000 acres) in 1984, seagrass areas in the bay increased to 28,300 hectares (≈ 70,000 acres) by 1997, a 79 percent increase. While declines were noted in 1994 and 1995, they were followed by increases in 1996 and 1997. The magnitude of seagrass recovery in the 1990s is significant, but it must be noted that today's total area represents only 14 percent of the seagrass meadows that existed in earlier times. Areas of bay seagrasses for the years between 1978 and 1997 are plotted in Figure 4.9. A computer-generated plot of seagrass meadows comparing areas in 1984 with those in 1995 is shown in Figure 4.10.

Seagrass recovery in the Potomac River has been dramatic. The river had no seagrasses in 1978, but advanced wastewater treatment at the Blue Plains Treatment Plant has had a substantial impact. Phosphorus loading from the river's main point source was reduced, water turbidity declined, light transmission to bottom waters increased, and a considerable regrowth of the river's seagrasses took place. Seagrass recovery

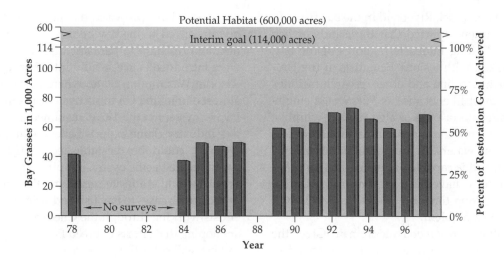

Figure 4.9 Areas of Chesapeake Bay seagrasses for years between 1978 and 1997.
(Source: Data from The Chesapeake Bay Program, Annapolis, Maryland, 1998.)

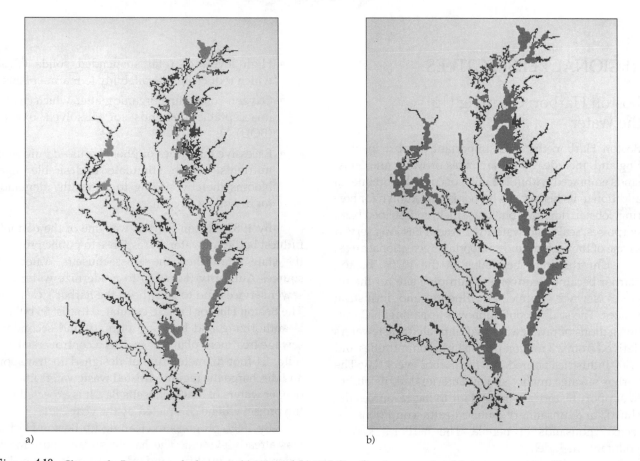

a)   b)

Figure 4.10 Chesapeake Bay seagrass bed areas in (a) 1984 and (b) 1995. Significant recovery in total seagrass areas is evident.
(Source: Chesapeake Bay Program, Annapolis, Maryland.)

has also occurred at other bay sites, particularly in the middle reaches and along the bay's eastern shores.

Despite several indications of recovery, a new problem is affecting the bay's living resources.

Beginning in 1992, damaging lesions on fish and outright fish kills were reported in Chesapeake rivers, including Maryland's Choptank, Patuxent, Manokin, and Pocomoke. Similar outbreaks were seen in

Virginia's Rappahannock River and in the bay itself. In 1997, bay watermen reported that thousands of fish had been killed along mideastern shores of the bay, their skin marked with lesions. Scientists at the University of North Carolina and other research facilities identified the most likely cause as *Pfiesteria*, a single-celled dinoflagellate (a type of alga having whiplike tails or flagella). *Pfiesteria* is a marine alga that prefers shallow estuarine waters. As a predator, it possesses a powerful toxin and is capable of causing fish kills. *Pfiesteria* is thought to have been present in the Chesapeake for a very long time, but no one knows why it has now become a problem.

Several other unresolved issues remain. Given the precipitous decline in Chesapeake Eastern oysters, fish-

ery experts have proposed that Japanese Pacific oysters be introduced to the bay because they are hardy, prolific, and resistant to parasitic diseases. Pacific oysters were introduced successfully into France in the 1970s. Also, in Washington State, where overharvesting had almost eliminated Olympia oysters, the introduction of Pacific oysters in the 1920s sustained a highly productive industry. Some experts believe that a major oyster industry could be reestablished in the Chesapeake based on the Pacific oyster, but others fear that the new species would eradicate remaining Eastern oysters and introduce diseases to bay fish and wildlife. For now, scientists working through the Chesapeake Bay Program are hoping that efforts to breed Eastern oysters resistant to MSX will solve the problem.

REGIONAL PERSPECTIVES

Boston Harbor: Cleaning Up the Water

Boston Harbor, the most important seaport in New England, includes 50 square miles of water and 30 islands embraced within 180 miles of coastal shoreline. In addition to its history, the harbor is well known for herring, lobster, flounder, bluefish, smelt, cod, striped bass, porpoises, seals, and waterfowl. And it has long served as one of the region's most important recreational areas.

Unfortunately, beginning in the 1970s, Boston Harbor became a convenient dumping site for the region's sewage, municipal effluents, and industrial wastes. The scale of pollution was appalling: 500 million gallons of raw sewage and untreated wastewaters derived from 2.3 million people in 43 communities and 5,500 industries poured into the harbor every day. The harbor's water quality slowly deteriorated, its ability to support historically important living resources declined, and unsanitary public health conditions affected thousands of people. The harbor's known pollutants include:

- Pathogenic organisms, as indicated by high levels of coliform bacteria.
- Toxic materials, including metals such as arsenic, cadmium, chromium, copper, lead, mercury, nickel, and zinc as well as organic compounds such as benzene, naphthalene, petroleum hydrocarbons, DDT, and PCBs.

- High levels of total suspended solids (TSS), which reduce light availability to marine plants.
- Oxygen-consuming organic matter, which creates unacceptable demands for dissolved oxygen (BOD).
- Excessive quantities of phosphorus and nitrogen nutrients, which stimulate undesirable algal blooms thereby adding to unhealthy demands for dissolved oxygen.

By 1985, Boston Harbor was one of the nation's filthiest bodies of water. In response to public protests, the state established the Massachusetts Water Resources Authority (MWRA) to modernize water and sewer services and to spearhead the harbor's cleanup. The Boston Harbor Project, estimated to cost $6 billion, is authorizing and building primary and secondary sewage treatment plants as well as a controversial 9.5-mile, 24-foot-diameter tunnel designed to transport and discharge the region's treated wastewaters into the deeper waters of Massachusetts Bay. It is expected that the project will be completed by the year 2001.

By 1999, progress in cleaning up Boston Harbor was already evident. The harbor's water now meets federal water quality standards most of the time. The number of summertime beach closings declined by 50 percent between 1995 and 1999. BOD and TSS levels have fallen steadily since 1992; liver tumors, once frequently seen in bottom fish such as flounder, are no longer found; and increasing numbers of porpoises and seals are returning to the harbor. But the restoration of Boston Harbor is costly. In many cases, Boston's water and sewer rates have risen drastically from $140 to $855 a year.

Louisiana: Wetlands Decline

Coastal wetlands are among the earth's most highly prized environments. They are not unlike estuaries in that they are inundated periodically by both freshwater and seawater. And, like estuaries, they serve as spawning grounds for innumerable species of finfish and shellfish, overwintering grounds for migrating birds, and critical habitat for waterfowl and aquatic wildlife. Taken together, estuaries and coastal wetlands are essential for at least 70 percent of America's seafood production, including shrimp, oysters, mussels, and clams.

Coastal wetlands also serve as biological filters by trapping, decomposing, and detoxifying pollutant runoff and discharges from agricultural sites, cities, manufacturing plants, mining operations, and petrochemical industries. In marshlands, where submerged vegetation consumes nutrient runoff, bacteria and other organisms biodegrade pesticides, and sediments adsorb heavy metals.

Louisiana's coastal wetlands account for close to 40 percent of total coastal marshes in the United States, but today, unfortunately, they have become one of the best examples of coastal wetland decline in the nation. Dramatic losses in wetland and marshland areas have occurred over the last three decades, along with adverse impacts on fish, shellfish, and wildlife. The causes of the decline are many, but the main reasons appear to be:

- Flood control projects on the Mississippi River using levees to keep the river within its banks have channelized the river, preventing normal, periodic re-channelizing of the delta at the mouth of the Gulf of Mexico. Formerly, re-channelizing in the delta resulted in continuous distribution of river sediments along the Louisiana coast, thereby creating and maintaining essential marshlands. But now the channelized river simply carries sediments out into the Gulf where they are lost to the sea.

- Dredging of what some people have regarded as unwanted and unneeded swampland has led to economic developments in Louisiana's coastal zone, including the creation of lucrative oil and gas industries. These developments, together with the release of hazardous industrial wastes of many kinds, have diminished the ecology of marshlands and the environmental services they provide.

- The increasing frequency of tropical storms, coupled with sea-level rise (phenomena many scientists believe are caused by global warming) are accelerating the pace of erosion in Louisiana's coastal marshes.

Major studies are underway to determine what actions and priorities are needed to reverse the loss of Louisiana's coastal wetlands.

Europe: The Black Sea Ecosystem

The Black Sea is an almost totally isolated inland marine ecosystem. Shaped like a butterfly, the sea occupies the southeastern corner of Europe. It receives freshwater inflow from precipitation and major rivers as well as seawater inflow through its connection with the Mediterranean Sea. Almost one-third of continental Europe's land area drains into the Black Sea through rivers such as the Danube, Dniester, Dnieper, and Southern Bug.

The Black Sea's coastal zone is densely populated, with close to 16 million permanent residents plus another four million visitors during the summer tourist season. Most of the region's people occupy seaside towns such as Kocaeli, Samsun, and Ordu in Turkey, and Odessa, Nikoleav, and Sevastopol in Ukraine. Sewage pollution originating from these urban areas adds to polluted runoff from the Black Sea's catchment basin, runoff carrying poorly treated sewage, and agricultural chemicals.

For example, the Danube River, the longest and most important river in central Europe, flows through parts of Germany, Austria, Hungary, Yugoslavia, Romania, Czechoslovakia, and Bulgaria. Its accumulated agricultural and municipal runoff is discharged to the Black Sea, adding large quantities of nitrogen and phosphorus nutrients each year. The cultural eutrophication problem created by added nutrients is reflected in massive algal blooms, diminished availability of light to submerged seagrasses in shallow areas, and dramatic declines in fish populations, especially sturgeon, bluefish, bonito, swordfish, tuna, and mackerel. Fishery resources are further strained by overexploitation of limited fish stocks. Taken together, eutrophication and overfishing are leading to catastrophic degradation of the Black Sea fishery.

Yet, despite well-documented declines in the Black Sea ecosystem, it is important to remember that the sea is historically very fertile. It has naturally high levels of nitrogen and phosphorus due to runoff from its extensive drainage basin and its isolation from the open ocean. The effect of this fertility is higher-than-typical algal populations and bacterial decomposer organisms, resulting in greatly diminished levels of dissolved oxygen below about 180 meters (≈ 590 feet). In fact, nearly 87 percent of the sea's total water volume is anoxic. The sea's contemporary problems derive from greater-than-normal loading of nutrients.

A strategy to save the Black Sea was developed in 1992. The Bucharest Convention, signed by representatives from Bulgaria, Georgia, Romania, Russia, Turkey, and Ukraine, created a basic framework with these goals:

- Control land-based sources of pollution.
- Enforce new rules regarding the dumping of municipal and industrial wastes.
- Engage in joint action in cases of oil pollution and other industrial accidents.

The Black Sea Environmental Programme was established in 1993. Black Sea countries were awarded $2 billion (U.S.) by the Global Environmental Facility to be managed through the World Bank and the U.N. Development Programme. This marks the beginning of coordinated efforts to reverse ecological degradation in the Black Sea.

KEY TERMS

advanced wastewater treatment

anadromous fish

anoxia

best management agricultural practices

biological oxygen demand

catadromous fish

crustaceans

cultural eutrophication

electromagnetic spectrum

epiphytes

filter feeders

heavy metals

hydrophytic plants

hypoxia

limiting nutrient principle

natural eutrophication

negative feedback

population crash

reaeration

salinity

turbidity

wetlands

DISCUSSION QUESTIONS

1. What physical, biological, and chemical factors make estuaries such as the Chesapeake so biologically productive? Identify and discuss as many as you can.

2. Many of the contaminants entering Chesapeake Bay and its tributaries come from industrial wastes that are piped directly to public wastewater treatment plants. Many of these plants do not have adequate technology to remove pollutants such as heavy metals. Should industries be required to pretreat their wastes before conveying them to wastewater treatment plants? Discuss your answer.

3. It appears that the first serious effort to control pollution entering Chesapeake Bay began on the Potomac River. Why? Discuss your answer.

4. Soil runoff in the Chesapeake Bay watershed occurs largely from agricultural lands that were once forested. Given this fact, would you advocate a major program of watershed reforestation? Discuss your answer.

5. Computer models can at best simulate environmental systems. Forecasts based on them are approximate and subject to uncertainty. Given this uncertainty, would you argue for major investments in advanced wastewater treatment plants and costly best management practices to improve Chesapeake Bay? Discuss your answer.

INDEPENDENT PROJECT

A research project involves a 40.5-hectare (≈ 100-acre) farm under cultivation in a temperate region of the world. The farm was once an undisturbed forest of mixed coniferous and deciduous trees, but 50 years ago the trees were cleared and traditional farming started. No soil erosion practices were adopted.

1. Calculate how much soil (in kilograms or pounds) is likely to have been lost through erosion over this 50-year time span.

2. Assume that farming stopped at this site, and natural revegetation led to a stand of young vine maples and Douglas firs in 10 years. After 20 years, a fairly dense young forest has grown up.

3. Estimate, using information provided in this case study, how long it would take for soil formation to replace the amount of soil lost earlier on the farm.

4. What steps could have been taken to reduce the amount of time needed to replace soil that was lost while this land was being farmed?

INTERNET WEBSITES

Visit our website at http://www.mhhe.com/environmentalscience/ for specific resources and Internet links on the following topics:

> Chesapeake Bay Homepage
>
> Chesapeake Bay Basin-Wide Information Site (BIOS)
>
> Animation of changes in Chesapeake Bay seagrasses since 1984
>
> State of Maryland - Sea Grant Program site
>
> Chesapeake Bay *Pfiesteria* site

SUGGESTED READINGS

Baker, Beth. 1992. Botcher of the bay or economic boom? *Bio Science* 42(10):744–747.

D'Elia, Christopher F. 1987. Nutrient enrichment of the Chesapeake Bay. *Environment* 29(2):6–11; 30–33.

Environmental Indicators. 1998. Annapolis, MD: Chesapeake Bay Program.

Horton, Tom. 1993. Chesapeake Bay—Hanging in the balance. *National Geographic* 183(6):2–35.

Horton, Tom, and William M. Eichbaum. 1991. *Turning the tide*. Annapolis, MD: Chesapeake Bay Foundation.

Leffler, Merrill. 1995. The new oyster wars: Battling disease in the lab and the bay. *Maryland Marine Notes, Summer 1995*. College Park, MD: Maryland Sea Grant College.

Murphey, Deirdre. 1990. *Contaminant levels in oysters and clams from the Chesapeake Bay 1981–1985*. Baltimore, MD: Department of the Environment, Water Quality Programs.

Nutrient reduction progress & future directions. 1997. Annapolis, MD: Chesapeake Bay Program.

Pfiesteria piscicida: A toxic dinoflagellate associated with fish lesions and fish kills in mid-Atlantic Coastal Waters. 1998. Washington, D. C.: U.S. Environmental Protection Agency, Office of Water.

The last wetlands. 1990. *Audubon Magazine* (July).

The Chesapeake Bay Program: The state of the Chesapeake Bay. 1995. Washington, D. C.: U.S. Environmental Protection Agency.

Virginia Institute of Marine Science. 1998. *Distribution of submerged aquatic vegetation in the Chesapeake Bay*. Gloucester Point, VA: College of William and Mary.

CHAPTER 5

Air Pollution: History, Legislation, and Trends

PhotoDisc, Inc./Vol. 31

... the smoke ... overhangs the town and descends in fine dust which blackens every

object; even snow can scarcely be called white in Pittsburgh.

Zadok Cramer's description of Pittsburgh in the
early 1800s from his book *The Navigator*

The point where the Allegheny and Monongahela Rivers meet in western Pennsylvania gained great importance in the early 1700s. Its proximity to rivers made it a crossroads for thousands of America's early settlers; its unique geography spurred rapid growth, industrial development, and the rise of a major city. Pittsburgh's early industries depended heavily on the region's rich deposits of bituminous coal. However, burning this coal released smoke, soot, and sulfur dioxide into the air, so that almost every day, clouds of smoke blocked the sun's light, turning the sky shades of gray and making the city a dismal place to live.

After World War II, the first effective efforts were made to control air pollution in the United States. The Clean Air Act of 1970 established national air quality standards and timetables for implementing them. Twenty years later, the Clean Air Act of 1990 reauthorized these standards and focused attention on remaining air pollution issues. When local and regional air quality regulations were finally adopted and enforced, air quality in Pittsburgh improved greatly. Today, the city is a model urban center where air pollution is no longer a major daily problem.

This chapter's case study traces the story of Pittsburgh's air quality. But in order to understand those events, we will begin by providing some basic information about the earth's atmosphere and the nature of air pollution.

BACKGROUND

The Nature of the Earth's Atmosphere

The earth's atmosphere is envisioned as four concentric layers: the troposphere or lower atmosphere, stratosphere, mesosphere, and thermosphere (Figure 5.1).

Troposphere

The **troposphere** extends from ground level to about 12 kilometers (≈ 7.4 miles) above the earth's surface. Over 90 percent of all atmospheric gases (by weight) are found in this layer—mainly nitrogen, oxygen, argon, water vapor, and carbon dioxide. The earth's lower atmosphere consists of about 78 percent nitrogen gas and 21 percent oxygen gas. The relative abundance of the most important chemical components in the earth's troposphere is given in Table 5.1. The troposphere also contains **aerosols**, very small solid and liquid particulates made up of sea salts, pollen, dust, water droplets, fogs, and mists. Atmospheric particulates vary in diameter from about 0.01 μm to 100 μm. (μm = micrometer, or micron; 1 μm equals one-millionth of a meter. It would take about 100 μm to equal the thickness of one page in this book.)

Air temperature in the troposphere declines with increasing elevation by approximately 6.5 °C per kilometer (≈ 3.5 °F per 1,000 feet). This decline, known as a

negative lapse rate, reaches a minimum temperature close to −40 °C (−40 °F) just below the stratosphere. Meteorological events (weather) occur only in the troposphere, not in higher atmospheric regions. The earth's biosphere is often defined as being made up of soil, water, and the troposphere.

Stratosphere

The **stratosphere** lies above the troposphere, beginning approximately 12 kilometers (≈ 7.4 miles) above the earth and extending to an elevation of 52 kilometers (≈ 32.4 miles) above the planet's surface. The temperature in the stratosphere rises with increasing elevation (a positive lapse rate), reaching a maximum of about 0 °C (32 °F) at the stratosphere's outer limit. The temperature rise is due mainly to the absorption of intense solar ultraviolet (UV) radiation by **ozone,** an important component of the stratosphere concentrated in a layer approximately 19 to 26 kilometers (≈ 12 to 16 miles) above the earth's surface (see Figure 5.1). The fact that intense UV radiation is absorbed (at least in part) by the stratosphere's ozone layer is critically important. Indeed, this shielding of the earth's biosphere from potentially damaging ultraviolet rays makes life on earth possible (see Chapter 13). The troposphere and stratosphere together account for about 99 percent of the planet's atmosphere by weight.

Mesosphere and Thermosphere

Two additional well-defined atmospheric layers lie above the stratosphere. The **mesosphere** begins at an elevation of about 52 kilometers (≈ 32.4 miles), where stratospheric temperatures reach a maximum. Temperatures gradually decline in the mesosphere to about −60 °C (≈ −76 °F) at 87 kilometers (≈ 54 miles) elevation. The **thermosphere** extends beyond the mesosphere. Consistent with its name, temperatures there are as high as 1,200 °C (≈ 2,200 °F) due to the absorption of solar energy by thinly dispersed oxygen and nitrogen molecules.

The Importance of Oxygen

It is believed that the earth's primitive atmosphere contained little or no oxygen. Over geologic time, photosynthetic plants produced oxygen, ultimately making it one of the principal components of the planet's gaseous envelope. Oxygen is of critical importance to living things for several reasons:

1. Almost all plants and animals exhibit **aerobic respiration,** a fundamental cellular-level biochemical process whereby energy stored in organic

TABLE 5.1 Principal Gases in the Troposphere and Their Relative Abundance in Dry Air

Gas	Formula	Percent
Nitrogen	N_2	78.084
Oxygen	O_2	20.9476
Argon	Ar	0.934
Carbon dioxide	CO_2	0.0360
Neon	Ne	0.001818
Helium	He	0.000524
Methane	CH_4	0.0002
Krypton	Kr	0.000114
Hydrogen	H_2	0.00005
Xenon	Xe	0.0000087

SOURCE: Data from *U.S. Standard Atmosphere,* National Oceanic and Atmospheric Administration (NOAA), National Aeronautics and Space Administration (NASA), and United States Air Force, 1976. Carbon dioxide has been updated to 1997.

food matter is released through metabolism. Oxygen is consumed and carbon dioxide is released.

2. Many geochemical processes, most notably weathering of minerals, depend on oxygen. Oxygen influences the chemistry of the planet, causing many elements such as iron, copper, and manganese to occur mainly as their oxides.

3. The decomposition of organic matter and the recycling of chemical nutrients generally involve oxygen.

4. Oxygen converts potentially harmful atmospheric gases to less harmful or harmless substances, although this process may occur slowly. For example, carbon monoxide is oxidized to carbon dioxide; methane is oxidized to water and carbon dioxide; and hydrogen sulfide is oxidized to water and sulfur dioxide.

Atmospheric Inversions

Air circulation and mixing occur continuously in the troposphere, making the lower atmosphere essentially homogeneous. That is, air samples collected at different locations in the troposphere all have essentially the same percentage distribution of gases. Atmospheric circulation is facilitated because air warmed near the earth's surface expands, becoming less dense (more buoyant) than cooler air at higher elevations. Warm air rises and is displaced by cooler air. At times, a warm

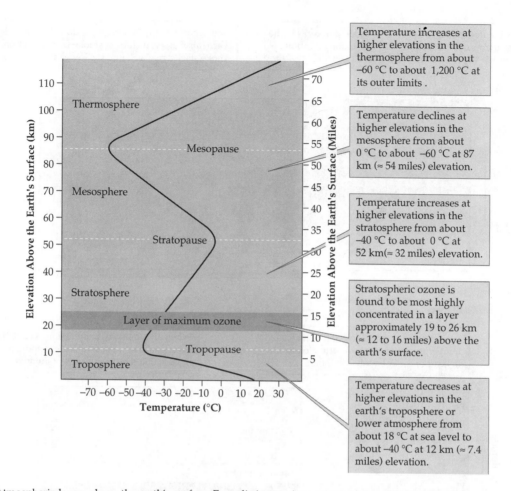

Figure 5.1 Atmospheric layers above the earth's surface. Four distinct regions exist in the earth's atmosphere: the troposphere or lower atmosphere, stratosphere, mesosphere, and thermosphere. Their boundaries are defined by temperature transitions.

air mass traps a cooler one below it, acting as a lid and thereby preventing normal circulation. This event is known as a **thermal inversion.** It commonly takes place in valleys and on the windward side of mountain ranges where an air mass can stagnate in the absence of wind. At times, inversions aggravate air pollution by trapping pollutants close to the ground where they build up (Figure 5.2).

One of the best-known examples of air pollution worsened by an inversion occurred in Donora, Pennsylvania, in 1948. Donora, a small town south of Pittsburgh in the Monongahela Valley, had steel mills and zinc smelters that burned high-sulfur coal. The town also had a manufacturing plant that produced sulfur dioxide to make sulfuric acid. On October 26, a thermal inversion occurred, trapping sulfur dioxide and carbon monoxide fumes in Donora for five days. Air pollution increased to toxic levels, and close to half the town's 14,000 people became severely ill. Twenty people died.

In December 1952, a similar atmospheric inversion caused a dense smog to stagnate for four days over London, England. The smog originated from coalfired steel mills, coal-burning power plants, and households that burned coal. More than 4,700 deaths were attributed to this smog event, primarily due to respiratory illnesses such as bronchitis, influenza, pneumonia, tuberculosis, emphysema, and asthma worsened by diminished air quality. London's Great Smog of 1952 led to England's first air pollution legislation, its Clean Air Act of 1956.

Similar episodes of air pollution have taken place elsewhere. In July 1970, stagnant, polluted air plagued a number of U.S. urban areas, including New York, Philadelphia, Baltimore, and Washington, D.C., and caused several public health alerts. The ability to forecast thermal inversions has led to better local and regional air pollution management practices.

Fundamentals of Air Pollution

Air pollution is largely an urban problem. Population centers like Pittsburgh are where most fuels are burned to generate electricity, power industrial production,

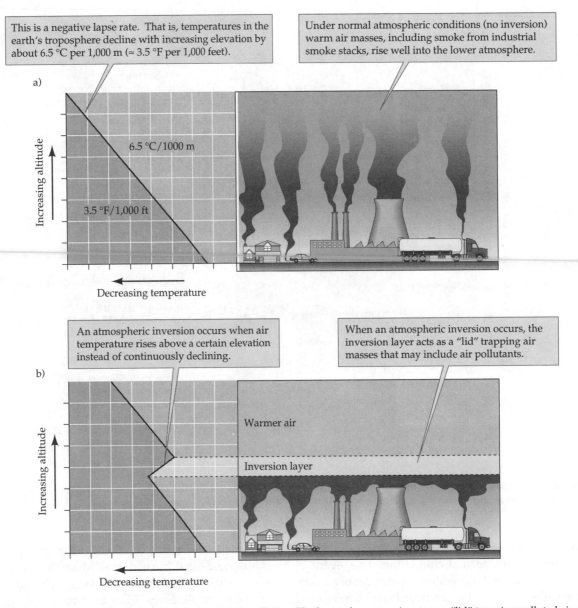

This is a negative lapse rate. That is, temperatures in the earth's troposphere decline with increasing elevation by about 6.5 °C per 1,000 m (≈ 3.5 °F per 1,000 feet).

Under normal atmospheric conditions (no inversion) warm air masses, including smoke from industrial smoke stacks, rise well into the lower atmosphere.

a)

Increasing altitude

6.5 °C/1000 m

3.5 °F/1,000 ft

Decreasing temperature

An atmospheric inversion occurs when air temperature rises above a certain elevation instead of continuously declining.

When an atmospheric inversion occurs, the inversion layer acts as a "lid" trapping air masses that may include air pollutants.

b)

Increasing altitude

Warmer air

Inversion layer

Decreasing temperature

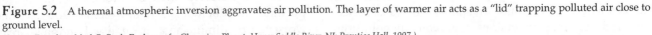

Figure 5.2 A thermal atmospheric inversion aggravates air pollution. The layer of warmer air acts as a "lid" trapping polluted air close to ground level.
(Source: Data from Mark B. Bush, Ecology of a Changing Planet, Upper Saddle River, NJ: Prentice Hall, 1997.)

and run transportation systems. Coal and petroleum are the fossil fuels most responsible for causing air pollution. They release the most common air pollutants—particulate matter, sulfur dioxide, and carbon monoxide. Of these, only particulate matter is visible. Sulfur dioxide, though invisible, is easily sensed because of its biting, pungent odor. Carbon monoxide cannot be seen or smelled but is highly toxic.

Particulate Matter

Burning fossil fuels can add visible particulate matter to the atmosphere—smoke and soot resulting from *incomplete combustion.*

Incomplete combustion of fossil fuels like coal occurs when oxygen is limited. It produces unburned particulate matter and carbon monoxide, whereas *complete combustion* produces carbon dioxide and other products.

The very small particles resulting from fossil fuel combustion remain suspended in the atmosphere for days, months, or even years, depending on their size, circulation patterns, and particle densities. Carbon particulates subjected to high temperatures in internal combustion engines, coal-fired power plants, and cigarettes are transformed in part into complex organic chemicals called **polynuclear aromatic hydrocarbons (PAHs).** PAHs are hydrocarbons containing benzene-

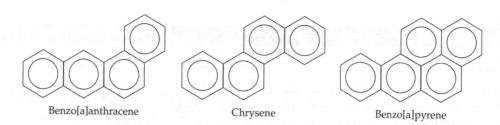

Benzo[a]anthracene   Chrysene   Benzo[a]pyrene

Figure 5.3 Three of the 16 polynuclear aromatic hydrocarbons that the U.S. Environmental Protection Agency has identified as probable human carcinogens: benzo[a]anthracene, chrysene, and benzo[a]pyrene. It is believed that benzo[a]pyrene, a component of cigarette smoke, plays a significant role in causing lung cancer.

like (six-membered) rings fused together. The U.S. Environmental Protection Agency (EPA) has identified 16 PAHs as probable carcinogenic (cancer-causing) agents in humans. Figure 5.3 shows the chemical structures of three of these hazardous PAHs: benzo[a]anthracene, chrysene, and benzo[a]pyrene. Medical authorities believe that benzo[a]pyrene, a component of cigarette smoke, plays a significant role in causing lung cancer.

Particulates in smoke can be 10 μm or less in diameter, making them a public health concern because they are **respirable.** That is, they escape capture by nasal hairs, respiratory tract mucus, and cilia, and invade air sacs in the lungs, where they cause tissue injury and increase the seriousness of pneumonia, bronchitis, and asthma. Respirable particulates in cigarette smoke are a leading cause of emphysema, and are also believed to cause lung cancer and cardiovascular disease.

Carbon particulates, because of their porosity, possess very large surface areas. This property enables them to **adsorb,** or bond, to the surface of metals such as cadmium, lead, zinc, and chromium, chemical elements identified as **heavy metals.** Heavy metals adsorbed on respirable particulates are also a public health concern. For example, air-borne lead associated with particulate matter is highly toxic to humans, especially children.

Sulfur Dioxide

Sulfur dioxide is a colorless, invisible gas with a biting, acrid, pungent smell. Volcanic eruptions release quantities of this gas that can affect downwind atmospheric conditions. Anthropogenic (human-related) sulfur dioxide emissions are due mainly to burning coal. For example, the combustion of one ton of bituminous coal releases about 82 kilograms (≈ 180 pounds) of sulfur dioxide to the atmosphere. Sulfur dioxide is also released, although to a lesser extent, when petroleum is burned. The principal environmental impacts of sulfur dioxide are:

- Photosynthesis in plants is impaired because the gas oxidizes plant chlorophyll, producing yellowed leaves and needles, a condition called **chlorosis**. At high levels, sulfur dioxide causes vegetative dieback of crops, shrubs, and trees.

- In animals and humans, sulfur dioxide irritates throat and nasal passages and exacerbates lung and respiratory tract illnesses.

- Atmospheric sulfur dioxide reacts with moisture and oxygen to form acidic mists, fogs, and acid rain.

Carbon Monoxide

Carbon monoxide occurs naturally in the earth's atmosphere at very low levels, about 0.15 parts per million (ppm). Most of it is formed through the oxidation of atmospheric methane, which is released from marshes and wetlands as plant matter decays. We will study the role of methane in global warming in Chapter 12.

Coal-based industries in Pittsburgh released considerable amounts of carbon monoxide to the air as a result of the incomplete combustion of coal. While not very harmful at low levels, carbon monoxide at higher levels combines with blood hemoglobin, forming carboxyhemoglobin, which diminishes oxygen transport in humans and animals. At 50 ppm, carbon monoxide impairs human reflexes and adversely affects vision. At 100 ppm, headache, fatigue, and shortness of breath occur. At 750 ppm or more, unconsciousness and death can result.

Carbon monoxide pollution is common in metropolitan areas where automotive traffic is concentrated. Incomplete combustion of gasoline and diesel fuel is the principal cause. Carbon monoxide levels are reduced by catalytic converters, devices installed in automobiles to help control air pollution, but unhealthy levels can still accumulate under certain circumstances—for example, in parking garages, tunnels, and urban areas where atmospheric inversions occur.

INSIGHT 5.1
Acid Rain

In theory, pure water has a pH of 7.0 and is neither acidic nor basic. A pH below 7.0 reflects acid conditions, while a pH greater than 7.0 reflects basic conditions (Figure 5.A). pH is therefore a measure of how acidic or basic an environment is.

Unpolluted rain is often slightly acidic (pH < 7.0) because atmospheric carbon dioxide gas is dissolved in it. Carbon dioxide reacts with water to form carbonic acid, which lowers rainwater's pH to about 5.6. However, this, is not considered acid rain. By definition, acid rain refers to precipitation having a pH lower than 5.6.

Acid rain occurs when rainwater contains acid-forming contaminants such as sulfur dioxide (SO_2) and oxides of nitrogen (NO_x). The principal source of sulfur dioxide is the combustion of sulfur-containing fossil fuels, particularly coal and petroleum. Once in the atmosphere, sulfur dioxide is converted to sulfuric acid (H_2SO_4), a very strong acid. Traces of sulfuric acid can cause rainfall pH to drop to 2.0 or even lower. Oxides of nitrogen produce a similar effect by forming strong acids such as nitric acid (HNO_3) in the atmosphere.

Acid rain falls periodically in Canada, southwestern England, Scandinavia, parts of Europe, and

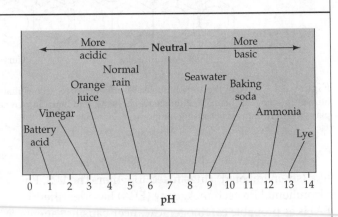

Figure 5.A   Typical pH values of common substances.

the northeastern United States. It erodes art and architectural works, corrodes metal structures such as bridges, and defaces tombstone inscriptions. Acid rain is also implicated in the decline of freshwater aquatic fish and wildlife. One of the best examples of this has occurred in New York's Adirondack State Park (see Regional Perspectives at the end of this chapter).

Acid Rain

More recently, air pollution has been linked to **acid rain,** precipitation that is more acidic than normal. Acid rain damages historic artworks, statues, and classic architecture and erodes names, dates, and markings on cemetery gravestones. It also impairs fish reproduction. Acid rain is discussed in more detail in Insight 5.1.

London Smog

When air is stagnant, during times of diminished air circulation or when an atmospheric inversion occurs, the pollutants we have just considered accumulate and may exert toxic effects on plant and animal life. Since this type of air pollution occurs most often in older industrial cities like London, it is called **London smog.**

L.A. Smog

In 1942, airline pilots were the first to describe **L.A. smog.** It appeared as a recurring yellow-brown haze over Los Angeles, California. The nature of L.A. smog

remained a mystery for years. Air particulates and sulfur dioxide were not responsible, but no one had a clue as to why the city's air was often irritating to the eyes, nose, and throat. In the early 1950s, Arie J. Haagen-Smit, a biochemist, trapped a sample of L.A. smog in a test tube and found that it contained a complex brew of chemicals, including hydrocarbons and peroxides. This suggested to him that automotive hydrocarbon emissions were involved in the formation of L.A. smog.

After considerable research, it was found that L.A. smog is caused by a combination of factors: atmospheric hydrocarbons and oxides of nitrogen (NO_x)—both from automotive exhaust—plus solar radiation. A number of potentially toxic chemicals, including ozone (O_3), are formed.

The low levels of ozone that occur naturally in the lower atmosphere cause little harm to humans, plants, or animals. They result mainly from lightning strikes that split oxygen molecules into highly energetic oxygen atoms. When an energetic oxygen atom reacts with oxygen, ozone is formed:

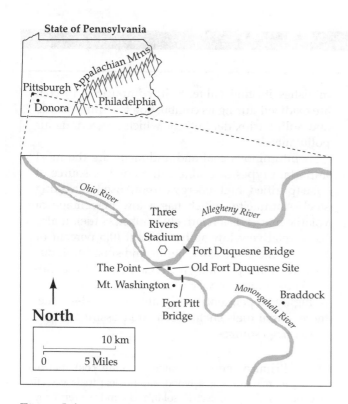

Figure 5.4 The confluence of the Allegheny and Monongahela Rivers, where Pittsburgh began.

Lightning

$$O_2 \rightarrow 2O$$

$$O + O_2 \rightarrow O_3$$

Ozone, depending on its concentration, can cause eye and nose irritation and is damaging to lung tissue. It is especially hazardous to the young and the elderly. Since ultraviolet radiation in sunlight is responsible for triggering this kind of pollution, L.A. smog is also called **photochemical smog.** The atmospheric hydrocarbons needed to generate L.A. smog originate from several sources:

1. A nearly empty automobile gas tank is nearly full of gasoline vapor. For this reason, gasoline vapor (mostly hydrocarbons) escapes from automobile gas tanks during refueling unless special devices are used.

2. Some unburned gasoline escapes from motor vehicle engines during normal driving. Catalytic converters, if they are working properly, help oxidize unburned hydrocarbons. Converters, however, do not maintain their initial efficiency.

3. Petroleum refineries, depending on their design and age, release hydrocarbons to the atmosphere.

4. Industries that use hydrocarbon solvents in their manufacturing processes may release hydrocarbon vapor to the air.

L.A. smog is found in cities where automotive density, intense sunlight, and frequent atmospheric inversions occur. The most serious events are reported in Los Angeles, San Diego, and Sacramento, California; Hartford, Connecticut; New York City; Philadelphia, Pennsylvania; Chicago, Illinois; Baltimore, Maryland; Houston, Texas; and Milwaukee, Wisconsin. Several approaches are being taken to limit photochemical smog. They include carpooling, shifting from private to public transportation, redesigning car engines to lessen pollution, and using reformulated gasoline to improve automobile engine combustion. Reformulated gasoline will be discussed later in this chapter.

CASE STUDY: CLEAN AIR IN PITTSBURGH

Pittsburgh's Origins

The site where two rivers, the Allegheny and the Monongahela, join to form the headwaters of the Ohio River gained early military and commercial importance because the Ohio is a water route westward and one of the major tributaries of the Mississippi. In 1754, the French built Fort Duquesne on this site to defend their claims to the Ohio Valley. Four years later, during the French and Indian War, the English captured the fort and renamed it Fort Pitt in honor of the British First Minister, William Pitt the Elder (Figure 5.4). By the early 1760s, the 200 people living near Fort Pitt founded Pittsburgh, which was to become one of America's first industrial cities. Until the 1840s, Pittsburgh was an agricultural and fur trading area where Indians, trappers, explorers, and pioneers established fur-trading encampments and staging areas for their exploits.

Besides geography, another factor shaped Pittsburgh's destiny: the presence of abundant coal (Insight 5.2). Rich deposits of bituminous or soft coal extended for miles in all directions. Nearby Coal Hill (now Mt. Washington) abounded in coal and became one of the area's main mining sites. In his book *The Navigator*, Zadok Cramer wrote that Pittsburgh had as many as 50 coal mines in its early days. They were dug horizontally in nearby hills, and the coal was extracted by wheelbarrows, loaded into horse-drawn wagons, and delivered to factories and private homes.

The abundance of coal meant that even the poor had cheap fuel to heat their homes and cook their food. Access to coal also spurred the development of

INSIGHT 5.2

The Geologic Evolution of Coal

Coal is a rocklike mineral that forms slowly over geologic time. Much of today's coal dates back to about 390 million years ago. Different kinds of coal, referred to as **coal ranks,** were produced by a process called *coalification*. The process begins when partially decomposed terrestrial plant matter buried in bogs, marshes, and swamps is transformed by microorganisms into **peat,** a spongy, water-saturated organic material. While peat is technically not "coal," it is dried and used as a source of energy in some countries.

Over long time periods, peat, deposited under layers of soil and sediment, was subjected to increasing pressure and temperature. Air was excluded, water expelled, and a series of geochemical processes gradually transformed peat first into **lignite** (brown coal) and then into subbituminous coal, bituminous coal (soft coal), and anthracite coal (hard coal). The plant origin of coal can sometimes be seen in coal deposits. For example, leaf and stem remains are found in lignite, the lowest-ranking type of coal, and fossilized impressions of plant fragments are recognizable in higher-rank coals.

Table 5.A compares four different ranks of coal. Fixed carbon is a measure of coal's value as a fuel—its energy content reported as kilocalories per kilogram of coal. Volatile matter is low-molecular-weight matter easily expelled to the air when coal is heated. It is a source of air pollution. Ash, the material left after coal's combustion, consists mainly of inorganic minerals. The presence of water in coal diminishes its fuel value. Sulfur impurities in coal are oxidized during its combustion to sulfur dioxide and sulfur trioxide, both of which are serious air pollutants.

Bituminous coal and anthracite are the most important types of coal. Bituminous is a source of coke (purified, high-energy carbon) used in making steel. Anthracite, which has a low percentage of volatile matter, is a relatively smokeless fuel. It also has a relatively low sulfur content (0.5 percent or less). On the other hand, lignite and subbituminous coal are highly polluting fuels. Germany's Ruhr Valley, known for its heavy industry and air pollution, is highly dependent on lignite, a prevalent fuel there. Fossil fuels such as coal are classified as *primary energy* sources.

Primary energy sources include coal, petroleum, natural gas, wood, and nuclear fuels as well as geothermal, hydroelectric, solar, and wind power. They are used to run automobiles, buses, trains, aircraft, and ships; to operate factory machinery; to generate electricity; and to power domestic and commercial heating and cooling systems.

Secondary energy is electricity generated using primary energy sources—mainly coal, hydroelectric systems, and natural gas. Other secondary energy sources include batteries and fuel cells. The U.S. Energy Information Agency has estimated that by the year 2000, 38 percent of the world's total primary energy use will involve conversion into electric power.

TABLE 5.A   Average Composition and Energy Value of Different Ranks of Coal*

Coal Rank	Fixed Carbon (%)	Volatile Matter (%)	Ash (%)	Water (%)	Sulfur (%)	Energy Value
Anthracite	88	1	10	1	≈ 0.5	8,300
Bituminous	65	20	10	2	≈ 3–5	7,200
Subbituminous	45	30	10	15	≈ 0.5	5,600
Lignite	30	30	10	30	≈ 0.5	3,900

SOURCE: Data from the Federal Energy Regulatory Commission, 1997.

*Energy Values are kilocalories per kilogram.

Pittsburgh's early industries: copper, brass, and tin factories, iron and steel mills, shipbuilding and glassmaking enterprises, cotton and woolen mills, breweries, and distilleries. Industrial growth became visible in the skies over Pittsburgh as smoke filled the air most of the time. The smoke came from the incomplete combustion of bituminous coal.

By the late 1840s, scores of Pittsburgh's factories lined the Monongahela River. While some used river water for power, most relied on coal burning steam engines to operate their equipment. Coal was also used to fuel railroad steam locomotives and steamboats that transported goods and people up and down local rivers. Many of America's early immigrants came to Pittsburgh to mine coal and work in factories. By 1860, Pittsburgh had grown to 50,000 people, and as many as 200,000 tons of coal were being burned annually. The economy flourished, and the prevailing philosophy became, "Where there's smoke there's money!" Indeed, the overall birth and progress of the Industrial Revolution—first in England, then in America—could be measured in terms of smoke-filled skies

By the time of the Civil War, iron-making had become Pittsburgh's most important enterprise, and the city supplied cannons and cannonballs to both the North and the South. Pittsburgh also became the nation's leading glassmaking center. Such developments spurred the building of railroads and canals linking Pittsburgh with other American cities.

Although iron was of great importance, it is a brittle metal. Steel, an alloy of iron, is far stronger and less brittle, and thus ideal for constructing railway rails, bridges, and tall buildings. In 1872, Andrew Carnegie, while traveling in England, learned about the Bessemer process for making steel. He brought the idea back to the United States and built one of its first Bessemer steel plants in Braddock, just east of Pittsburgh. Steelmaking became big business in 1901 when J. P. Morgan acquired Carnegie's steel interests and established the United States Steel Company (now USX). By then, Pittsburgh had emerged as the nation's steel capital and one of the country's major industrial centers (Insight 5.3).

Air Pollution Concerns

By the 1860s, Pittsburgh had earned its reputation as the dirtiest, grimiest, grubbiest city in America. Bituminous coal was burning night and day to run trains, steamboats, and factories as well as to heat homes and cook meals. Its continuous combustion fouled the air and darkened the skies. The result of burning so much

coal was perhaps best described in an 1868 *Atlantic Monthly* article as "hell with the lid taken off."

The problem was worsened by frequent atmospheric inversions that trapped polluted air close to the ground, with the following consequences:

- Air pollution limited visibility. Indeed, areas of downtown Pittsburgh often could barely be seen from only short distances away. Streetlights burned 24 hours a day.
- Smoke and soot fell from the sky, depositing black grime on buildings, houses, artworks, and laundry on clotheslines.
- Eye, throat, and respiratory irritation occurred frequently, especially in susceptible individuals.
- Tree growth was commonly stunted and vegetation visibly damaged.

Strategies to Improve Air Quality

Ultimately, air pollution was recognized as more than just an aesthetic problem; public health and ecological issues were also involved. Thus, Pittsburgh began trying to improve its air quality.

In London, air pollution had been monitored by measuring how much particulate fallout, called **dustfall,** collected on the ground each day. Inches of dustfall became the yardstick of air pollution. Tall smokestacks were built there to abate air pollution, but they failed most of the time. This type of air pollution abatement strategy was also attempted in Pittsburgh. The stacks lifted pollutants above ground level, allowing air currents to carry them away. But although tall stacks were sometimes effective and prevailing winds blew smoke away from the city, they failed to solve the city's air pollution problems (Figure 5.5).

Pittsburgh's first air pollution ordinance was adopted in 1868. It attempted to control the city's appalling problems by substituting smokeless fuels such as coke and anthracite coal for bituminous coal. It directed that home-heating stoves must burn coal more efficiently. But high costs associated with implementing the ordinance led to its failure. There was no relief on the horizon for a seemingly unending air pollution problem.

In the 1880s, cheap natural gas discovered in Murraysville, Pennsylvania, was piped to Pittsburgh's homes and industries. Since natural gas produces no smoke when it burns, a dramatic improvement in air quality took place almost overnight. For a time, Pittsburgh experienced the best of two worlds: industrial activity and pollution-free skies. But unfortunately,

INSIGHT 5.3

Making Iron and Steel

Smelting iron ore to make iron metal was known in Egypt as early as 3,000 B.C. The process involves heating a mixture of iron ore and coal. Carbon in coal chemically reduces the ore producing iron. In the following chemical equation, C represents carbon; Fe_2O_3 is one type of iron ore; CO_2 is carbon dioxide gas; and Fe is iron metal:

$$3C + 2Fe_2O_3 \rightarrow 3CO_2 + 4Fe$$

Using a furnace to make iron dates back to the Middle Ages. The process was gradually improved through many centuries by using higher grades of iron ore, refining the ore prior to smelting, and using **coke,** a purified form of carbon, instead of coal. Iron made in this way is called pig iron.

The blast furnace concept that Andrew Carnegie brought back from England involves heating iron ore with coke and limestone. Coke is made in coke ovens by heating coal in the absence of air to remove volatile matter. Coke burns with a hotter, cleaner flame than coal. Limestone is used to form a glasslike slag to remove impurities. Figure 5.B shows a model of a blast furnace.

A large blast furnace can make 200 tons of pig iron a day. Pig iron contains 5 percent or more carbon and other impurities, causing it to be brittle. Steel, which generally contains 1.7 percent or less carbon, can be made directly from iron ore or indirectly from pig iron using a blast furnace. The high temperatures required for this process ($\approx 1,650\,°C$ or $3,000\,°F$) are achieved using coke as the fuel.

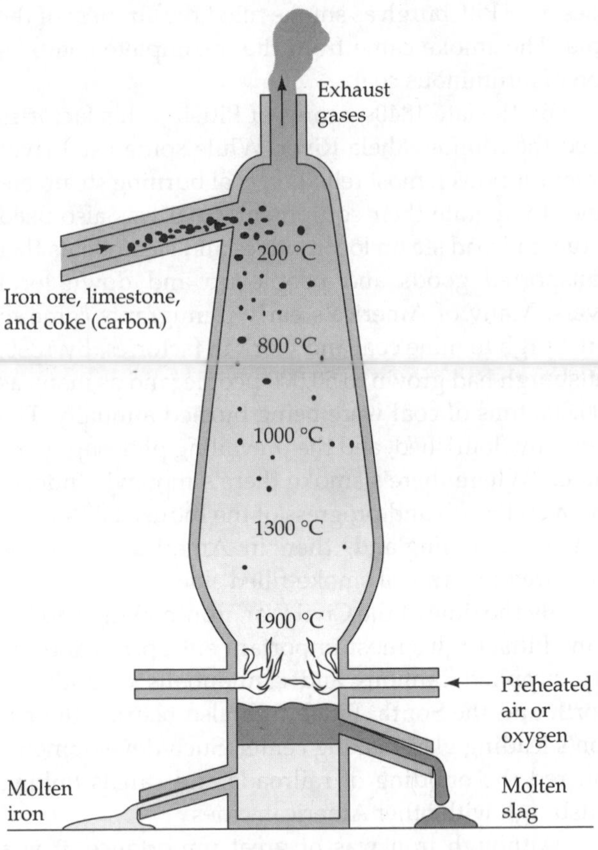

Figure 5.B   A blast furnace. Iron ore, limestone, and coke are added at the top of a heated furnace. Preheated air or oxygen is used to support combustion. Pig iron is formed from the molten iron removed at the base of the furnace. Impurities are removed in the molten slag.

---

local gas supplies lasted only six years. After that, Pittsburgh's industries resumed burning coal, and smoke-filled skies returned as well.

During World War II, transcontinental pipelines were built, connecting oil and natural gas wells in Texas and Oklahoma to eastern states to meet the needs of wartime industry. The availability of cheap natural gas changed the patterns of fuel use in Pittsburgh. By 1950, a considerable fraction of all domestic heating and industrial production had shifted from coal to natural gas. Coal-fired steam locomotives were retired, and diesel-electric locomotives took their place. These locomotives burn diesel oil to run electric generators that power the locomotive, and burning diesel oil produces less smoke than burning coal. The transition to natural gas as the city's principal fuel led to spectacular improvements in Pittsburgh's air quality that are reflected today in a modern, clean city (Figure 5.6).

National Legislation and Regulations

Not until the 1960s did clean air become a national priority. The Clean Air Act of 1963 was the first federal air pollution legislation in the United States. It gave the government jurisdiction over interstate air pollution, but only at the request of affected state governments. The act turned out to be weak because states were more concerned about protecting industrial interests and their tax base than bettering air quality.

The Air Quality Act of 1967 accomplished little more than its 1963 predecessor. It directed the

Figure 5.6 Pittsburgh in the 1990s.
(Photo courtesy of the Carnegie Library of Pittsburgh, Pittsburgh, Pennsylvania.)

Figure 5.5 Pittsburgh's tall smokestacks before 1940.
(Source: The Carnegie Library of Pittsburgh, Pittsburgh, Pennsylvania.)

Department of Health, Education, and Welfare to create 91 Air Quality Control Regions and required individual states to establish air quality standards. However, few Air Quality Control Regions were set up, and effective standards were never established. Individual states granted air pollution exemptions in order to keep industries and attract new ones.

The Clean Air Act of 1970

The first effective air quality legislation in the United States was the Clean Air Act of 1970. By that time, severe air pollution incidents were becoming more frequent, especially in urban areas. The 1970 legislation identified two types of air pollution sources:

1. *Stationary sources*   Coal-burning power plants, iron and steel mills, metal smelters, and factories.

2. *Mobile sources*   Automobiles, buses, trucks, and aircraft.

The Clean Air Act directed the newly established U.S. Environmental Protection Agency to develop **ambient air quality standards,** maximum allowable levels of specific pollutants to be enforced by federal laws and regulations. The standards would apply to outdoor air in urban areas. Individual states had to submit plans and timetables detailing how and when the new standards would be implemented.

National Ambient Air Quality Standards   The EPA has established six national ambient air quality standards. Each one applies to a specific contaminant called a **criteria air pollutant.** Two sets of standards are identified: primary standards to protect public health, and secondary standards to protect the general welfare (which includes effects of air pollutants on soils, crops, vegetation, buildings, animals, wildlife, visibility, property, and economic values). In addition, the agency has set up 247 air quality regions so that monitoring can be carried out systematically and efficiently. Table 5.2 lists the six EPA criteria air pollutants and the standards in effect in 1998.

All of the contaminants identified in Table 5.2, except lead, occur naturally in the atmosphere. Air pollution is not defined in terms of the mere presence of one or more criteria pollutants but in terms of concentrations that exceed established standards over specified time periods. From a legal and environmental management point of view, the definition of pollution includes accepted standards or threshold levels above which one or more contaminants exist for a defined period of time.

Automotive Air Pollution   The 1970 Clean Air Act targeted automotive air pollution as an important air quality issue. The act mandated 90 percent reduction in sulfur dioxide and carbon monoxide emissions by 1975 and 90 percent reduction in oxides of nitrogen by 1976 (Insight 5.4). The goals were based on 1970 emission levels. The automobile industry argued that the cost of achieving these reductions could not be justified in

INSIGHT 5.4
Oxides of Nitrogen

Atmospheric nitrogen gas, N_2, is chemically very stable and unreactive due to the triple bond that joins its two nitrogen atoms: $N \equiv N$. But given sufficient energy, usually in the form of heat, nitrogen reacts with oxygen to form various oxides that are collectively referred to as oxides of nitrogen (NO_x). This reaction occurs in internal combustion engines, no matter what fuel is being used, making it difficult to reduce oxides of nitrogen levels in automotive engine exhausts.

TABLE 5.2　EPA National Ambient Air Quality Standards[a] for the Six Criteria Air Pollutants

Criteria Air Pollutant	Monitoring Time Period[b]	Primary Standard	ppm	Secondary Standard	ppm
Particulates (PM-10)[c]	Annual	50 μg/m³	—	50 μg/m³	—
	24 hours	150 μg/m³	—	150 μg/m³	—
Particulates (PM-2.5)[d]	Annual	15 μg/m³	—	15 μg/m³	—
	24 hours	65 μg/m³	—	65 μg/m³	—
Sulfur dioxide	Annual	80 μg/m³	0.03	—	—
	24 hours	365 μg/m³	0.14	—	—
	3 hours	—	—	1,300 μg/m³	0.50
Carbon monoxide	8 hours	10 mg/m³	9	—	—
	1 hour	40 mg/m³	35	—	—
Nitrogen dioxide	Annual	100 μg/m³	0.053	100 μg/m³	0.053
Ozone	1 hour[e]	235 μg/m³	0.12	235 μg/m³	0.12
	8 hours[f]	157 μg/m³	0.08	157 μg/m³	0.08
Lead[g]	Quarterly	1.5 μg/m³	—	1.5 μg/m³	—

SOURCE: Data from The U.S. Environmental Protection Agency, 1998.

[a]Some standards are set in terms of micrograms per cubic meter (μg/m³), some in terms of milligrams per cubic meter (mg/m³), and others in terms of parts per million (ppm) — the ratio of a pollutant's volume to one million volumes of air.

[b]The monitoring time period refers to average air pollutant levels during this time.

[c]PM-10 refers to air particulates with diameters of 10 μm or less.

[d]PM-2.5 refers to air particulates with diameters of 2.5 μm or less. This new regulation will become effective in the year 2005.

[e]The ozone 1-hour average is the current regulation.

[f]The ozone 8-hour average is a new regulation effective in the year 2004.

[g]Hydrocarbons originally appeared as one of the criteria pollutants; they were later replaced by lead.

terms of existing technology. The government delayed compliance deadlines several times.

While political and economic arguments led to delays in Clean Air Act mandates, several new cost-effective technologies were developed and implemented to reduce automotive air pollution, including:

1. Exhaust gas recirculation to lower oxides of nitrogen levels in automotive tailpipe emissions.

2. Improved catalytic converters to reduce tailpipe emissions of hydrocarbons and carbon monoxide by oxidizing them to water and carbon dioxide.

3. Reduction in levels of lead in regular gasoline to protect catalytic converters, which are poisoned by lead.

4. Reengineered automobile engines to operate with less lead in gasoline.

Tetraethyl lead was commonly added to gasoline to improve engine performance and increase power. But its discharge to the atmosphere (tailpipe emissions) caused lead air pollution. In addition, lead is a catalyst poison—that is, it kills a catalyst's ability to function. Beginning in the 1970s, lead levels in regular gasoline were reduced from about 1 gram to 0.1 gram per gallon. The EPA later required that only unleaded gasolines be available.

The Clean Air Act of 1990

By 1990, considerable progress had been made in meeting 1970 Clean Air Act mandates, but a number of urban centers failed repeatedly to meet standards for particulates, carbon monoxide, and ozone. Although strengthening air pollution regulations to solve the remaining problems was opposed by coal-burning electric utilities, labor unions, coal and oil industries, automobile manufacturers, and many politicians, the tide of public support favored cleaner air, less smog, reduced health risks, and control of acid rain.

These considerations led to the Clean Air Act of 1990. Among other issues, the act focuses on acid rain and requires the use of low-sulfur coal, flue-gas desulfurization (scrubbers), and clean coal technologies at the nation's largest coal-burning power plants. Market-based emission control strategies are also encouraged. This is discussed later in this chapter.

Furthermore, all regions of the United States are now required to meet the one hour ground level ozone health standard of 0.12 ppm, except Los Angeles, Baltimore, and New York , which are presently unable to meet this standard (see Table 5.3). Strict controls on mobile sources of air pollution, especially cars, trucks, and buses, are now in effect. Automotive fuels have been reformulated to contain oxygenated compounds such as ethanol (ethyl alcohol) to burn cleaner and achieve mandated reductions in oxides of nitrogen and hydrocarbons, both of which contribute to ozone formation.

Reformulated Gasoline   The 1990 Clean Air Act requires that "clean gasolines" be marketed in urban regions of the United States that fail to meet standards for carbon monoxide, unburned hydrocarbons, oxides of nitrogen, and ozone. The reformulated gasolines that appeared on the market in 1991 contain oxygen-based additives such as ethanol and methyl tertiary butyl ether (MTBE) referred to as oxygenates. But in 1999, the EPA found that MTBE was showing up in groundwater sources of drinking water; as a result, it was decided to phase out this particular oxygenate. Reformulated gasolines must have a minimum oxygen content of 2.7 percent by weight. It was expected that the development of reformulated gasolines would increase the cost of gasoline significantly, but this has not happened.

The use of reformulated gasolines is mandated year-round in 10 areas of the United States: Los Angeles, San Diego, and Sacramento in California; Hartford, Connecticut; New York City; Philadelphia, Pennsylva-

nia; Chicago, Illinois; Baltimore, Maryland; Houston, Texas; and Milwaukee, Wisconsin. Other regions that exceed the ozone standard are voluntarily participating in this program.

Shaping Effective Clean Air Strategies

Command and Control Strategies   Air pollution abatement under the Clean Air Act of 1970 depended mainly on conventional approaches to solving environmental problems called **command and control strategies.** Using this approach, the EPA established uniform standards that every industry, utility, and individual production facility had to comply with.

However, over time, it was found that regulating air pollution this way wasn't cost effective. Some industries were better able to control certain air pollutants at a lower cost than other industries, and companies operating several production units sometimes found it less costly to control pollution at some sites and more costly at other sites. Yet the law required each unit to meet the same standards. Command and control strategies sometimes exacted high costs on manufacturing firms—costs that consumers ultimately had to bear. Also, efforts directed toward meeting uniform standards diverted resources that could have been used to develop newer, better, and more cost-effective pollution abatement practices. More practical strategies were developed under the Clean Air Act of 1990.

Market-Based Strategies   Regulations that offer flexibility in the way individual industries and manufacturing firms can meet environmental standards are called **market-based pollution control strategies.** Tradable pollution permits are an example. Under the 1990 Clean Air Act, sulfur dioxide emissions in the United States are to be reduced by 10 million tons below 1980 levels. At the same time, oxides of nitrogen emissions are to be reduced by 2 million tons. To achieve these objectives, the EPA issued permit allowances for sulfur dioxide and oxides of nitrogen emission to 111 electric power utilities. Each permit allows fixed amounts of these two pollutants to be emitted to the air annually. Unused portions of a permit allowance can be traded, sold, or leased to other utilities. They can even be banked for future use. These 111 electric utilities are among the largest U.S. power producers. Since electric power generation is based mainly on burning coal, these utilities are major sources of sulfur dioxide and oxides of nitrogen. By the year 2000, almost all U.S. electric utilities were included under these provisions of the 1990 Clean Air Act. The EPA estimates that this approach to regulating air quality will save the nation about $1 billion annually.

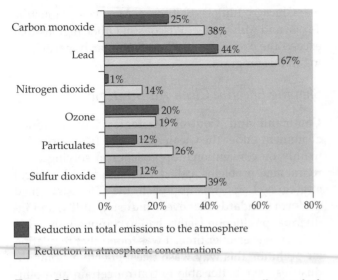

Carbon monoxide — 25% / 38%
Lead — 44% / 67%
Nitrogen dioxide — 1% / 14%
Ozone — 20% / 19%
Particulates — 12% / 26%
Sulfur dioxide — 12% / 39%

■ Reduction in total emissions to the atmosphere

□ Reduction in atmospheric concentrations

Figure 5.7 Percent reductions in EPA priority air pollutants, both total emissions and prevailing ambient atmospheric concentrations, between 1988 and 1997.
(Source: Data from EPA National Air Quality and Emissions Trends Report, 1997.)

Monitoring U.S. Air Quality Today

By 1997, considerable progress had been made in reducing air pollution in the United States. This progress was reflected primarily in declining levels and decreased total emissions of the EPA's criteria air pollutants. The EPA, other federal agencies, state and local governments, and private organizations monitor these pollutants in urban, suburban, and selected rural areas. Figure 5.7 summarizes percent reductions in both total emissions and atmospheric concentrations of criteria air pollutants over the 10-year period between 1988 and 1997.

The data show that U.S. air quality has continued to improve over this period. The most dramatic reduction is in atmospheric lead. In 1997, lead total emissions were 44 percent lower and atmospheric concentrations 67 percent lower than in 1988. If lead data for the 20-year period from 1978 to 1997 are considered, a 98 percent decline in total emissions and a 97 percent decline in atmospheric levels can be demonstrated. This was achieved by mandating lower levels of lead in gasoline. Today, the principal sources of atmospheric lead are no longer automotive; they are metal smelters and battery manufacturing plants.

Levels and emissions of the other criteria air pollutants have also been reduced, but to varying degrees. The air pollutant most difficult to control is oxides of nitrogen, which is reported in terms of nitrogen dioxide (NO_2). Over the 10-year period from 1988 to 1997, total emissions fell only 1 percent, while atmospheric levels fell 14 percent.

The decline in carbon monoxide is due to automotive catalytic converters and reformulated gasolines. Diminished levels of sulfur dioxide are the result of a number of EPA-mandated programs, including:

1. Flue-gas desulfurization (scrubbers) at large coal-fired electric power generating stations. Sulfur dioxide emissions are trapped in a slurry of limestone and water. Scrubbers are about 95 percent effective.

2. Fluidized-bed combustion. This involves burning a mixture of pulverized coal and limestone on a bed of sand while air is pumped through the combustion mixture. Limestone combines with sulfur dioxide in the presence of air to form calcium sulfate, a solid precipitate.

3. Clean coal technology. Coal is ground to a powder and suspended in water. Sulfur impurities in coal, such as iron pyrite (FeS_2), float to the surface and are removed.

Nonattainment Areas

If a particular region of the country fails to meet one or more of the EPA criteria air pollutant standards, it is designated as a **nonattainment area.** In September 1998, there were 130 U.S. nonattainment areas. Their formal identification means that the EPA can promulgate special air pollution reduction measures along with a timetable for reaching compliance. Table 5.3 lists the numbers of U.S. nonattainment areas in 1991 and 1998 according to specific air pollutant and the size of the 1998 population affected.

Air Toxics

Air toxics are hazardous air pollutants known to cause or suspected of causing cancer or other serious human health problems or ecosystem damage. In 1998, the EPA listed 188 specific air toxic chemicals or groups of similar chemicals in its National Toxics Inventory (NTI). Knowledge of their toxicity is based on both human and animal experimental data. The inventory includes chemicals and materials such as:

- *Asbestos*   A mineral found in older heating systems, ceiling tiles, insulation products, automobile brake and clutch linings, and cement, paint, and plastic tile products. Asbestos causes fibrosis in lung tissues and possibly lung cancer in humans.

- *Benzene*   An industrial solvent and component of gasoline. Benzene has been implicated as a cause of leukemia.

INSIGHT 5.5
Radon

Radon is a naturally occurring, invisible, odorless, radioactive gas formed by the radioactive decay of an isotope of radium. Radon's concentration in the earth's atmosphere is low, and its radioactivity is normally not considered harmful. But although radon is not a public health hazard in open environments, significant levels can build up in closed environments such as groundwaters or home basements.

When radon is concentrated in relatively closed environments, breathing it can cause health problems. Its radioactive decay products emit alpha particles, which can penetrate lung tissues. Alpha particle emissions are known to cause lung cancer, depending on the degree and duration of exposure. The EPA standard for in-home radon exposure is 4 picocuries per liter of air (average annual exposure). (A picocurie is a unit of radioactivity exposure.)

TABLE 5.3 Numbers of U.S. Nonattainment Areas in 1991 and 1998 by Pollutant and Population

Air Pollutant	Number of 1991 Areas	Number of 1998 Areas	1998 Population Affected (1,000s)
Carbon monoxide	42	20	34,047
Lead	12	10	1,375
Nitrogen dioxide	1	0	0
Ozone	100	38	99,824
Particulate matter	70	77	29,890
Sulfur dioxide	51	34	4,695

SOURCE: Data from the U.S. Environmental Protection Agency, 1997.

- *Vinyl chloride* A toxic chemical that may be released from polyvinyl chloride pipe. Exposure is believed to cause lung and liver cancer.

- *Beryllium* An uncommon metal used in specialty alloys and electronic components. It is known to be extremely toxic, causing lung, liver, and kidney damage.

- *Mercury* A chemical element whose compounds are used as industrial catalysts, agricultural pesticides, and batteries. Exposure to methyl mercury, a mercury compound produced from mercury by common bacteria, causes severe central nervous system damage.

- *Radionuclides* Radioactive substances that release ionizing radiation capable, in some cases, of inducing tumors and cancers.

- *Arsenic* Together with some of its chemical derivatives, arsenic is believed to be toxic and, in some cases, carcinogenic.

- *Coke oven emissions* Gases given off by coal when it is converted into coke for steel production. They include a mixture of highly toxic chemicals, including PAHs, some of which cause respiratory diseases and cancer.

- *Radon* A naturally occurring radioactive gas formed through the radioactive decay of radium (Insight 5.5).

Monitoring Global Air Quality

Carbon Dioxide

Carbon dioxide, a natural component of the earth's atmosphere, is not one of the EPA's criteria air pollutants, but it is especially significant from a global perspective. Levels of carbon dioxide in today's atmosphere are about 360 ppmv (parts per million by volume). That's about 0.036 percent. While this concentration might seem relatively low, atmospheric carbon dioxide levels have increased by almost 30 percent since the start of the Industrial Revolution. The increase is due almost totally to human influence.

Carbon dioxide is not a criteria air pollutant because at prevailing atmospheric levels it does not adversely affect human health. Nevertheless, it is increasingly evident that carbon dioxide should be considered an air pollutant. The international scientific community believes rising levels of atmospheric carbon dioxide are accelerating climate change through global warming, sea level rise, and more frequent droughts, floods, and severe storm events. A considerable body of evidence supports this view (see Chapter 12).

Methane

Like carbon dioxide, methane is a natural component of the air. Its current level is about 2 ppmv (0.0002 percent) (see Table 5.1). Although it is a minor component of the atmosphere, methane levels are rising, and these increases are significant because methane, like carbon dioxide, is a **greenhouse gas.** That is, it contributes to global warming by trapping heat energy that would otherwise escape from the earth and radiate into outer space. Rising levels of methane are due mainly to two anthropogenic factors:

1. Methanogenic soil bacteria in oxygen-deprived wetland ecosystems produce methane. While this is a natural process, wetland areas devoted to wet rice agriculture have expanded greatly since the start of green revolution agriculture. This is adding greater quantities of methane to the atmosphere each year.

2. Cattle ranching, aimed chiefly at beef and hamburger production, has expanded greatly world-wide. Cattle, as a part of their digestive process, produce considerable quantities of methane, releasing it at both ends of their bodies.

The Chlorofluorocarbons (CFCs)

The industrial chemicals known as **chlorofluorocarbons** (CFCs) are used as refrigerants in air conditioners, food coolers, and freezers. Unavoidably, CFCs find their way into the air and slowly rise to the stratosphere. There, intense solar radiation splits off some of their chlorine atoms, a process that initiates ozone-destroying chain reactions. We now know that the CFCs and certain other chemicals periodically deplete stratospheric ozone, especially above the earth's north and south poles (see Chapter 13).

Since stratospheric ozone shields the earth from damaging ultraviolet radiation, its depletion is a serious problem, particularly above Antarctica between October and December each year. It is also observed above the North Pole and at temperate latitudes above Siberia, Canada, and the United States.

The CFCs are global air pollutants. As a result of the 1987 Montreal Protocol and subsequent international agreements, CFC production in industrialized nations was banned starting in January 1996. Worldwide production is to cease in 2002. Ozone-safe CFC substitutes have been developed and are being used in new refrigeration equipment. While a black market in CFCs exists, it appears that these chemicals will totally disappear from the earth's environment in the future. The Montreal Protocol is the first example of international decision-making aimed at solving a global environmental problem.

REGIONAL PERSPECTIVES

New York: Acid Rain in Adirondack State Park

In the 1930s, as part of a statewide lake study, the average pH of the roughly 300 higher-elevation lakes in New York's Adirondack State Park was found to be 6.4, a fairly typical lake pH. All the lakes studied had healthy populations of lake trout, brown trout, yellow perch, and smallmouth bass. Indeed, the Adirondack lakes were known as superb sites for sport fishing.

By the 1970s, however, the park's higher-elevation lakes had become 25 times more acidic, with average pH values close to 5.0. The effect on lake fisheries was dramatic: At least 90 percent of the lakes had become barren—totally devoid of any fish. Research has shown that lake acidification affects fish in two ways:

1. Fish eggs fail to hatch in waters as acidic as pH 5.0.
2. Fish gills in adult fish react to aluminum compounds leached from watershed soils by acid rain runoff. The mucus formed within the gills prevents the uptake of dissolved oxygen, which leads to asphyxiation.

The cause of lake acidification is related to acid rain triggered regionally by the atmospheric transport of acid-forming air pollutants, particularly sulfur dioxide and oxides of nitrogen. In the case of the Adirondack lakes, plumes of acid-forming air pollutants originating in the Ohio Valley from coal-burning electric power utilities and heavy industries are transported by prevailing winds to the northeast where they precipitate as acid rain.

Mexico City: Urban Air Pollution

Today, insidious recurring episodes of atmospheric pollution are common in a growing number of urban centers worldwide, and urban centers are growing in number and population density. It is presently estimated that there are no fewer than 20 "megacities" in the world, each populated by 10 million or more people, including: Tokyo, Japan; Beijing and Shanghai, China; Rio de Janiero, Brazil; Seoul, Korea; New Delhi, India; Cairo, Egypt; and Mexico City, Mexico.

Mexico City is Mexico's capital as well as its leading industrial and cultural center. In Mexico City, the pace of industrial activity and the density of automotive traffic add massive amounts of particulate matter, carbon monoxide, sulfur dioxide, oxides of nitrogen, and hydrocarbons to the atmosphere every day. The pollution is worsened by the fact that Mexico City lies within a valley, so it is surrounded on all sides by mountains that periodically trap air pollution until changing weather patterns create air currents powerful enough to dilute or remove them from overhead.

Like other megacities, Mexico City's growth is phenomenal. Its population increased from about 3 million in 1950 to an estimated 19 million in 1999. Its ancient Aztec heritage has given way to a modern metropolis noted for skyscrapers, a subway system, and millions of automobiles. Deforestation, the often inevitable consequence of urban growth, highway construction, and industrial expansion, has exposed the region's soils to wind erosion, which is a major source of airborne particulate matter that sometimes limits visibility to less than one mile.

L.A. smog commonly prevails in Mexico City, and its cause is the same as that found in southern California—photochemical reactions triggered by sunlight acting on atmospheric oxides of nitrogen and hydrocarbons from automotive sources. This produces ozone and other eye, throat, and respiratory irritants along with a number of toxic substances. Ozone levels in Mexico City are often as high as 0.45 ppm, almost four times the U.S. ozone standard, creating unhealthy conditions approximately 330 days each year. Atmospheric thermal inversions prevail above Mexico City from November through March, especially during the early morning hours, thereby exacerbating air pollution problems.

Chesapeake Bay: Atmospheric Deposition of Nutrients

The eutrophication of Chesapeake Bay is discussed in detail in Chapter 4. Studies have shown that nutrient runoff from the bay's extensive multi-state watershed is impacting its living resources, especially its submerged seagrass meadows, the bay's first trophic level. Here, sunlight penetrating shallow estuarine waters and driving seagrass photosynthesis accounts for most of the Chesapeake's primary production.

While the lion's share of nitrogen- and phosphorus-based nutrients entering bay waters is transported through watershed runoff, it has been discovered that nutrients also enter the estuary through both wet and dry atmospheric deposition—the fallout of airborne particulate matter associated with water droplets, rainfall, and dry matter precipitated by gravity. Watershed nutrient runoff, together with atmospheric nutrient deposition, has accelerated bay eutrophication by stimulating algal blooms that block sunlight from submerged seagrasses.

The origins of airborne nutrient deposition to Chesapeake Bay include wind-driven exposed agricultural soils and applied fertilizers, atmospheric transport of soils eroded from deforested areas, soils carried aloft from urban and suburban construction sites, and oxides of nitrogen from automotive and industrial sources, some of which are converted into nitrates, a major nitrogen-based nutrient. The Ohio Valley, one of the country's principal industrial regions, is a major source of airborne nutrient deposition to Chesapeake Bay. According to some estimates, 25 percent of total nitrogen-based nutrients entering the bay originate from atmospheric deposition. It is less certain how much of the bay's total phosphorus-based nutrients are added through atmospheric transport.

KEY TERMS

acid rain

adsorb

aerobic respiration

aerosols

air toxics

ambient air quality
standards

chlorofluorocarbons

chlorosis

coal rank

coke

command and control
strategy

criteria air pollutant

dustfall

greenhouse gas

heavy metals

incomplete combustion

L.A. smog

lignite

London smog

market-based pollution
control strategy

mesosphere

negative lapse rate

nonattainment area

ozone

peat

photochemical smog

polynuclear aromatic
hydrocarbons (PAHs)

primary energy

radon

respirable

secondary energy

stratosphere

thermal inversion

thermosphere

troposphere

DISCUSSION QUESTIONS

1. In the summer of 1988, great fires occurred in Yellowstone National Park. Assuming that the fires were of natural origin, which park officials believe to be the case, should the ash, smoke, and air particulates produced by the fire be regarded as air pollution? Discuss your views.

2. Determine the truth of the following statement: There is no pollution if there are no environmental laws or regulations that establish specific environmental quality standards. Discuss your answer.

3. Given the fact that the United States possesses very large deposits of coal, much of it having a high sulfur content, is it simply a matter of common sense that this coal should be mined and burned? Discuss your answer.

4. Do market-based air pollution control strategies in the Clean Air Act of 1990 give major industries too much flexibility? Are certain industries simply being given the right to pollute? Do you think that the older command and control approach applied uniformly to all industries and their production plants was better? Explain your answers.

5. To solve some of California's most serious air pollution problems, it was suggested that there should be a limit on the number of cars a family may own. Is this an acceptable solution to air quality problems? If not, what would be a better way to solve these problems?

INDEPENDENT PROJECT

Choose a city in your state known to have periodic air pollution. What is the nature of these problems? Is it London or L.A. smog? Contact a regional or state air pollution office or authority and ask for recent data documenting levels of ozone, sulfur dioxide, carbon monoxide, and particulates. Plot this information on a graph and note whether there is a trend. Compare your data with current air quality standards. Is there compliance or noncompliance with these standards? What is your assessment of the air quality where you live?

INTERNET WEBSITES

Visit our website at http://www.mhhe.com/environmentalscience/ for specific resources and Internet links on the following topics:

EPA Office of Air & Radiation

EPA air quality resources

California air quality data

Fuel ethanol and air quality

Indoor air quality primer

SUGGESTED READINGS

Air quality management plan: South Coast Air Basin. 1989. El Monte, CA: South Coast Air Quality Management District.

Brown, Lester R., Hal Kane, and Ed Ayers. 1993. *Vital signs 1993: The trends that are shaping our future.* New York; London: The Worldwatch Institute.

Cramer, Zadok. 1814. *The navigator; containing directions for navigating the Monongahela, Allegheny, Ohio, and Mississippi Rivers.* Pittsburgh: Cramer, Spear, and Eichbaum.

Davidson, Cliff I. 1979. Air pollution in Pittsburgh: A historical perspective. *Journal of the Air Pollution Control Association* 29(10):1035–41.

Hockenstein, Jeremy B., Robert N. Stavins, and Bradley W. Whitehead. 1997. Crafting the next generation of market-based environmental tools. *Environment* 39(4):13–20; 30–33.

Kubasek, Nancy K., and Gary S. Silverman. 1994. *Environmental law.* Upper Saddle River, NJ: Prentice Hall.

Menzie, Charles A., Bonnie B. Potocki, and Joseph Santodonato. 1992. Exposure to carcinogenic PAHs in the environment. *Environmental Science and Technology* 26(7): 1278–1284.

Simon, Joel. 1997. *Endangered Mexico—An environment on the edge*. San Francisco: Sierra Club Books.

Smith, Zachary A. 1992. *The environmental policy paradox*. Englewood Cliffs, NJ: Prentice Hall.

Tarr, Joel A. 1996. *The search for the ultimate sink: Urban pollution in historical perspective*. Akron, OH: The University of Akron Press.

U.S.E.P.A. Office of Air Quality Planning and Standards. 1997. *National air quality and emissions trends report*.

Wayne, Richard P. 1991. *Chemistry of atmospheres*. Oxford: Clarendon Press.

PART III

Human Populations, Agricultural Soils, and Food Supply

Corel/Indigenous People

Chapter 6
Population Principles and Demography

Chapter 7
Green Revolution Agriculture

Chapter 8
Sustainable Agriculture

Population Principles and Demography

PhotoDisc, Inc./Vol. 60

Every nation has an economic policy and a foreign policy. The time has come to speak more openly of a population policy. What, in the judgment of its informed citizenry, is the optimal population?

from *The Diversity of Life* by Harvard biologist,
Edward O. Wilson

World population today is a staggering statistic, with an estimated 6 billion humans living on the earth in the year 2000, according to the U.S. Census Bureau. And at an average annual growth rate of about 1.3 percent, close to 78 million people are being added every year. Population in some countries is doubling in less than one person's life span. If the current rate of growth continues, the number of human beings on the earth will double by 2053, for a total of 12 billion.

World population growth is creating environmental, economic, social, and political problems never before encountered. We will begin this chapter by exploring the basic principles of population growth, using plants and animals in undisturbed environments as examples. Then we will turn our attention to the dynamics of human populations and focus on China as a classic case of uncontrolled population growth and attempts to limit further expansion.

BACKGROUND

General Principles of Population Growth

Understanding how plant and animal populations function in nature lends important perspective to human population dynamics. The basic principles to understand are: biotic potential, exponential population growth, environmental resistance, and ecosystem carrying capacity.

Biotic Potential

Plants and animals have reproductive strategies that enable them to produce more offspring than needed to maintain their numbers. Some species produce large numbers of offspring; others are more limited. The maximum rate at which a given species can increase in number, assuming no environmental limits to its reproduction or survival, is called its **biotic potential**. Mathematically, biotic potential is related to a particular population's ability to expand in number, a function symbolized by r, which stands for a population's **intrinsic capacity for increase**.

For example, consider arctic lemming, a critically important food source in arctic food webs, especially for predators such as arctic foxes, snowy owls, jaegers, and weasels (Figure 6.1). Lemming have a high biotic potential. Each year, during the spring, summer, and fall, a female may give birth to as many as nine pups after only 20 days gestation (the development period before birth). Theoretically, if there were no environmental limits to their multiplication, 100 lemming could produce more than 28,000 young in a single growing season.

Orca whales typify species with a small biotic potential. Orcas are most often found in northeastern Pacific Ocean waters and northwestern coastal estuaries such as Puget Sound. Commonly called "killer

INSIGHT 6.1

Population Size and Doubling Time

Mathematically, biotic potential is related to two factors: a species' intrinsic capacity for increase, symbolized by r, and its initial population size, symbolized by N_0. Population size, N, at any future time, t, can be calculated using the equation below, where e is the base of the natural logarithm system.

$$N = N_0 e^{rt}$$

Consider the case of marine diatoms. Their intrinsic capacity for increase, r, is reported to be 0.38/day. This means that the population size of marine diatoms can increase by 38 percent in one day. Starting with one diatom ($N_0 = 1$), the predicted size of a population after 31 days would be 130,614 diatoms.

Doubling time is another important population statistic. It is defined as the time required for the population of a particular species to double in size, assuming that its numbers are increasing. It is calculated as follows:

Doubling time = 69 ÷ Percent rate of growth

In this equation, the number 69 is derived from a growth rate equation based on the number 0.69, which is the approximate value of the natural logarithm of 2. Since percent rate of growth is 100 times r, the doubling time of the marine diatom population = 69 ÷ 38 = 1.8 days = 43 hours.

TABLE 6.1 Intrinsic Capacity for Increase for Selected Species

Species	r-value
E. coli (bacteria)	0.0347/minute
Paramecium (protozoan)	0.05/hour
Diatoms	0.38/day
Lemming	0.47/week
Domestic dog	0.07/week
Rabbit	≈1.5/year
Orca whale	≈0.05/year
Gray whale	≈0.025/year
African elephant	≈0.02/year

SOURCE: Data based on T. Fenchel, "Intrinsic Rate of Natural Increase: The Relationship with Body Size" in *Oecologia* 14, 317–26, 1974.

Figure 6.1 Arctic lemming, showing their white winter coat and their summer brown coat.
(Photo courtesy of Hilmar Maier.)

whales," though they are actually highly social and almost always harmless to humans, orcas have a life expectancy of about 100 years. They do not mature sexually until close to 12 years of age, and females average only one birth every 10 years. These reproductive characteristics limit their ability to increase in number. African elephants also have limited biotic potential. They are reckoned to be the slowest breeder of all known animals. Although they may live 90 years, mating does not occur until they are about 30 years old. Then, after 22 months gestation, a female gives birth to a single calf. These facts, together with recent onslaughts of elephant poaching, explains why the number of African elephants is declining and why they are on the U.S. endangered species list.

Table 6.1 lists intrinsic capacity for increase (r values) for selected species.

Exponential Population Growth

Whether its biotic potential is high or low, the population of a particular species can exhibit **exponential population growth** unless there are factors that limit its growth, such as disease or lack of food. Exponential population growth is defined as the repeated doubling of a species' population size over constant time intervals (Insight 6.1).

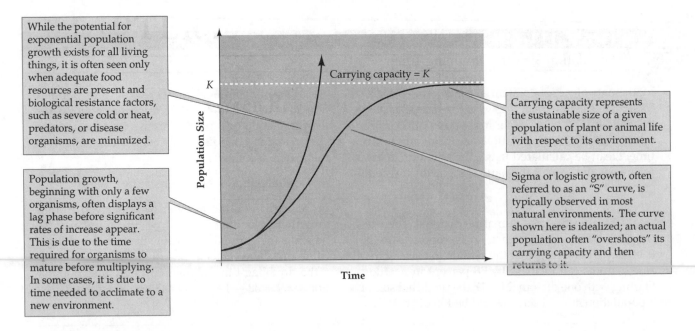

While the potential for exponential population growth exists for all living things, it is often seen only when adequate food resources are present and biological resistance factors, such as severe cold or heat, predators, or disease organisms, are minimized.

Population growth, beginning with only a few organisms, often displays a lag phase before significant rates of increase appear. This is due to the time required for organisms to mature before multiplying. In some cases, it is due to time needed to acclimate to a new environment.

Carrying capacity = *K*

Carrying capacity represents the sustainable size of a given population of plant or animal life with respect to its environment.

Sigma or logistic growth, often referred to as an "S" curve, is typically observed in most natural environments. The curve shown here is idealized; an actual population often "overshoots" its carrying capacity and then returns to it.

Figure 6.2   Population growth among marine diatoms. While exponential growth is possible, sigma or logistic growth is more typical reflecting algal population expansion which approaches and is limit by an ecosystem's carrying capacity, *K*.

An exponential growth curve resembles the letter J and is sometimes called a **J-curve** (Figure 6.2). A striking example of exponential growth is often seen among marine diatoms, a species having a large *r* value.

A **lag phase** is a common element in population growth curves. It represents an interval of little population increase—the time needed by organisms to reach maturity and/or to acclimate to a particular environment. The lag phase is often followed by exponential growth, which in most cases eventually gives way to slower rates of growth, as we will see later.

Environmental Resistance

Every plant and animal species on earth would exhibit exponential growth if there were no limits to the survival and reproduction of their progeny, limits such as crowding, disease, competition, predation, and inadequate food resources. Such limiting factors are referred to as **environmental resistance**. While some populations increase and decrease seasonally or oscillate in response to cyclic changes in food supply or predator populations, environmental resistance allows undisturbed populations to neither consistently increase nor decrease in size over long time intervals.

Let's reconsider marine diatoms. Oceanographers estimate that up to five million diatoms can occupy one cubic centimeter of seawater in antarctic waters. While diatom populations may multiply exponentially at times, their numbers cannot increase indefinitely. Competition for space, light, and nutrients limits their growth. Their numbers are also constrained because they are a food source for zooplankton such as marine copepods (see Figure 1.11). Diatom populations represent a balance between biotic potential and the forces of environmental resistance that limit their growth and reproduction.

Density-Dependent Factors   A population's **density** refers to the number of organisms of the same species in a defined habitat. Factors related to density that can limit population size are called **density-dependent factors**. High density, or overcrowding, in an area increases competition for food and exposes organisms to predators and disease-causing organisms, so population size is likely to be lowered. For example, the density of a lemming population appears to be linked to the abundance of tundra vegetation. In years of lush, plentiful grasses, sedges, and herbs, the population becomes larger. But then overgrazing occurs, triggering mass migrations as lemming search for new food sources (Figure 6.3).

Density-Independent Factors   Other factors that affect population size are **density-independent;** these include temperature extremes, changes in weather, and catastrophic events such as flooding, earthquakes, and volcanic eruptions. Blue heron populations in England and Wales exemplify the importance of temperature as a density-independent factor. Their numbers decline during very cold winters when lakes and ponds freeze,

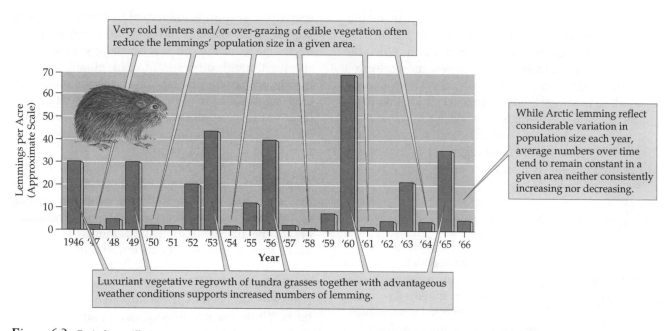

Very cold winters and/or over-grazing of edible vegetation often reduce the lemmings' population size in a given area.

While Arctic lemming reflect considerable variation in population size each year, average numbers over time tend to remain constant in a given area neither consistently increasing nor decreasing.

Luxuriant vegetative regrowth of tundra grasses together with advantageous weather conditions supports increased numbers of lemming.

Figure 6.3 Periodic oscillations among arctic lemming at Point Barrow, Alaska, between 1946 and 1966.

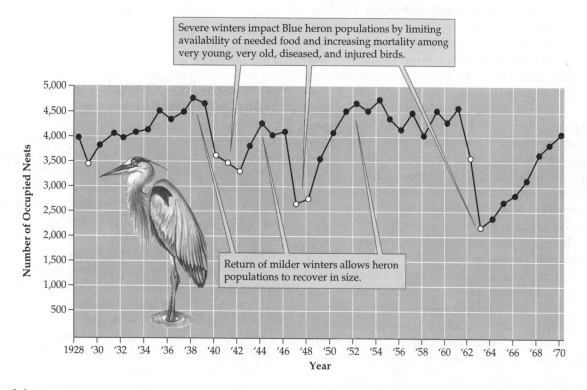

Severe winters impact Blue heron populations by limiting availability of needed food and increasing mortality among very young, very old, diseased, and injured birds.

Return of milder winters allows heron populations to recover in size.

Figure 6.4 The impact of severe winters on blue heron populations in England and Wales, an example of a density-independent factor affecting a wildlife species' population size.

making it difficult for them to catch the fish they need to eat to survive (Figure 6.4).

Another example is the olive ridley turtle, one of eight species of sea turtles (all threatened or endangered) that migrate annually to sandy beaches, such as those along Costa Rica's Pacific coast, to lay their eggs. Every year during October nights, thousands of females, each about 76 centimeters (≈ 30 inches) long, haul themselves ashore at high tide, dig nests in the sand, and lay an estimated 30 million eggs.

Although their biotic potential is very high, many turtle eggs are eaten by coyotes or illegally poached by humans; less than 10 percent of them hatch. When the baby turtles make their dash for the sea, most are eaten by birds, crabs, and small mammals. Those that do make it to the water's edge face sharks and other predators next. On the average, only one in a hundred newborn turtles survives, matures, and returns to the beaches the next year to lay its eggs. Yet that is enough to assure their survival as a species.

Ecosystem Carrying Capacity

The type of population growth most typically seen is known as **sigma or logistic growth,** which leads to a steady state or relatively constant population size called *carrying capacity*. When graphed, the sigma growth curve looks like the letter S (see Figure 6.2).

The **carrying capacity** for a plant or animal in a particular ecosystem is the maximum population size of the species that the ecosystem can sustain over long periods of time. It is the population size or density reached when a species' biotic potential is in balance with the environmental forces that limit its growth and reproduction. As an example, the sigma curve in Figure 6.2 shows how diatom populations generally behave in the real world. After an initial lag phase, population density rises exponentially. But the competition among diatoms for limited resources also increases. In the case of antarctic marine diatoms, population density levels off at approximately 5 million diatoms/centimeter^3. This is the carrying capacity, symbolized by K, of antarctic waters for diatoms.

An ecosystem's carrying capacity for a particular species is not necessarily constant. For example, if a species' biotic potential is limited by lack of food, an increase in the food supply will expand the carrying capacity. Conversely, carrying capacity can be reduced if a species' habitat or food supply is impaired. This has happened to China's giant panda bears, which dwell on mountain slopes in Szechwan Province and subsist almost entirely on bamboo shoots growing in the forests throughout their territory. Although the pandas were able to adjust to the cycle of various bamboo species periodically dying back, they could not adjust to bamboo deforestation caused by expanding human settlement. It is now estimated that fewer than a thousand pandas are left in China.

K-Strategists and r-Strategists   Higher animals, such as orca whales, monkeys, chimpanzees, and elephants, generally exhibit low r values. They possess complex social patterns, are strongly competitive, have stable population densities, and use resources efficiently. Such organisms are called **K-strategists** because each species, through processes of self-regulation, tends to maintain a population size at or near environmental carrying capacity, K. Male and female K-strategists introduced to an appropriate environment would most likely display sigma (logistic) population growth and level off at a population equal to K.

On the other hand, lichen, mosses, bacteria, fungi, weeds, insects, and rodents possess high r values, indicating that they reproduce rapidly. Called **r-strategists**, they often function as pioneer or colonizer organisms in disturbed environments. In 1990, 10 years after Mount St. Helens erupted in Washington State, 11 r-strategist plant species had colonized on a mud flow southwest of the volcano's crater where plant life had been totally obliterated by the volcanic blast. While the r values of most plant and animal species lie somewhere between very low and high extremes, the extremes themselves are instructive and describe very different kinds of living things (see Table 6.1).

Demography: Principles Governing Human Populations

The principles of plant and animal populations discussed so far are relatively simple, but population issues become much more complex once the focus turns to humans. One of the striking characteristics of *Homo sapiens* is its ability to influence the environment: Age-old hunting and gathering practices have been superseded by planned agricultural systems and rapid transportation of foods to consumers far away. Modern dwellings protect humans against predators, harsh winters, and storms. New knowledge has led to improved nutrition, better sanitation, and effective remedies against disease. In short, people are now able to manipulate the factors that would otherwise naturally limit the population growth of their species. As a result, an immense rising tide of human population is taking place in many parts of the world.

The study of human population trends is called **demography.** Demographers analyze the factors that influence human population growth, such as total fertility rates, rates of natural increase, migration patterns, population age structures, demographic transitions, and environmental factors. These data are used to project future population trends and to analyze alternative scenarios aimed at solving population-related environmental problems.

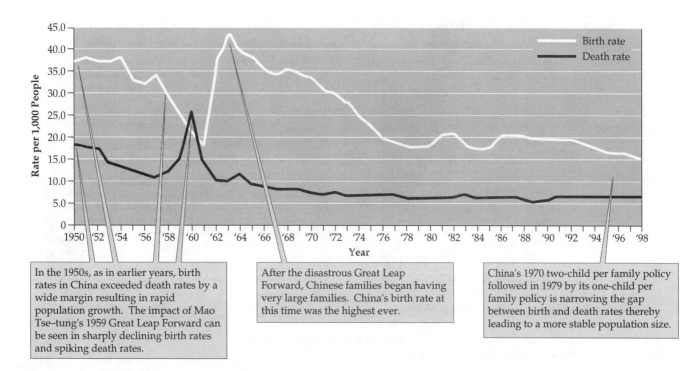

In the 1950s, as in earlier years, birth rates in China exceeded death rates by a wide margin resulting in rapid population growth. The impact of Mao Tse–tung's 1959 Great Leap Forward can be seen in sharply declining birth rates and spiking death rates.

After the disastrous Great Leap Forward, Chinese families began having very large families. China's birth rate at this time was the highest ever.

China's 1970 two-child per family policy followed in 1979 by its one-child per family policy is narrowing the gap between birth and death rates thereby leading to a more stable population size.

Figure 6.5 Crude birth rates and death rates in China between 1950 and 1998.
(Source: Data from the U.S. Bureau of the Census, International Data Base.)

Total Fertility Rate

The average number of children that women in a given city, region, or country give birth to during their childbearing years is referred to as the **total fertility rate.** Population growth is influenced more powerfully by total fertility rates than by any other demographic variable. This is due to a built-in multiplier effect: One child may produce three or more children, and each of these may produce three or more, and so on. Even if death rates remain constant, the total population size rises. The worldwide average total fertility rate in 2000 is estimated to be 2.78.

Rate of Natural Increase

The annual rate at which a human population increases (not including immigration or emigration) is called its **rate of natural increase**. It is defined as a country or region's birth rate minus its death rate:

Rate of natural increase = Birth rate − Death rate

A country's total population may increase, remain relatively constant (zero population growth), or decline. Therefore, rates of natural increase may be positive, zero, or negative.

Birth rate, more accurately called **crude birth rate,** is the number of annual live births per 1,000

members of a particular population. It may be converted into a percentage by dividing by 10. As an example, consider birth rates in China since 1950 (Figure 6.5). In 1957, the crude birth rate was about 35 per 1,000 people (3.5 percent). Since China's population at that time was 624 million, close to 22 million children were born that year.

But birth rates alone do not determine how fast a population grows. Death rates are also important. **Death rate**, more accurately called **crude death rate,** is the number of annual deaths per 1,000 members of a population. It too can be converted into a percentage by dividing by 10. For example, in 1957, the average death rate in China was close to 10 per 1,000 (1.0 percent) (see Figure 6.5). The country's rate of natural increase that year was therefore 3.5 − 1.0, or 2.5 percent.

Migration Patterns and Growth Rates

The movement of people across geopolitical boundaries into and out of different countries obviously affects populations but is often difficult to document. A country's **growth rate** is the rate of natural increase in a given year, plus the number of people immigrating into the country minus the number of people emigrating out of the country. If death rates fall and net immigration remains constant, even a constant fertility rate will result in faster population growth. But if death rates rise, rates

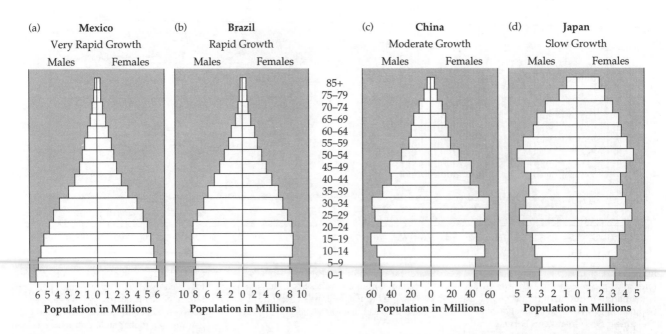

Figure 6.6   Projected Year 2000 population age structures for Mexico, Brazil, China, and Japan.
(Source: Data from the U.S. Bureau of the Census, International Data Base.

TABLE 6.2   Year 2000 Population Statistics for Selected Countries

COUNTRY	Total Population (Millions)	Birth Rate, per 1,000	Death Rate, per 1,000	Total Fertility Rate	Natural Increase (%)	Annual Growth Rate (%)	Doubling Time (Years)
China	1,256	14.5	7.0	1.79	0.76	0.71	97
India	1,018	24.9	8.3	3.11	1.65	1.65	42
United States	275	14.2	8.8	2.07	0.54	0.84	82
Brazil	174	19.9	9.1	2.23	1.09	1.08	64
Russia	146	9.7	15.0	1.33	−0.53	−0.34	—
Pakistan	141	32.6	10.2	4.56	2.24	2.15	32
Japan	126	10.6	8.3	1.50	0.23	0.20	345
Nigeria	117	41.4	13.0	5.95	2.84	2.87	24
Mexico	102	24.5	4.8	2.79	1.97	1.69	41
Germany	82	8.5	10.8	1.27	−0.23	−0.02	—
Ethiopia	61	44.0	21.6	6.75	2.23	2.11	33
France	59	11.1	9.2	1.58	0.19	0.23	300
United Kingdom	59	11.8	10.6	1.72	0.12	0.22	313
Italy	57	9.4	10.4	1.25	−0.10	−0.09	—
Canada	31	11.6	7.3	1.64	0.43	1.02	68
Kenya	29	29.9	15.0	3.69	1.49	1.46	47
Sweden	8.9	11.7	11.0	1.78	0.07	0.27	259

SOURCE: Data from the U.S. Bureau of the Census: International Data Base: http://www.census.gov/ftp/pub/ipc/www/idbnew.html

of population growth will decline; it is even possible that a country's total population size will decline.

Although social upheaval, job opportunities, political oppression, war, epidemics, and many other factors continuously influence demographic statistics, they generally have little effect on total world population. The most noteworthy exception was the Black Death (bubonic plague) of the fourteenth century when world population actually declined.

Population Age Structures

A country's future population size is determined largely by its **age structure,** the current relative proportion of its population in different age groups. The proportion of individuals between 15 and 44 years of age is a major determiner of a country's future population because these individuals have the greatest reproductive potential. Age structure diagrams are used to depict the relative size of different age groups in a population. Figure 6.6 shows age structure diagrams that compare projected year 2000 age structures for Mexico, Brazil, China, and Japan. The wider the base of an age structure diagram, the larger the fraction of young people entering their reproductive years and the more likely it is that the country will experience rapid population growth.

Table 6.2 compares population growth statistics for selected countries in the year 2000, ending in the far right-hand column with their doubling time, the number of years it will take, all other factors remaining constant, for the population to double in size (see Insight 6.1). These statistics show that Mexico is one of the fastest-growing countries in the world. Its annual growth rate of 1.69 percent translates into a doubling time of about 41 years (69 ÷ 1.69). This rapid growth is reflected in the country's age structure diagram (see Figure 6.6a), which has a very wide base relative to the rest of its population. Brazil is also experiencing rapid growth, though individuals below age 15 make up a somewhat smaller fraction of its total population compared to Mexico.

In China, the proportion of people under age 15 is declining due to the country's coercive population control policies (see Case Study). In 2000, annual population growth in China fell to 0.71 percent, a rate somewhat less than that of the United States. (But keep in mind that, even at this modest rate of growth, China is adding close to 9 million more people each year.) The annual growth rate in Japan is very small, about 0.20 percent, a pace reflected in the relatively narrow base of the country's age structure diagram (see Figure 6.6d). At this rate, Japan's doubling time is 345 years.

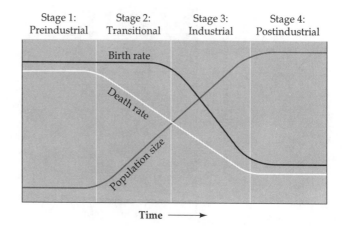

Figure 6.7   The four stages of a demographic transition. Most of today's industrialized nations are in Stages 3 or 4, while most developing societies are in Stage 2.

Population growth in the United States is currently close to 0.84 percent, a rate considerably greater than the rate of natural increase (0.54 percent). This is due to a high incidence of immigration into the United States. Presently, the doubling time of the U.S. population is 82 years. Canada's year 2000 growth is slightly greater than 1 percent, a rate that will cause the country's population to double in 68 years.

Today, populations in about 32 industrialized societies are neither increasing nor decreasing significantly and have, for all practical purposes, achieved zero population growth. Included among these countries are France, Germany, Italy, Japan, the United Kingdom, and Sweden.

Demographic Transitions

Demographers recognize a series of four stages that occur as a society matures (Figure 6.7). These demographic transitions are among the most important of all population growth patterns and are often linked to industrial development:

Stage 1, Preindustrial: Birth rates and death rates are high. As a result, the population grows slowly. Total fertility rates are relatively high.

Stage 2, Transitional: Birth rates remain high, but death rates fall due to improved sanitation, nutrition, and health care. As a result, the population rapidly increases in size.

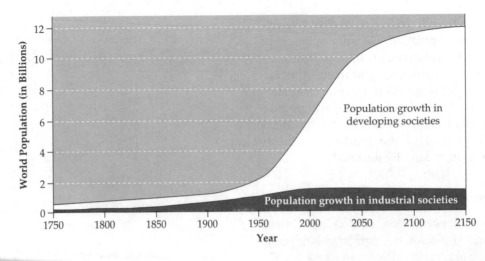

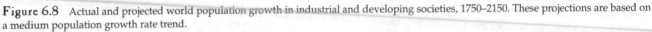

Figure 6.8 Actual and projected world population growth in industrial and developing societies, 1750–2150. These projections are based on a medium population growth rate trend.
(Sources: Data from Population Reference Bureau and United Nations Population Division.)

Stage 3, Industrial: People become more educated and affluent; birth rates fall; the gap between birth rates and death rates narrows; and a country's population growth slows.

Stage 4, Postindustrial: Birth rates and death rates are low, small families are preferred, and the total population size stabilizes.

A look back at the demographic trends in Great Britain illustrates these four stages. In 1775, birth rates were as high as 39 per 1,000 (3.9 percent). Death rates were also high—about 34 per 1,000 (3.4 percent). The rate of natural increase was therefore quite low—about 5 per 1,000 (0.5 percent). This was Stage 1 in Britain's demographic transition, the preindustrial stage. As food supplies were increased, drinking water made safe, and means of sanitation improved, death rates gradually fell. However, birth rates declined only slightly. As the gap between birth and death rates widened, rates of natural increase surged and population growth increased rapidly. This was Britain's transitional demographic stage, or Stage 2. During the country's Industrial Revolution, birth rates continued to fall and total population began to level off, typical of Stage 3. Today, modern Britain has entered its postindustrial stage, Stage 4, in which both birth and death rates are low and total population has stabilized.

Demographic transitions similar to those that occurred in Great Britain have taken place in other European countries as well. In most industrialized societies today, the gap between birth rates and death rates has narrowed to the point where the average annual rate of natural increase is about 0.3 percent. Thus, most of today's industrialized nations are in Stages 3 or 4 of their demographic transitions.

But in developing societies, where the gap between birth and death rates remains fairly wide, average annual rates of natural increase are close to 2.0 percent. Thus, most developing societies are in Stage 2. Their populations are expanding for the same reasons that industrial nations grew so rapidly—declining death rates. While birth rates in developing nations are beginning to decline, the gap between birth and death rates has not yet narrowed significantly. Many developing nations, including China, Mexico, India, and Brazil, entered Stage 2 of their demographic transitions in the early 1900s. Before then, their birth rates averaged about 42 per 1,000 (4.2 percent), and death rates were close to 37 per 1,000 (3.7 percent). Rates of natural increase were about 5 per 1,000 (0.5 percent), the same as those prevailing in 1775 Great Britain.

The differences between the rates of population growth in today's industrialized and developing societies is striking. When that difference is projected into the future, the outcome is even more remarkable (Figure 6.8). Demographers estimate that between now and the year 2150, the lion's share of increased population growth worldwide will take place in the world's developing societies, while population growth in many industrial nations is likely to decline or even become negative.

Environmental Factors

Human population growth or decline within regions or individual countries is also related to characteristics of the natural environment. Among these, the availability of safe water, nonpolluting sanitary facilities, clean air, and adequate arable land are the most important. Some of these factors are discussed in the following case study.

CASE STUDY: POPULATION GROWTH IN CHINA

Chinese officials reported that on February 15, 1995, China's population reached 1.2 billion people when a baby boy was born to a Beijing couple. The day was proclaimed Population Day. Peng Peiyun, the State Minister for Family Planning, and He Luli, Beijing's Vice-Mayor, came to the hospital to congratulate the happy couple. Minister Peng pointed out that if annual population growth in China had continued at its 1973 annual rate of 2.79 percent, the 1.2-billionth citizen would have been born in 1986—nine years earlier. But, as a result of the family planning policies delaying this birth, there were 300 million fewer people in China that day.

It is difficult to comprehend the sheer magnitude of a population the size of China's. Nowhere else on earth is population growth more profoundly evident. In 1982, China had grown to one billion people—the first nation to reach that size. It is estimated that at the start of the year 2000, China's population was increasing by 9 million each year and that close to 22 percent of the world's people lived there. Based on these statistics, it is projected that by 2050, as many as 450 million more Chinese people will have been added.

Historical Influences on China's Population

Although ancient China was known for its poetry, porcelain, calligraphy, silk, tea, and important scientific discoveries, crop production was valued above almost all other endeavors. While less than 14 percent of China's land area is **arable** (suitable for raising crops), close to 70 percent of its people occupy these lands. They are mostly peasant farmers who till the fertile plains and valleys carved out long ago by the Yellow, Yangtze, and Pearl Rivers (Figure 6.9). Rice is grown in southern and central China, while wheat is favored in the north. Before the Communist Revolution, Chinese tenant farmers were compelled to cultivate farms owned by wealthy landlords. Little mechanization

Figure 6.9 A map of China

existed, and crops were planted, cultivated, and harvested mostly by hand. As the demand for rice increased, more people had to be working in the fields—not only farmers but their wives and children as well, since rice is a labor-intensive crop (Insight 6.2). The best chance a Chinese peasant had to improve his lot in life was to raise a large family and grow as much rice as possible. So larger families meant greater prosperity, and the culture placed high value on fertility, particularly in rural areas, where most families had four, five, or more children. Two other factors contributed to high fertility rates in China:

1. High rates of infant mortality, together with recurring epidemics, encouraged couples to bear many children to replace those who had died or were likely to die.

2. Chinese families desired sons over daughters as a type of "social security." That is, sons would continue the family line and perform the traditional rituals, honoring their ancestors and taking care of their parents in old age; daughters married into their husbands' families and no longer provided help at home. As a result, many families kept having more children in order to get boys.

The Policies of Mao Tse-tung

Mao Tse-tung's Communist Party seized power in China in 1949. The new government redistributed agricultural land taking it from the control of wealthy landlords and allocating it to individual farmers who became owners of small, independent farms, most of

INSIGHT 6.2
Wet Rice Agriculture

Cultivating rice is an example of labor-intensive agriculture as it is practiced in China and many other Asian nations. In wet rice agriculture, rice seeds are sown in seedbeds, and seedlings are later transplanted by hand, one by one, into rice paddies that are flooded naturally during monsoon seasons or irrigated if necessary. Standing water is required as rice grows. Then, as the rice ripens, the paddies are drained. The crop is then harvested and threshed manually. This is how wet rice agriculture

has been carried out for over 2,000 years. It remains a labor-intensive practice today.

Growing rice requires only a minor fraction of available arable land relative to the numbers of people that can be fed. Today's improved green revolution rice varieties have higher levels of protein than those found in earlier kinds of rice, making rice an important element in the diets of many people around the world.

them only one to three acres in size. Since harvests could now be sold on the open market, farmers began to produce more grain year by year. Spurred by this success, Mao became determined to build China into a powerful and prosperous nation through a centrally planned economic system. To accomplish this, agricultural practices were changed. Small farms were collectivized into larger ones where as many as 10 farm families worked together. Farmers were directed where, when, and what crops to plant.

In 1958, larger collectives called communes were formed, each one incorporating as many as 4,000 farm families. Individual farmers lost their property rights and became tenant farmers once again. They received hourly wages but had no stake in the land they farmed.

In 1959, Mao Tse-tung launched China's Great Leap Forward (GLF). The GLF was an ambitious program aimed at industrializing China. Mao was determined to advance the country's infrastructure by building iron-making blast furnaces, new steel mills, highways, huge dams, and massive power projects. Millions of people were enlisted in this endeavor, including many farmers. But with fewer farmers working the land, agricultural production faltered. Food shortages followed, and by the early 1960s China had insufficient grain to feed its people. Widespread famine broke out despite efforts to import food. It is estimated that between 1959 and 1961, 30 million Chinese people died of starvation. Also during the GLF, total fertility rates fell from 5.8 to 3.3, and the country's population declined. The low birth rates and very high death rates that prevailed at that time can be seen today in China's 40–44-year-old cohort in Figure 6.6.

Once the GLF had ended in disaster, China's agricultural communes were dismantled, and a new strategy, the Household Responsibility System, was

adopted. While individual farmers did not regain land ownership, they were granted limited rights to the land they farmed. Later, the government reclaimed all agricultural land and reallocated it to farm families based on family size. Clearly, this practice encouraged more children, a strategy Mao Tse-tung strongly supported. Mao believed that a large population would mean greater production on farms and in factories. He also thought population growth would give China a strategic advantage in the event of war.

The aftermath of the GLF was evident by 1963 when China's total fertility climbed to 7.5 and its birth rate soared to 43 per 1,000 (4.3 percent), the highest ever recorded there (see Figure 6.5). Close to 30 million children were being added to the population each year, partly a reflection of the people's desire to reestablish large families. While death rates remained low (10 per 1,000, or 1.0 percent), rates of natural increase rose to 33 per 1,000 (3.3 percent), a dramatically high growth rate for any country. If that rate had continued, China's population of 700 million would have doubled to 1.4 billion people by 1984. But this did not happen.

Deng Xiaoping and Industrialization

Following Mao Tse-tung's death in 1976, industrialization in China accelerated dramatically under the leadership of his successor, Deng Xiaoping. Deng boldly established five special economic zones and 14 open coastal cities to expand economic trade with foreign countries, attract financial investment, and import advanced technologies. Thus, Deng established a new political climate in China: "one country, two systems," in which free, market-based economic practices were encouraged.

Shanghai became one of the special economic zones, and as a result the city has been transformed into a booming industrial center with expanding job markets. The same thing has happened in other special economic zones. Western-style free-market economies have blossomed, and China has started to achieve its great leap forward on the world stage.

Industrialization is attracting many young people from rural areas to China's new, revitalized cities because of exciting opportunities for interesting, high-paying jobs. Young families are experiencing a way of life their parents never knew—an affluent lifestyle that is changing their ideas and ideals regarding family planning. In many cases, couples are deciding not to have any children at all. In other cases, childbearing is simply delayed and often limited by choice to one child.

Impacts of China's Population Growth

Arable land areas in China, small to begin with, are shrinking as new cities, highways, power projects, and factories are constructed. By necessity, therefore, China is shifting from an exporter of grains and agricultural products to a major importer of wheat, rice, and other produce.

China's expanding population size, coupled with its rapid urbanization, has also created critical environmental problems. Air pollution is one of the most serious of these problems, especially in cities where high-sulfur soft coal is burned to fuel factories, produce basic metals, operate trains, and heat homes. Coal currently accounts for about 75 percent of China's total energy demand. Its combustion, which is often inefficient and incomplete, produces emissions containing carbon monoxide, sulfur dioxide, and air particulates (see Chapter 5).

Population growth in rural areas is causing damage to forests as more and more trees are cut for household heating and cooking. Deforestation leads to land degradation because tree roots hold the ground in place, conserving soils and minimizing water- and wind-driven soil erosion. It is estimated that approximately 20 percent of China's agricultural land has been lost to soil erosion since 1957.

In China, as in many countries, the supply of clean water for agricultural and human use is not keeping pace with population growth. Water shortages occur frequently in urban areas of China, and safe drinking water is often not available. Groundwater aquifers, used to irrigate crops in northern China, are being gradually depleted, creating the possibility that agricultural production may decline. Domestic sewage and industrial wastewaters, often poorly treated prior to their discharge, are contaminating

some of China's principal rivers so that many of the country's waterways are now too polluted to be used for crop irrigation. The stage is set for possible widespread water-borne diseases.

China is the second leading contributor of carbon dioxide gas to the earth's atmosphere (the United States being first). Carbon dioxide, formed when fossil fuels such as coal are burned, has a long residence time in the atmosphere and is building up to higher and higher levels each year. This is believed to be the principal cause of global warming (see Chapter 12). Computer models project that as this occurs, the world's climate will change, altering prevailing temperatures, rainfall patterns, and growing seasons and likely resulting in major environmental impacts such as flooding and drought. Although global problems such as these are not due solely to China's growth and development, nevertheless China and several other developing nations are major contributors.

Population Control in China

Coercive Family Planning Policies

Two Children—One Family   By 1970, the seriousness of China's population problem was clear to the country's leaders. Therefore, the Chinese Communist Party (CCP) developed a family planning strategy aimed at limiting population growth. Its slogan was:

> One child is not too few,
> two children are just right,
> and three children are too many.

Limiting families to two children became China's official population plan in 1970. It was based on the concept of **replacement level fertility**, the number of children needed to replace their parents but not add to population growth. Replacement level fertility averages 2.1 children per family. By the end of the 1970s, China's rate of natural increase was about 2.2 percent. Close to 20 million children were being added annually, and total population was rapidly approaching one billion people. Computer projections based on this information convinced Chinese leaders that the two-child per family policy would fail to control population growth soon enough to avert unmanageable growth. They believed the survival of China's next generation was at risk unless more restrictive approaches to birth control were adopted.

One Child—One Family   In 1979, China changed its population policy to One Child—One Family

Compliance was forced through a combination of rewards and punishments. Here are the main elements of the One Child—One Family policy as it was initially set forth:

- Chinese women, many of whom marry at age 16, are encouraged to delay marriage until age 25 for those living in cities and age 23 for those in rural areas. The rule was relaxed somewhat in 1980 by allowing all women to marry at age 20.
- Specific birth quotas were established in villages, towns, and cities. Pregnancies must be pre-authorized by government officials who, if they approve, issue a birth coupon prior to each conception.
- Government-run birth control clinics supply free contraceptives and terminate unauthorized pregnancies.
- Women are granted six months of maternity leave at full pay following the birth of their first child. An only child receives free medical care, unlimited educational opportunity, and guaranteed employment upon graduation. Children whose births were unauthorized are not entitled to these benefits.
- Rural families with only one child pay less tax and are granted land for home construction.

There is considerable uncertainty in projecting the size of China's future population. Planners hope that by the year 2000, the One Child—One Family policy, though it has been made less stringent, will have succeeded in lowering birth rates below 10 per 1,000. Computer models project that at that rate, the country's total population will level off at 1.3 billion. However, demography experts project China's year 2000 birth rate at 14.5 per 1,000 (see Table 6.2), since many Chinese families are not obeying the one-child policy, especially in remote, rural areas where not all children are counted during an official census. United Nations demographers believe that birth rates in China are likely to rise as centralized government control weakens. If this happens, the country's population may surpass 1.5 billion by 2025.

Human Rights Issues

The United Nations' policy is that every country has a sovereign right to develop and implement its own population policy. In China, sterilization of women, mandatory abortions, and infanticide are common means of enforcing the one-child policy. These coercive measures are employed mainly in urban areas, where in 1999, at least 30 percent of China's people lived. The

Figure 6.10 Chinese people in Tienanmen Square, Beijing, prior to the June 4, 1989 massacre.
(Photo courtesy of Ruth Robins Collins.)

traditional preference for sons in Chinese families has led to sex-selective abortions and abandonment of girl babies. For example, if a pregnant woman's ultrasound test reveals a female fetus, an abortion is likely to be performed even though gender-specific abortions are officially illegal in China. Typically, if a couple's first child is a girl, she may well be killed (infanticide) or given up (abandoned) to a state-run orphanage. In some cases, after giving birth to their first child, one of the parents faces mandatory sterilization.

China's one-child policy has proved to be highly controversial, with many citizens resisting its enforcement, especially in rural agricultural areas. In the 1980s, a more tolerant family planning practice appeared when government officials began to look favorably on rural families having a second child. This gave farm families another chance to have a son if their first child was a girl. It also meant there would be more agricultural workers. Nevertheless, strict reproductive limits are still being enforced in many parts of China.

Sex-selective abortion, infanticide, and child abandonment are causing an increase in the ratio of boys to girls in China. Normally, 106 boy babies are born to every 100 girl babies, a universal statistic in all countries of the world. This was true in China too until 1979, the year the one-child policy was adopted. After 1979, the male-to-female sex ratio began to increase. In 1992, the ratio was reported as 119 boys to 100 girls. If this is accurate, 1.1 million girl babies (16 percent of all girls born that year) were missing or unaccounted for. (More recent statistics are not available, but 14 million abortions were officially reported by the Chinese government in 1991.)

Many world leaders regard China's population policies as unduly harsh and reflecting little regard for

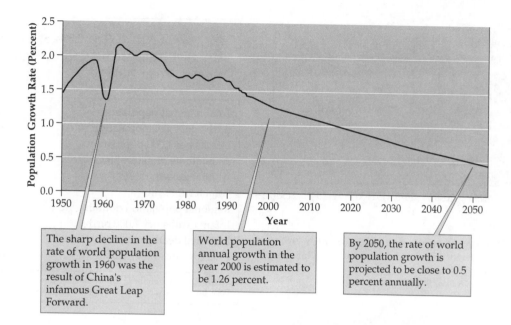

Figure 6.11 Actual and projected rates of worldwide population growth: 1950–2050. *(Source: Data from the U.S. Bureau of the Census, International Data Base.)*

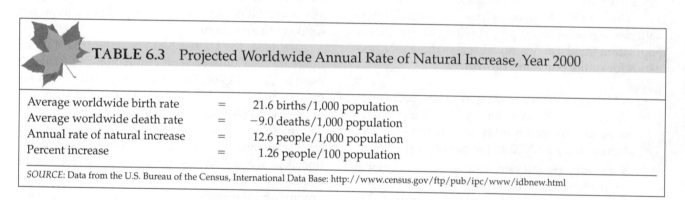

TABLE 6.3   Projected Worldwide Annual Rate of Natural Increase, Year 2000

Average worldwide birth rate	=	21.6 births/1,000 population
Average worldwide death rate	=	−9.0 deaths/1,000 population
Annual rate of natural increase	=	12.6 people/1,000 population
Percent increase	=	1.26 people/100 population

SOURCE: Data from the U.S. Bureau of the Census, International Data Base: http://www.census.gov/ftp/pub/ipc/www/idbnew.html

fundamental human rights, especially reproductive rights. A glaring example is the widespread existence of state-run orphanages. In June 1995, Channel 4 television in Britain broadcast a documentary called "The Dying Rooms" that showed infants and children in Chinese orphanages being allowed to starve to death. The government considered these children surplus population.

The 1989 massacre in China's Tienanmen Square is symbolic of human rights issues in China (Figure 6.10). China's coercive tactics and their harmful effects on children are affecting international affairs, including private investment, trade, and tourism.

FUTURE WORLD POPULATION TRENDS

Thomas Malthus, a noted British social philosopher, believed that human populations exhibit exponential growth. He thought growth in food supply was at best limited to linear increases proportional to expanding cultivated land. He envisioned a dismal "doomsday" outcome: worldwide famine, starvation, and death. In 1798, Malthus published *An Essay on the Principle of Population* in which he wrote: "The power of population is indefinitely greater than the power of the Earth to produce subsistence for man."

Not surprisingly, Malthus became the world's best-known pessimist. Was he correct? Does global disaster lay ahead?

As Malthus predicted, the rapidly rising numbers of humans, particularly in developing countries, have raised questions about the ability of planetary resources to feed all those people. But in recent decades, population growth worldwide has declined from a high of 2.2 percent in 1963 to 1.26 percent in 2000, and it is projected to continue declining (Table 6.3 and Figure 6.11). But although this indicates world population may ultimately stabilize, 1.26 percent growth at the present time means that almost 76 million people

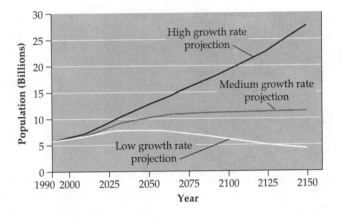

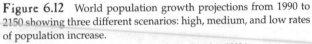

Figure 6.12 World population growth projections from 1990 to 2150 showing three different scenarios: high, medium, and low rates of population increase.
(Source: Data from the United Nations Population Division, 1992.)

TABLE 6.4 Sub-Saharan Africa: Basic Demographic Statistics, Year 2000	
Total regional population	652 million
Births per 1,000 population	40.3
Deaths per 1,000 population	15.5
Annual rate of natural increase	2.48%
Annual growth rate	2.49%
Total fertility rate	5.63
Regional population doubling time	28 years
Infant deaths per 1,000 births	100
Maternal deaths per 100,000 births	980
Literacy rate: male/female	67%/47%

SOURCE: Data from the U.S. Bureau of the Census, International Data Base: http://www.census.gov/ftp/pub/ipc/www/idbnew.html

are being added to the planet each year—an annual increase equal to four Mexico Cities!

The United Nations is one of several international agencies that projects long-range global population trends. Figure 6.12 displays three U.N. scenarios for future population growth:

1. A high growth rate, leading to a total global population of 28 billion people by 2150. This scenario assumes that total fertility rates worldwide will decline to 2.5 by 2050 and remain at that level.

2. A medium growth rate, leading to a total global population of 11.5 billion people by 2150. This scenario assumes that total fertility rates worldwide will decline to 2.1 (replacement level fertility) by 2050 and remain at that level.

3. A low growth rate, leading to a total global population of 4.3 billion people by 2150. This scenario assumes that total fertility rates worldwide will stabilize at 1.7 by 2025 and remain close to that level.

No one knows which of these scenarios is most likely. If total fertility rates were to remain at year 2000's level of 2.78, world population could soar to 694 billion by the year 2150. Perhaps more importantly, 99.7 percent of this immense number of people would

be living in what are today's developing countries. Even if more modest population growth occurs globally, about 90 percent of 2150's world population will reside in these nations.

In 1994, the United Nations organized its third International Conference on Population and Development in Cairo, Egypt. The agenda included analyses of growth and societal impacts, inequitable distribution of energy, uneven access to food, safe water, medicine, and material goods, and environmental dilemmas such as disease, malnutrition, and air and water pollution. A plan of action was agreed upon that focused on principles known to have brought about sustainable growth and development in other nations. The most important of these principles are:

- Enhance the status of women by expanding their opportunities for education, empowering them to choose when and whom they will marry, and widening their access to birth control information and contraceptives.

- Improve women's welfare by increasing their access to health-care facilities and modern medical treatment.

- Reduce poverty by establishing job-training centers and creating new economic opportunities for families.

REGIONAL PERSPECTIVES

Sub-Saharan Africa: Family Planning Efforts

The 49 African nations south of the Sahara Desert are referred to as Sub-Saharan Africa (SSA). In 2000, approximately 652 million people lived in this region, generally as members of large families. Average birth rates surpassed 40 per 1,000, while death rates were close to 15 per 1,000. The region's average total fertility rate of 5.6 accounts for its large families and explains why annual population growth is as high as 2.5 percent. If this rate of growth continues, SSA will double in population within 28 years, making it all but impossible for most of the nations to supply their people with food, safe drinking water, sanitary services, health care, and educational facilities, much less create economic opportunities and spur needed developments. Table 6.4 summarizes some basic SSA statistics.

SSA total fertility is among the highest in the world. For example, Nigeria's rate of growth is about 2.9 percent per year, a rate that could cause its population to double in only 24 years. Many factors lie behind runaway population growth in this part of the world:

- Like most men and women in developing nations, Africans in SSA typically want many children. It is part of a long-standing cultural heritage to have as many as six to eight children.

- In part, the desire for many children comes from knowing that one child in six is not likely to see his or her fifth birthday. This is mostly due to inadequate immunization, unsafe drinking water, and poor sanitation.

- Without a national pension or social security system in SSA nations, families want many children to ensure that someone will take care of them in their later years.

As SSA nations have become more democratic, they have welcomed international efforts to help them solve critical problems. Family planning is high on the list and is getting a lot of attention. In 1967, Kenya, recognizing that population growth was a major obstacle to its development, became the first SSA nation to adopt a national family planning program. One of the principal aims of Kenya's program is to advance the education of women. The problem is that across most of SSA, girls do not attend school at all. In a few countries, girls go to primary school, but very few ever reach secondary school. Boys fare somewhat better, but illiteracy rates in SSA are very high; at least one-third of men and one-half of women cannot read or write.

For women, illiteracy includes failure to understand and practice effective birth control, inability to find and keep a job, and difficulty in learning better farming practices, a responsibility that falls mostly on the rural women.

By 1980, family planning in Kenya had lowered fertility rates from 6 to 5.4. Desired numbers of children among both men and women declined from 7 to 4. In large measure, these accomplishments stem from improving the status of women, lessening the pressure on women to marry early, and controlling the onslaught of childhood diseases through immunization for measles, tetanus, whooping cough, tuberculosis, polio, and diphtheria.

By the late 1990s, 25 SSA countries had adopted family planning programs. Kenya's community-based program continues to serve as a model. It offers family counseling, education about contraception, and birth control supplies. It is the beginning of a success story in SSA.

Kerala, India: Achieving a Stable Population

Kerala, a state on the southwestern edge of India's subcontinent, is a lush, idyllic place with a dense, rural population. Lying along the eastern edge of the Arabian Sea, Kerala is a tourist's paradise. Its coastal areas are known for palm trees, rice fields, and coconut trees, and its river deltas support abundant fish and wildlife. Uplands are home to intensive agriculture—coffee and tea plantations, rubber trees, sugarcane, cashew nuts, and spices such as pepper, cardamom, cinnamon, nutmeg, mace, ginger, and turmeric. Tropical forests occupy mountainous terrain further to the east.

Kerala, a small welfare state in a very large country, has 29 million people, a population almost equal to that of Canada. It is known for a high quality of life that is based not on high income or material goods but on a deep-rooted sense of community and a long-established respect for family life. Per capita income is low, perhaps $300 U.S. per year, and the economy grows only very slowly.

Kerala was not always this way. Like the rest of India, especially before its independence from Britain in 1947, Kerala was mired in the evils of a highly restrictive caste system. The many people regarded as untouchables had no access to places of worship and could not even use public roads, schools, hospitals, or parks. The struggle for freedom started in the early 1800s and was led by women. But it was not until 1957, with the democratic election of a communist government, that real reform began in Kerala. The caste system was eliminated, land reform returned land to the tiller, food was distributed equitably, health care was advanced, and female education became the hallmark of Kerala's governmental policies.

Kerala has worked consistently to improve the status of women. Perhaps the best indicator is that 93 percent of high school-age girls attend school. It is estimated that 90 percent of women in Kerala can read and write, while only 39 percent are literate across the rest of India (1995 data). Female education in Kerala has helped lower total fertility to 1.9, while women in India as a whole have an average of 3.1 children during their childbearing years.

The average life span in Kerala is 72 years, indicative of a progressive, healthy society. Both birth rates and death rates have been significantly lowered, and the region's age structure distribution is such that population growth is stable. Kerala has, in effect, entered a postindustrial (Stage 4) demographic transition. Few other developing nations, if any, have been able to accomplish this.

Canada: Typical Postindustrial Population Patterns

Statistics describing Canada's population are typical of many developed nations. For this reason, it is instructive to include an analysis of Canadian demographics as a regional perspective.

Canada's low birth rates and low death rates today (see Table 6.2) place the country in Stage 4 of its demographic transition—the postindustrial or, as some have called it, the "information age." Canada's year 2000 population is estimated at 31 million, and its annual growth rate at 1.02 percent, one of the lowest in Canadian history and having a doubling time of 68 years. Two factors contribute to the country's low growth rate: falling rates of natural increase and immigration. Natural increase stands at 4.3 per 1,000 or 0.43 percent per year. Annual net immigration adds 0.59 percent, the sum of these equaling the country's growth rate of 1.02 percent.

Annual population growth in Canada (1.02 percent) is greater than in the United States (0.84 percent) mainly because Canada admits proportionally greater numbers of immigrants than the United States does. But this is changing as the Canadian government puts tighter limits on immigration. A somewhat different perspective emerges in some of Canada's provinces where population growth rates are down in many cases. For example, Newfoundland on the country's east coast is experiencing negative population "growth" owing to an extremely low fertility rate (1.31 children per woman, the lowest ever recorded in a

Canadian province). On the west coast, British Columbia's population is increasing at a rate more than two times the national average. In part, a westward migration of Canadians is responsible for some of the demographic patterns evident in the nation's provinces.

Falling numbers of births in Canada reflect declining total fertility rates, which in recent years have dropped from 1.69 to 1.64, a decrease of almost 3 percent. It is believed that this trend is related to increased education and training of women, leading to their greater participation in the workforce. Childbearing is being delayed, and as a result family size is turning out to be smaller than couples had originally desired.

KEY TERMS

age structure	environmental resistance
arable	
biotic potential	exponential population growth
birth rate	growth rate
carrying capacity	intrinsic capaity for increase
crude birth rate	
crude death rate	J-curve
death rate	K-strategists
demography	lag phase
density	rate of natural increase
density-dependent factors	replacement level fertility
density-independent factors	r-strategists
doubling time	sigma or logistic growth
	total fertility rate

DISCUSSION QUESTIONS

1. Choose a plant or animal species not discussed in this chapter and analyze its biotic potential—how it reproduces, how many offspring or seeds it produces, and what factors influence its fertility. Is this species an r- or K-strategist? Assume an r value for this species, and calculate its doubling time.

2. Continue your discussion of the species you selected in question one by analyzing environmental resistance factors that affect its population size. Which factor do you think is the most important? Why? What are the typical numbers of this species in areas you are familiar with?

3. Considering the population dilemma China faces if fertility rates are not further reduced, do you

think the Chinese government is justified in enforcing its one-child policy? Why or why not? Explain the reasons for your viewpoint.

4. Why did the demographic transition in the industrial world take place as rapidly as it did (in less than 125 years), while the end of the demographic transition in developing societies is not yet in sight? Please elaborate.

5. Was Thomas Malthus correct in predicting that ultimately famine, starvation, and death would overtake the world's population? Discuss your answer.

6. It is important to be able to project the future population size of a country. Given the discussion of U.N. population projections in this chapter, what single strategy would most influence future world population size? Explain and discuss your answer.

INDEPENDENT PROJECT

Choose one of the countries listed in Table 6.2 (not already discussed in this chapter) and research its history, culture, and population trends. Using the information you collected, discuss the factors you think account for current birth rates, death rates, and population trends. What population problems do you see developing and what policies would you recommend to deal with them? What environmental problems exist or are likely to occur in this country? What laws and regulations might be adopted to minimize or perhaps prevent these problems?

INTERNET WEBSITES

Visit our website at http://www.mhhe.com/environmentalscience/ for specific resources and Internet links on the following topics:

Welcome to China

China—An information database

Chinese science, culture, and art

U.S. Census Bureau World and U.S. population clocks

Population data on individual countries

SUGGESTED READINGS

Aird, John S. 1996. China's war on children. *The American Enterprise* (March/April):58–61.

Brown, L. R., et al. 1996. *State of the world*. New York and London: The Worldwatch Institute.

Brown, L. R. 1995. Who will feed China? *The Worldwatch Environmental Alert Series*. New York; London: W. W. Norton.

Ding, Jian. 1995. China's population hits 1.2 billion. *Beijing Review* (March): 6–11.

Dumas, J., and Alain Bélanger. 1995. Report on the demographic situation in Canada. *Statistics Canada*, ISSN 0715-9293.

Ellis, William S. 1994. Shanghai—Where China's past and future meet. *National Geographic* 185:(3)2–35.

Jisen, Ma. 1996. *1.2 Billion—Retrospect and prospect of population in China*. UNESCO ISSI 148/1996: 261–67.

Lutz, Wolfgang. 1994. The future of world population. Population Reference Bureau. *Population Bulletin* 49(1).

Martin, Roberta, ed. 1996. *Central themes for a unit on China*. East Asian Curriculum Project, Columbia University, New York (March).

Orrell, Keith, and Pat Wilson, eds. 1994. China struggles with the population equation. *Geographical* (June): 43–45.

Population Reference Bureau. 1992. *The UN long-range population projections—What they tell us*.

Prosterman, Roy L., Tim Hanstad, and Li Ping. 1996. Can China feed itself? *Scientific American* (November): 90–96.

Rosen, James E., and Shanti R. Conly. 1998. *Africa's population challenge: Accelerating progress in reproductive health*. Population Action International, Washington, D.C.

Tien, H. Yuan, et al. 1992. China's demographic dilemmas. Population Reference Bureau. *Population Bulletin* 47(1).

Xin, Wang. 1995. Population vs. development: Challenge of the new century. *Beijing Review* (May): 12–15.

Zhenhua, Xie. 1966. A new industrial civilization. *Our Planet* 7(6): 12–13.

CHAPTER 7

Green Revolution Agriculture

Permission to reprint, courtesy of Case Corporation, Racine, Wisconsin

Let's get our priorities in perspective. . . . We must feed ourselves and protect ourselves against the health hazards of the world. To do that, we must have agricultural chemicals. Without them, the world population will starve.

Norman E. Borlaug

As human populations expand, the demand for food is accelerating worldwide. Today, malnutrition, hunger, and famine affect at least one billion people. These adversities stem from poverty, lack of arable land and water, and in some cases political agendas. But since the 1960s, agricultural innovations known as the "green revolution" have proven immensely valuable in producing more food for the world, especially in developing nations where populations are increasing most rapidly.

The green revolution was initiated in the 1940s when the Rockefeller Foundation in New York City funded an agricultural research program in Mexico to avert almost certain starvation for millions of people. The work of a team of plant scientists ultimately led to new varieties of disease-resistant, high-yielding wheat and, in time, to improved varieties of maize, rice and other crops. This chapter explores the nature of green revolution agriculture, its spread to developing and industrialized nations, environmental problems it has created, and recent bioengineering strategies that have evolved to continue to address the problem of world hunger.

BACKGROUND

Fundamentals of Agriculture

Scholars believe that agriculture as a purposeful human pursuit began about 10,000 BP (years before the present time) when indigenous people in many parts of the world turned from hunting and gathering toward nurturing wild plants as sources of food. Early agriculture depended solely on wild plant varieties, but over time farmers selected particular wild plants that prospered in specific microclimates, resisted unfavorable weather, and were able to survive disease and attacks by plant pests. Later, through careful selection and cultivation, these superior crop varieties were developed and cultivated. We will begin our discussion of agriculture with wild crop varieties.

Wild Crop Varieties and Subsistence Agriculture

Some scholars believe that wild einkorn wheat, emmer wheat, and barley were first harvested in Iraq and Iran some 9,000 years ago. At the same time, beans, peas, gourds, water chestnuts, and pumpkins grew wild in northeastern Mexico. In Ethiopia, farmers culled coffee, wheat, and sorghum, an important cereal crop—all growing as wild plants on nearby hills, in cool moist valleys, and in forested areas. In the Andes Mountains of southern Peru, wild potatoes are still an important food staple, especially for the region's poor.

Wild crop varieties such as these were the first sources of grains, seeds, fruits, and tubers for most of the world's people as they abandoned their nomadic hunting-and-gathering lifestyles. Harvesting naturally occurring wild plants allowed poor people almost everywhere to settle down in stable communities and eke out a living through what today we would call **subsistence agriculture.**

Looking back at wild plants as food sources may lead us to think that such plants have little value today. After all, modern varieties of wheat, maize, barley, rice, potatoes, and other crops produce immensely greater yields, feeding directly or indirectly not only the farm families that grow them but as many as 100 or more others. However, wild crop varieties shelter and maintain genetic diversity, which is displayed in hundreds if not thousands of slightly different varieties within each kind of crop. The genetic diversity of wild plants assures their survival, giving some resistance to specific diseases, others the ability to withstand prolonged drought, and still others the capacity to survive attack by fungal, viral, and bacterial infections. For example, late blight, the fungus that attacked Ireland's potato crop in the 1840s causing a million deaths and forcing another million people to migrate to other countries, is today controlled with agricultural fungicides. But a fungicide-resistant form of late blight broke out in the 1980s and now causes billion-dollar losses in global potato production. Fortunately, genetic resistance to this type of blight has been found in a wild potato variety high in the Andes. Using modern plant cross-breeding techniques, it will be possible to add this resistance to today's modern potato plant.

Landraces: Traditional Crop Varieties

Centuries of wild plant harvesting by subsistence farmers the world over led slowly to the selection of preferred wild crop varieties, which were then cultivated instead of simply harvesting wild crops. The relatively small number of favored crops they chose included maize, wheat, rice, beans, rye, barley, and potatoes. Crop preferences were based on desired traits such as taste, texture, smell, and color, as well as resistance to diseases, pests, droughts, heat, and early frost. Seeds from these carefully selected and propagated plants were handed down from generation to generation, becoming traditional or folk crop varieties known as **landraces.**

In many countries today, farmers plant traditional landrace seeds to grow their grains. Landraces possess most of the genetic heritage common to wild plant varieties. Yields from landrace crop varieties are consistent, though often meager, but farmers do not have to apply chemical fertilizers and pesticides, which in most cases they cannot afford anyway. In Ethiopia, where 80 percent of the people are farmers and the size of an average farm is about 2 hectares ($\approx$ 5 acres), using costly fertilizers and pesticides is not an option. Therefore, planting landrace sorghum and maize in Ethiopia essentially guarantees predictable harvests. The same is true in the Andes where landrace potato seeds produce 20 to 30 somewhat different kinds of potatoes, some of which always survive adverse conditions.

Wild crop varieties and traditional landraces possess rich and complex endowments of genetic diversity—biological DNA borne of evolutionary change and selected over time to withstand a variety of environmental stresses. Thus, landrace crops do not need chemical inputs such as pesticides to survive. They capitalize instead on their genetic biodiversity.

Crop Cultivars: Superior Crop Varieties

Although plant breeding has ancient roots, the systematic improvement of plants can be traced in part to Gregor Mendel, an Austrian botanist, who published his studies of plant heredity in 1866. Mendel, working with pea plants, found that traits such as seed pod texture and the color of flowers were inheritable and could be manipulated when plants with one particular trait were crossbred with plants having a different trait. Later, scientists who repeated Mendel's research identified the basic units of biological inheritance: *genes.*

A **gene** is a discrete biological unit responsible for transmitting hereditary information from parents to offspring. Genes, which occur in cell chromosomes, determine the specific characteristics of all organisms. Most genes are composed of the chemical deoxyribonucleic acid, or DNA.

Mendel did not invent **crossbreeding,** the art and science of cultivating desired plant varieties. It was practiced by farmers in China and other civilizations as long ago as 7,000 BP. Crossbreeding is accomplished through careful cross-pollination, whereby pollen from the anthers of one plant is purposefully transferred to the stigma of another plant of the same or closely similar species. The goal of crossbreeding in agriculture is to develop new plant varieties that display desired traits such as preferred color, texture, appearance, and taste; resistance to drought, disease, and early frost; shorter growing season and increased yields; and higher nutritional values. Plant breeders try to recognize specific genetic differences among plants of the same species and develop superior crop varieties called *cultivars* (CULTIvated VARieties).

A **cultivar** is an improved crop variety consistently cultivated in a particular region. Cultivars generally display highly desirable genetic traits such as high yield, appealing taste and color, disease resistance, short growing season, ability to utilize added chemical fertilizers to produce greater crop yields, and capacity to withstand drought, storms, and early frost.

The most important crop cultivars, green revolution varieties of wheat, maize, and rice were first developed by Norman Borlaug and other plant scientists through crop crossbreeding research that began in Mexico. But although green revolution cultivars are striking in their ability to produce very high yields, they reflect genetic uniformity rather than genetic diversity. Their uniformity is crossbred into them so that plants will grow and mature at the same rate, respond to fertilizer in the same way, and produce essentially the same yield at a predictable time. Thus, green revolution agriculture is a godsend to the hungry of the world, but it comes with a high price and numerous questions for the future, as we will see in the following case study.

CASE STUDY: ORIGINS, ACHIEVEMENTS, AND ENVIRONMENTAL ISSUES OF THE GREEN REVOLUTION

In 1943, in response to increasing demand for food in Mexico and other developing countries, the Office of Special Studies of the Rockefeller Foundation in New York City, working with the Mexican Ministry of Agriculture, initiated a cooperative research program. The foundation sent a team of plant scientists to work with Mexican scientists and farmers on ways to improve the yields of crops, particularly wheat and maize. It turned out to be a slow process requiring several generations of plant breeding before the first new kinds of high-yielding wheat could be introduced into Mexican agriculture in the 1960s.

Breeding High-Yielding Wheat

Mexico

Wheat had been introduced into Mexico by the Spaniards in the 1600s, but yields were typically low. The traditional wheat plants were commonly devastated by crop diseases called *rusts,* and their long, thin stalks (standing about 125 to 150 centimeters [≈ 50 to 60 inches] high) were weak and tended to blow down in windstorms, an effect called **lodging.** Fertilization and irrigation of these wheat varieties simply led to taller plants, making lodging an even more serious problem.

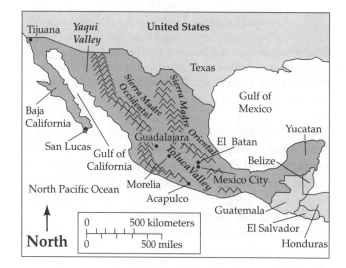

Figure 7.1 Mexico, showing the Yaqui and Toluca Valleys where green revolution wheat was first developed.

Norman Borlaug, one of the plant scientists assigned to the project, began a series of experiments in the Toluca Valley of central Mexico and the Yaqui Valley of Sonora, a state in northwestern Mexico (Figure 7.1). His work involved crossbreeding wheat cultivars from many different parts of the world.

In 1955, Borlaug began to develop short-stature, semi-dwarf types of wheat that expend less energy growing tall stalks and thus free up more energy to produce higher yields. To achieve this goal, Borlaug crossbred Norin 10—a dwarf variety of wheat from Japan, where growing dwarf wheat has long been an art form—with Mexican and other wheat cultivars. By 1962, the first successful semi-dwarf wheat varieties, Pitic 62 and Penjamo 62, had been produced, tested, and made available to Mexican farmers. Semi-dwarf wheat has the following characteristics:

- Short-stature plants—about 50 to 60 centimeters (≈ 20 to 24 inches) high, which minimizes lodging.
- Resistance to rust diseases common to wheat.
- Production of higher yields if chemical inputs such as fertilizers and pesticides are used.

Because wheat is by nature a *self-pollinating* plant, the desired traits of a particular wheat variety are uniformly inherited in the seeds each plant produces. Farmers who grow a particular wheat variety can save seed from one growing season to the next, knowing that the seeds will produce wheat having the same characteristics as last season's crop. This makes it cheap and easy for farmers to grow a particular variety of wheat year after year. Seed from other major crops, particularly rice, soybeans, and cotton, can also be saved and planted in this way.

Self-pollination is the transfer of pollen by pollinating insects or by hand from the anthers (pollen-producing organs) of a flowering plant to the stigma (pollen receptor) of the same plant. Fertilization leads to the development of new seeds. Self-pollination is common in wheat, barley, and oats.

When farmers in Mexico's Toluca and Yaqui Valleys began to cultivate semi-dwarf wheat, their *annual yields* increased dramatically. In 1961, before the new varieties were introduced, traditional wheat yields averaged 1,675 kilograms/hectare (≈ 24 bushels/acre) annually. At that time, Mexico was importing half of the wheat it needed. But by 1973, wheat yields had doubled to 3,350 kilograms/hectare (≈ 48 bushels/acre), and the country had become self-sufficient in wheat. By 1987, wheat yields were close to 4,500 kilograms/hectare (≈ 64 bushels/acre). As *total harvests* climbed to record levels, Mexico became an exporter of wheat. Since then, however, total wheat output has declined slightly due to droughts in northern Mexico and increased production of fruits and vegetables for export, products that command higher prices than wheat in many countries (Figure 7.2a).

Two types of statistics are particularly useful in discussing agricultural production: **annual yields,** usually given as kilograms per hectare or bushels per acre, and **total harvests,** typically reported as metric tons per year. Total harvest figures reflect how much food a given country produces each year for consumption, storage, and export to other countries. Table 7.1 highlights 1961, pre-green revolution yield and harvest data for wheat, maize, and rice in selected countries. This information will serve as a yardstick later as we assess the accomplishments of green revolution agriculture.

The development of semi-dwarf wheat in Mexico led to the establishment of the International Maize and Wheat Improvement Center in 1963. The center, known by its Spanish acronym CIMMYT, is in El Batan, 50 kilometers (≈ 30 miles) north of Mexico City. Today, it is an internationally funded, nonprofit, research and training center that works to improve the productivity, profitability, and sustainability of wheat, maize, and other crops in developing nations. Some of the newer varieties of wheat created by CIMMYT are shown in Figure 7.3.

Pakistan and India

In 1964, Borlaug was invited by the government of Pakistan to initiate wheat breeding programs there. This work was supported by the Ford Foundation in New York. By 1967, he had successfully crossed native

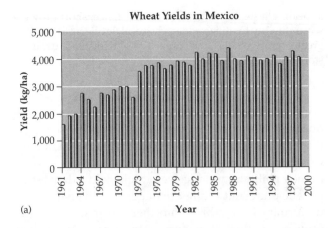

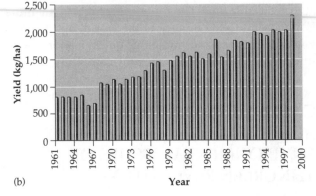

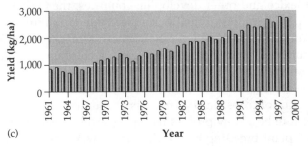

Figure 7.2 Annual yields of wheat in Mexico, Pakistan, and India between 1961 and 1998.
(Source: Data from the United Nations Food and Agriculture Organization (UNFAO), electronic data sheet, Worldwide Web.)

cultivars with Mexican semi-dwarf wheat, which led to new wheat varieties well suited to the Indian subcontinent's climates. By 1968, wheat yields rose from traditional levels of about 820 kilograms/hectare (≈ 12 bushels/acre) to about 1,000 kilograms/hectare (≈ 15 bushels/acre), an increase of about 25 percent. Since then, wheat yields and total harvests have continued to increase, and in 1998 Pakistani wheat yields exceeded 2,250 kilograms/hectare (≈ 32 bushels/acre) (see Figure 7.2b).

Meanwhile, Mexican semi-dwarf wheat, introduced into India as an experiment in 1962, performed

TABLE 7.1   Typical 1961 Pre-Green Revolution Crop Yields and Total Harvests of Wheat, Maize, and Rice[a]

Selected Country	Wheat		Maize		Rice	
	Yield	Harvest	Yield	Harvest	Yield	Harvest
Mexico	1,675	1.40	993	6.25	2,275	0.33
Pakistan	822	3.81	1,030	0.49	1,391	1.69
India	851	11.00	957	4.31	1,541	53.49
Philippines	—	—	628	1.27	1,391	1.69
Egypt	2,469	1.44	2,401	1.62	5,053	1.14
South Africa	773	0.87	1,285	5.29	1,587	0.01

SOURCE: Data from the United Nations Food and Agriculture Organization (UNFAO).

[a]Yields are given in kilograms/hectare/year. Harvests are reported in million metric tons per year. (One metric ton = 1,000 kilograms = 2,204 pounds = 1.1 tons.)

Figure 7.3   Wheat varieties under development at CIMMYT. The green plants on the left are a tall variety of wheat. To the right is an early-maturing dwarf wheat. Further to the right are semi-dwarf wheat varieties.

(Reprinted with permission by CIMMYT, El Batan, Mexico.)

so well that the Indian government tried to convince farmers to switch to semi-dwarf wheat. Because they were reluctant to do so, national programs were organized to publicly demonstrate the productivity of semi-dwarf wheat. Hundreds of experimental plots were set up in different agricultural areas. Half of each plot was planted in traditional Indian wheat; the other half sown with a semi-dwarf variety. By the following year, the outcome was clear: Traditional Indian wheat averaged 850 kilograms/hectare (≈ 12 bushels/acre), while semi-dwarf wheat averaged 2,500 kilograms/hectare (≈ 36 bushels/acre). The results sparked a revolution in Indian wheat agriculture (see Figure 7.2c). Between 1961 and 1998, total annual

wheat harvests in India rose from 11 million metric tons to 66 million metric tons.

Producing New Varieties of Maize

Maize is not only the principal food source in Mexico—it serves as an icon whose image pervades much of Mexican culture. Hundreds of maize landraces exist in Mexico, each one adapted to a particular microclimate where environmental factors such as elevation, prevailing temperatures, and duration of daily sunlight influence growth. For example, highland maize landraces prosper at cold, high elevations often exceeding 2,000 meters (≈ 6,600 feet). Other varieties grow only at warmer, mid and low elevations. Most maize landraces are intolerant to cold, susceptible to disease, display poor grain quality, and produce minimal harvests. They are also predisposed to lodging.

These were the problems that led CIMMYT plant breeders to develop new wheat varieties. But crossbreeding varieties of maize to produce new varieties with desired characteristics such as large grain size, short growing season, and disease resistance was mostly unsuccessful. The spectacular increases in wheat yields common in Mexico, Pakistan, and India could not at first be duplicated in the case of maize. The lack of success in breeding high-yielding maize was due to the fact that maize is a *cross-pollinating* plant. As a result, desirable traits and characteristics that were crossbred into new maize varieties were lost in just one growing season due to pollen transferred by wind or insects from unimproved varieties of maize to new cultivars.

Figure 7.4 An improved variety of maize growing in the Toluca Valley of central Mexico.
(Reprinted with permission by CIMMYT, El Batan, Mexico.)

Cross-pollination is the transfer of pollen by wind, by pollinating insects, or by hand from the anthers (pollen-producing organs) of one flowering plant to the stigma (pollen receptor) of a different plant of the same species. Fertilization leads to the development of new seeds.

The best way around this problem is to develop maize *hybrids* grown each year from seeds supplied by seed producers. Although hybrid crop seeds must be bought each year, farmers are essentially guaranteed a bountiful harvest (Figure 7.4). At the time Borlaug and his fellow plant scientists were working in Mexico, developing nations did not have well-developed seed industries. This did not come about until the 1980s. Today, both cross-pollinating and hybrid varieties of maize are planted in Mexico. Trends in annual maize yields in Mexico, mainland China, and Sub-Saharan Africa are shown in Figure 7.5.

Hybrids are plants grown from seeds produced through controlled cross-pollination of inbred, genetically uniform, pure-breeding plant varieties. Hybrid plants display not only uniformity but also "hybrid vigor," which includes high yields, disease resistance, drought tolerance, and hardiness. Nowadays, hybrid seeds are produced by seed companies that maintain selected, true-breeding, parent crop varieties. However, seeds saved from a hybrid crop will not grow the same crop the next year. Genetic traits bred into the original seeds segregate out in the next growing season, yielding a mix of varietal traits. Therefore, it is not worthwhile for farmers to save seed from hybrid crops for the next planting. Instead, new hybrid seed must be bought from an established seed industry or government agency.

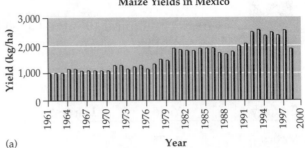

(a)

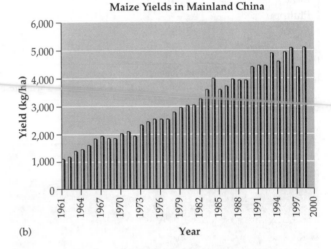

(b)

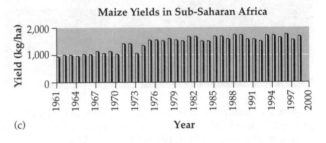

(c)

Figure 7.5 Annual yields of maize in Mexico, Mainland China, and Sub-Saharan Africa between 1961 and 1998.
(Source: Data from the United Nations Food and Agriculture Organization (UNFAO), electronic data sheet, Worldwide Web.)

Creating Improved Varieties of Rice

Rice is as important in some developing countries as wheat and maize are in Mexico. In Asia, where it is the principal crop, twice as much rice as wheat is typically grown. In 1960, the International Rice Research Institute (IRRI) was established in the Philippines by the Ford and Rockefeller Foundations to expand rice production there. Crossbreeding rice landraces from different regions led to the release of IR8 in 1965, the first high-yielding, short-stature rice variety having superior resistance to disease and insect attack. Since

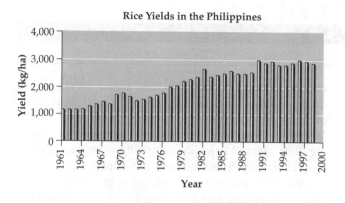

Figure 7.6 Annual rice yields in the Philippines between 1961 and 1998.
(Source: Data from the United Nations Food and Agriculture Organization (UNFAO), electronic data sheet, Worldwide Web.)

Figure 7.7 Oroville Dam in northern California stores water to irrigate crops further to the south.
(Photo courtesy of David C. Neilsen.)

then, new varieties have been introduced into many countries. By 1982, IRRI rice varieties were being grown in 85 percent of Philippine rice paddies. Both annual yields and total harvests doubled, and the country was transformed from a net importer to a net exporter of rice. Figure 7.6 depicts annual trends in Philippine rice yields between 1961 and 1998.

Green Revolution Infrastructure

Besides developing new varieties of wheat, maize, and rice, the green revolution affected the practice of agriculture in other ways, most notably in the increased reliance on monocultures, mechanized farm implements, crop irrigation, and chemical fertilizers and pesticides.

Monocultures

One of the main objectives of plant breeding is to develop crop varieties that have uniform growth characteristics—that is, individual plants that grow at the same rate, display common biological traits, and reach maturity at the same time. Green revolution varieties of wheat, maize, and rice possess this uniformity, making it economical and efficient to grow a single crop, or perhaps a single crop variety, over hundreds if not thousands of acres. Crops grown in this way are referred to as crop **monocultures.** Monocultures capitalize on crop uniformity, justify the purchase or rental of specialized farm implements, and minimize the time required to carry out field operations. Monocultures also allow farmers to apply uniform management strategies across large farmed areas. Strategies such as these are economically much more efficient than growing many different crops in one area.

Mechanization

Crop uniformity makes large-scale, mechanized farming efficient and cost-effective. Indeed, mechanized agriculture is required on large farms to realize the full potential of green revolution agriculture. Today, crop seeding, fertilizing, cultivating, pesticide application, and harvesting are often carried out using multipurpose farm implements—cultivators, planters, harrows, sprayers, and harvesters.

In some cases, mechanical implements have been designed to conform to the shape and characteristics of individual crops. One example is raspberry harvesters engineered to engage plants ready for harvest and shake mature berries into wooden boxes as they navigate down row after row in monoculture-planted berry fields. Wheat combines used from Washington and Oregon eastward to the American Great Plains cut and thresh crops, separate grains from chaff, blow chaff into the air, and drop hay to the ground.

Irrigation

High green revolution crop yields sometimes depend on irrigation. In semi-arid and dry regions, where yields are often meager due to lack of rain, irrigation can generate two and sometimes three harvests a year. In the year 2000, 40 percent of the world's food supply came from irrigated croplands. When monocultures, mechanization, and irrigation are combined, the results can be spectacular. It is not surprising, therefore, that immense quantities of water are used in modern agriculture. One example is California's Central Valley Project, which delivers Feather River water stored

INSIGHT 7.1
The Ogallala Aquifer

An **aquifer** is a naturally occurring water resource associated with underground sand, gravel, and porous rock formations. Aquifers usually contain freshwater and are often excellent water sources for drinking and agricultural irrigation. North America's largest aquifer, the Ogallala or High Plains Aquifer, stretches about 960 kilometers (≈ 600 miles) north and south under parts of Wyoming, Nebraska, Colorado, Kansas, Oklahoma, and Texas. Its volume is estimated to equal that of Lake Huron. The Ogallala comprises wetlands and water deposits up to 90 meters (≈ 300 feet) deep.

Withdrawing irrigation water from the Ogallala began in the 1940s. Increasingly greater quanti-

ties were pumped, and by 1980 water levels had dropped an average of 3 meters (≈ 10 feet). In parts of Texas, the decline was as much as 30 meters (≈ 100 feet). Bountiful harvests on farms above the Ogallala seemed to justify extracting so much water. However, concerned farmers have initiated water conservation measures by adopting minimum irrigation strategies and adopting no-till and low-till farming practices (See Chapter 8). By 1993, these efforts were resulting in much slower water level declines in the Ogallala. In parts of west Texas, northeast Oklahoma, and southeast Wyoming, aquifer levels had risen by as much as 0.6 meters (≈ 2 feet).

behind Oroville Dam via 900 kilometers (≈ 560 miles) of aqueducts to San Joaquin Valley farms (Figure 7.7). Another example is the Ogallala aquifer, an extensive underground water supply beneath much of the American Great Plains. Prodigious quantities of water are pumped each year from the Ogallala to irrigate crops on hundreds of midwestern farms (Insight 7.1).

Chemical Fertilizers

High agricultural yields are possible if crops are grown in fertile, nutrient-rich soil. While animal and human manure were once the most important fertilizers, commercial chemical fertilizers are now the principal source of nutrients used in green revolution agriculture, mainly because they are cheap and easy to apply. Chemical fertilizers consist of minerals containing nitrogen, phosphorus, and potassium, the three nutrients needed by crops in relatively large amounts. While earlier crop varieties lacked the genetic information needed to utilize added nutrients, green revolution varieties of wheat, maize, and rice were developed to produce higher yields in response to increased fertilization.

The success of green revolution crop varieties is linked to their ability to capitalize on the application of commercial fertilizers (see Figures 7.2, 7.5, and 7.6). Between 1961 and 1984, annual applications of commercial fertilizers on Mexico's farms increased eightfold, from about 16 million tons to 126 million tons a year. In response, wheat harvests rose by more than 300 percent, and maize harvests increased by more than 200 percent. Using commercial fertilizers, Indian wheat harvests increased by more than 400 percent, and Philippine rice harvests doubled.

Between 1961 and 1984, commercial fertilizer use resulted in worldwide grain increases of almost 4 percent yearly. Each additional ton of fertilizer produced as much as nine more tons of grain. But after 1984, applying more fertilizer resulted in diminishing returns, and worldwide grain yields increased less than 1 percent a year. Farmers attempted to raise their yields by using more fertilizers, but it did not work.

Unfortunately, the expanded use of chemical fertilizers has led to agricultural fields becoming non-point sources of nutrient runoff, causing eutrophication in rivers and lakes—massive algal blooms, undesirable weed growth, sagging levels of dissolved oxygen, and fish kills (see Chapter 4). In some cases, chemical fertilizers containing nitrate can pollute groundwater sources of drinking water (see Chapter 3). Nevertheless, commercial fertilizers continue to be one of the main chemical inputs in modern agriculture. In the 1990s, their annual use worldwide added up to more than 127 billion kilograms (≈ 140 million tons), worth close to $50 billion. Another factor that affects soil fertility is erosion by wind and water, against which special measures have been taken (Insight 7.2).

Chemical Pesticides

Throughout agriculture's long history, farmers have had to cope with pests and blights: weeds that compete with crops for space, light, nutrients, and moisture; insects that attack crops, often eating their leaves, stalks, and seeds; viruses that cause disease and decimate crop yields; and molds that infiltrate plant cells, altering plant biology. Often as much as 30 percent of farmers' crop yields have been lost each year to

INSIGHT 7.2

Soil Erosion

All soils are subject to **erosion,** the wearing away of land by natural processes such as wind and water, resulting in the loss of topsoil. Rates of soil erosion are influenced by cover vegetation, soil type, climate, weather, rainfall, and topography. Agricultural soils in particular may be highly susceptible to erosion. For example, while soil erosion on grazing lands is typically less than 3 metric tons/hectare/year (≈ 1.3 tons/acre/year), cultivated cropland can lose as much as 200 metric tons/hectare/year (≈ 89.0 tons/acre/year). This is because grazing lands are not subjected to intense cultivation.

The U.S. Natural Resources Conservation Service (NRCS), an agency of the U.S. Department of Agriculture (USDA) formerly called the Soil Conservation Service, conducts periodic national inventories of U.S. agricultural soils. In 1997, the NRCS reported that water and wind-related soil erosion in the United States averaged 11.6 metric tons/hectare/year (≈ 5.2 tons/acre/year). While this was 42 percent less erosion than in 1982, it was greater than the average rate at which new soil is formed—typically 11.2 metric tons/hectare/year (≈ 5.0 tons/acre/year), or the equivalent of about one inch every 500 years (Table 7.A).

Almost half of U.S. cropland soil erosion occurs in just five states: Texas, Minnesota, Iowa, Montana, and Kansas. Here, erosion exceeds 25 metric tons/hectare/year (≈ 11 tons/acre/year) in some areas. This is due in part to farming on *highly erodable land.* **Highly erodable land** (HEL) has a high potential for soil erosion due to its slope, lack of vegetative cover, and deficiency of organic matter. In 1997, about 30 percent of U.S. cropland (≈ 380 million acres) was classified as highly erodable. Beginning in 1985, the USDA Conservation Reserve Program has encouraged farmers to remove such lands from production and plant grass instead. Texas, one of the states where this is being done, has reduced HEL erosion from 13 tons per acre per year to 2 tons per acre per year. Nationally, HEL erosion has been cut by 47 percent since 1982.

TABLE 7.A   U.S. Rates of Water- and Wind-driven Soil Erosion Compared with Average Rate of Soil Formation

Year	1982		1997	
Type of Soil Erosion	Metric Tons/ Hectare/Year	Tons/Acre/ Year	Metric Tons/ Hectare/Year	Tons/Acre/ Year
Water-related erosion	10.1	4.5	6.7	3.0
Wind-related erosion	7.9	3.5	4.9	2.2
Total soil erosion	18.0	8.0	11.6	5.2
New soil formation	—	—	11.2	5.0

SOURCE: Data from the U.S. Department of Agriculture, Natural Resources Conservation Service.

agricultural pests, reducing harvests in some cases to subsistence yields. Green revolution agriculture has not eliminated these problems. On the contrary, new crop varieties are generally more susceptible to disease and pest attack because they do not have the genetic diversity of parent landraces and cultivars—biological endowments that confer disease and pest resistance. As a result, most green revolution crop varieties depend on the use of chemical pesticides. However, pesticides must be evaluated for their level of toxicity before they are applied to crops intended for human consumption (Insight 7.3).

In terms of weight, pesticides make up the second largest chemical input to modern agriculture (the largest being chemical fertilizers). During the 1990s, close to 2.3 billion kilograms (2.5 million tons) of pesticides, worth more than $30 billion, were applied to crops worldwide each year. The term *pesticide* includes a number of different chemical agents, each designed to attack a specific kind of pest (Table 7.2). Of these,

INSIGHT 7.3
Pesticide Toxicity

One measure of a chemical's toxicity is its LD_{50} value, the concentration that on average kills 50 percent of a test population of animals such as rats. LD_{50} is usually reported as milligrams of a chemical per kilogram of a test animal's weight. The smaller the value, the more toxic the chemical. LD_{50} values for selected pesticides are given in Table 7.B.

Chemicals with LD_{50} values greater than 500 mg/kg in rats or mice are generally considered safe in agriculture. However, this is not true in all cases. Chemicals with LD_{50} values less than 500 mg/kg may be toxic, especially if the LD_{50} is less than 100 mg/kg. For example, parathion and methyl parathion both have low LD_{50} values and are extremely hazardous pesticides. Their use has led to fish kills and even human deaths.

Malathion, an organo-phosphorus insecticide with an LD_{50} of 1,000 mg/kg, is much less toxic than parathion. It can be safely used to control aphids on vegetables, fruits, and forage crops. It can also control fleas on dogs and cats and kill body lice on humans. This is why malathion has been used to control outbreaks of Mediterranean fruit flies (medflies) in California.

TABLE 7.B   Pesticide Toxicity Data for Selected Herbicides and Insecticides

Pesticide	LD_{50} (Rats, mg/kg)
Parathion	13
Methyl parathion	24
DDT	115
2,4-D	370
MCPP	500
Kelthane	575
MCPA	700
Malathion	1,000
Atrazine	1,750
Simizine	5,000
Maneb	6,750
Captan	9,000
Benlate	9,590

SOURCE: Data from *The Merck Index*, Twelfth Edition, 1996.

herbicides, insecticides, and fungicides will be discussed here.

Herbicides Agricultural soils are ideal environments for weed growth, especially when they are cultivated, fertilized, and irrigated. When weeds compete with crops for limited space, nutrients, moisture, and sunlight, crop yields are reduced. Weeds were once simply rooted out of the ground using a hoe, but this is not practical on large farms today, so farmers have come to rely on chemical herbicides.

2,4-D, the first important chemical herbicide, became available in 1942. Its name is chemical shorthand for 2,4-dichlorophenoxyacetic acid. 2,4-D has been almost universally adopted on farms around the world because it is selectively toxic toward common broadleaf weeds. It is perhaps best known in domestic lawn and garden products—for example, for killing dandelions. In time, additional chemical herbicides were developed to free farmers from the tedious manual work and costly mechanical operations necessary to manage weed problems.

2,4-D is a member of a group of herbicides known collectively as phenoxy herbicides. They kill

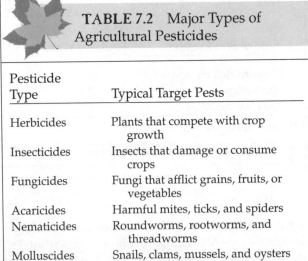

TABLE 7.2   Major Types of Agricultural Pesticides

Pesticide Type	Typical Target Pests
Herbicides	Plants that compete with crop growth
Insecticides	Insects that damage or consume crops
Fungicides	Fungi that afflict grains, fruits, or vegetables
Acaricides	Harmful mites, ticks, and spiders
Nematicides	Roundworms, rootworms, and threadworms
Molluscides	Snails, clams, mussels, and oysters
Termiticides	Wood-consuming termites
Rodenticides	Rats and mice
Avicides	Birds
Bactericides	Bacteria that cause plant or animal disease

SOURCE: Data from *Pesticide Safety Handbook*, Saskatchewan Agriculture and Food, 1990.

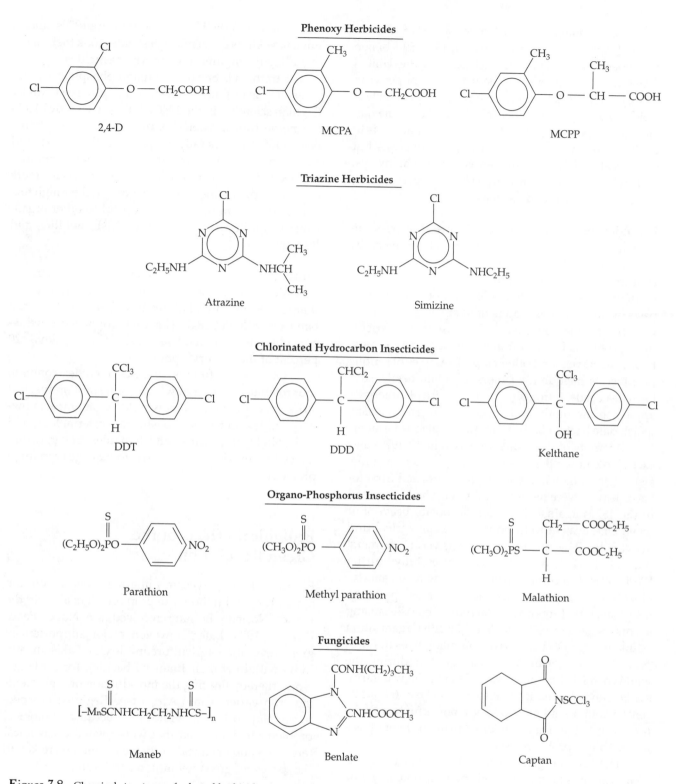

Figure 7.8 Chemical structures of selected herbicides, insecticides, and fungicides.

weeds by mimicking natural plant growth regulators called auxins, thereby triggering unsustainable, accelerated cell division. The chemical structures of selected phenoxy herbicides are shown in Figure 7.8.

Atrazine and simizine, examples of triazines, are effective at controlling weeds in cornfields by inhibiting photosynthesis. 2,4-D, atrazine, and several other chemical herbicides were on the market before the 1960s, but their use increased sharply when green revolution agriculture began. They made weed control more efficient and less costly on large farms.

The Economic Research Service of the U.S. Department of Agriculture estimates that in 1999, $9.1 billion was spent on agricultural pesticides in the United States, a cost that increases about 4.5 percent per year. Of this total, $5.8 billion (64 percent) was spent on herbicides. Indeed, chemical herbicides make up the major share of pesticides sold today. Their main use is to control weeds affecting wheat, corn, and soybeans, but they are also used to eliminate weeds along highways, under electric power transmission lines, on golf courses, and in home lawns and gardens.

Insecticides More than 750,000 insect species are known to exist, but less than 1 percent of them are pests. Bees, for example, pollinate many crops, flowers, and fruit trees and in addition make honey. The silkworm moth produces silk. Lady beetles control several insect pests by keeping their populations small.

But pest insects frequently cause major agricultural problems by consuming and damaging grains, fruits, vegetables, and other crops. Examples of serious pest insects include the Colorado potato beetle, soybean flea beetle, European corn borer, rice planthopper, apple codling moth, and cotton boll worm. Locusts, which date back to Biblical times, still plague farmers.

A variety of chemicals were used in the 1920s and earlier to combat pest insects. Many, such as cyanide, arsenic, lead, mercury, and nicotine extracted from tobacco leaves, were highly toxic. Arsenic had been used in the 1870s against the Colorado potato beetle. But when DDT was introduced in 1946 as an agricultural insecticide, it seemed like the answer to farmers' prayers. DDT had been used successfully during World War II to control malaria, yellow fever, and typhus among military personnel—diseases spread by insect *vectors* (organisms carrying disease-causing agents). In fact, in 1948, the World Health Organization estimated that DDT had saved 5 million lives in European, Mediterranean, and Asian theaters during the war. When DDT became available after the war, farmers quickly adopted it because it was cheap (about 22 cents a pound) and proved to be a powerful *broad-spectrum* insecticide. That is, it killed many different pests not just a few problem insects.

DDT is an **organic** chemical, a substance whose composition is based on the element carbon. Its name stands for *d*ichloro-*d*iphenyl-*t*richloroethane. Because of its success as a pesticide, new chemicals similar to DDT were synthesized and became known as the **chlorinated hydrocarbon pesticides.** Their use in the 1940s and 1950s controlled a great variety of agricultural pests and increased crop yields significantly. But because of adverse environmental effects, which we will discuss later, most chlorinated hydrocarbon pesticides were banned in the United States in the 1970s, and in time, new kinds of chemical pesticides took their place, including the **organo-phosphorus pesticides.**

Parathion is one of the organo-phosphorus pesticides. It was originally developed as a nerve toxin by German scientists in the 1940s. It kills insects by blocking nerve transmission. Because organo-phosphorus pesticides are generally more toxic than DDT and other chlorinated hydrocarbons, smaller quantities are needed to control insects, making their use more efficient and, in some cases, cheaper. But their high toxicity makes them potentially harmful to other organisms, including beneficial insects, fish, wildlife, and humans.

Fungicides One of the best examples of a harmful agricultural fungus is potato blight, the legendary pest that caused repeated failures in Ireland's potato crop between 1845 and 1847. This catastrophe led to widespread famine in Ireland, seriously affecting almost 20 percent of the country's population.

Other crop fungus diseases include: common root rot and stem rust in wheat; stem smut on rye; leaf stripe on barley; vine downy mildew on grapes; powdery mildew in beans, squash, and watermelons; and early blight on potatoes and tomatoes. Several commercial fungicides are now used to manage crop fungi problems.

Ecological Effects of the Green Revolution

For the role he played in initiating green revolution agriculture and in honor of crop breeders all over the world, Norman Borlaug received the Nobel Peace Prize in 1970. Later, however, major supporters of green revolution agriculture in Mexico, Pakistan, and India withdrew their financial backing for Borlaug's work, thereby limiting the introduction of high-yielding crop varieties into Africa (see Regional Perspectives). Support was withdrawn because ecological problems had surfaced that were related to chemical fertilizers and chemical pesticides, which were key to the success of green revolution agriculture.

Resistance to Pesticides

Because monocultures are highly susceptible to plant disease, pest attack, and weed growth, green revolution agriculture depends heavily on chemical pesticides. Now, after more than 50 years of using them, some environmental impacts have become clear.

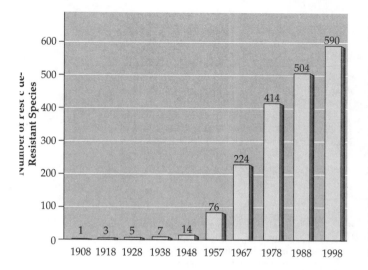

Figure 7.9 Trend toward increasing pest resistance to pesticides 1908–1998. Note that 1998 data are a projection by the author.
(Source: Data from George P. Georghiou, "Overview of Insecticide Resistance" in Managing Resistance to Agrochemicals, Maurice B. Green, et al. (eds.), pp. 18–41, American Chemical Society Symposium Series 421, 1990.)

Over time, weeds, insects, and disease-causing pests have developed increasing resistance to chemical pesticides. Individual pests that are by chance genetically more resistant survive and reproduce; those that are less resistant die out, a process known as **natural selection.** After several generations, a pesticide-resistant pest population often develops, forcing farmers to apply higher pesticide levels or switch to newer, generally more expensive chemicals.

An interesting example of pest resistance to DDT occurred in cotton boll weevils migrating from cotton fields in Mexico to Texas, California, and the Mississippi Valley. Female weevils attack cotton by boring holes into cotton bolls to lay their eggs, from which larvae then hatch and consume the cotton. At first, as little as 30 milligrams of DDT killed 1 kilogram of boll weevils, but by 1965, close to 1 kilogram of DDT (33,000 times more) was needed to achieve the same result. Cotton boll weevils had become remarkably resistant to DDT, and cotton fields were being devastated. It is estimated that in 1998 at least 590 pest species had developed resistance to one or more chemical pesticides (Figure 7.9).

Harm to Nontarget Organisms

Broad-spectrum pesticides not only kill pests (**target organisms**); they sometimes kill organisms never intended to be harmed, including beneficial insects such as plant pollinators or those that eat insect pests. Therefore, when beneficial insects are killed, insect pests often multiply, and species that were not previously a problem become pests (see Regional Perspective: Indonesia).

Pollinators, which include insects and several other animals, provide pollinating services for more than 75 percent of the staple crops that feed mankind. Honeybees are among the most important pollinators worldwide, but bee populations have been unintentionally decimated through repeated applications of pesticides such as parathion, methyl parathion, malathion, and diazinon. This is a prime example of harm to a **nontarget organism.** Now, many growers are forced to rent honeybee colonies from commercial beekeepers. In the United States, where this is now happening more than a million times a year, the annual cost to farmers is more than $13 million.

Pesticide Bioaccumulation and Biomagnification

When DDT was introduced as an agricultural pesticide, no harmful effects on fish and wildlife were anticipated. But by the 1950s, studies showed that DDT and a number of similar pesticides persist for many years in the environment and move through food webs as a result of *bioaccumulation* and *biomagnification*, causing toxic effects in fish and wildlife.

Bioaccumulation is the uptake and storage of an environmental contaminant such as a pesticide in an organism's tissues. Bioaccumulation of a pesticide can be toxic to organisms, depending on the pesticide, how much is applied, and how much accumulates.

Biomagnification is the concentration of a contaminant along a food chain or food web, resulting in higher and higher contaminant levels in higher-level consumers.

A classic case of bioaccumulation and biomagnification occurred in the 1950s in Long Island Sound, an estuary off the coast of New York State. DDT had been used to control offshore mosquitoes, but traces of the pesticide—approximately 0.000003 ppm (parts per million)—ended up in the water. Even though this is an extremely low concentration, the DDT bioaccumulated in algae and other plankton up to levels of 0.04 ppm, concentrations about 13,000 times higher than in the water itself. Then minnows feeding on plankton biomagnified DDT to 0.5 ppm. Fish that consumed the minnows had DDT levels of up to 2 ppm, and ospreys eating those fish had DDT levels as high as 25 ppm. In this case, DDT levels in ospreys feeding on Long Island Sound fish had been concentrated 8 million times over pesticide levels in the water (Figure 7.10).

In 1962, Rachel Carson, a U.S. Fish and Wildlife Service biologist, published the book *Silent Spring,*

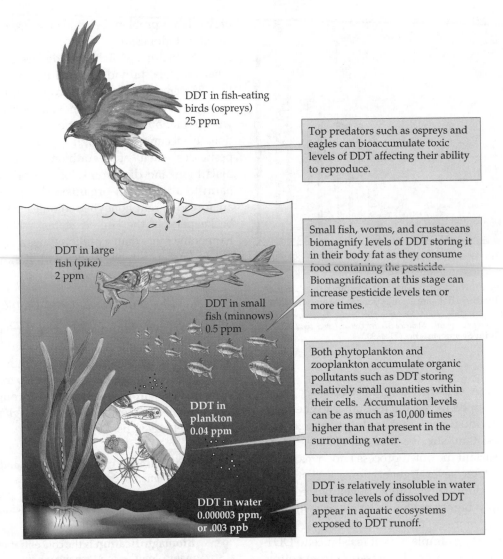

DDT in fish-eating
birds (ospreys)
25 ppm

Top predators such as ospreys and
eagles can bioaccumulate toxic
levels of DDT affecting their ability
to reproduce.

DDT in large
fish (pike)
2 ppm

Small fish, worms, and crustaceans
biomagnify levels of DDT storing it
in their body fat as they consume
food containing the pesticide.
Biomagnification at this stage can
increase pesticide levels ten or
more times.

DDT in small
fish (minnows)
0.5 ppm

DDT in
plankton
0.04 ppm

Both phytoplankton and
zooplankton accumulate organic
pollutants such as DDT storing
relatively small quantities within
their cells. Accumulation levels
can be as much as 10,000 times
higher than that present in the
surrounding water.

DDT in water
0.000003 ppm,
or .003 ppb

DDT is relatively insoluble in water
but trace levels of dissolved DDT
appear in aquatic ecosystems
exposed to DDT runoff.

Figure 7.10   Biomagnification of DDT in a Long Island Sound, New York, food web in the 1950s.

which focused on some of these problems. She described how DDT is taken up by plankton (bioaccumulation) and then appears at higher levels in birds (biomagnification). Carson also eloquently described the effects of DDT on the North American falcon, pelican, osprey, and bald eagle populations. She explained how DDT interferes with calcium metabolism in birds, producing thinner eggshells than normal. Thin-shelled eggs are easily crushed in nests, resulting in fewer hatchlings and smaller bird populations. The title of Carson's book suggested that if the use of DDT were to continue, the springtime sounds of birds would someday no longer be heard.

In 1972, the Environmental Protection Agency (EPA) banned DDT and similar chlorinated hydrocarbon pesticides, in the United States, although limited use of these pesticides may be granted by the EPA in cases of serious outbreaks. Since then, some U.S. bird populations have begun to recover. For example, nesting pairs of bald eagles in the lower 48 states increased from 40 in 1963 to more than 4,000 in 1996. In 1996, the bald eagle was removed from the "endangered" species list and upgraded to "threatened" in most states.

Pesticide Residues in Food

Because crops are treated with chemical pesticides, pesticide residues—traces of pesticides or chemicals derived from them—are sometimes detected in food products. In 1989, the Natural Resources Defense Council reported that residues of the plant growth regulator Alar discovered on U.S. apples could cause cancer in children who ate them or drank apple juice prepared from them. The problem centered on a **metabolite** or chemical breakdown product called

UDMH, which forms when Alar decomposes. Since UDMH causes cancer in laboratory mice, it was believed that Alar-treated apples posed a serious health risk. Consequently, they were removed from school lunch programs and grocery store shelves. Later, studies by Dr. Bruce Ames at the University of California at Berkeley and scientists at the EPA cast considerable doubt on the issue of health risks due to Alar. Nevertheless, Uniroyal Chemical Company, the manufacturer of Alar, suspended sales of the pesticide in the United States.

In March 1999, Consumers Union, the publisher of *Consumer Reports*, analyzed U.S. Department of Agriculture data on pesticide residues in 27 fruits and vegetables sold in U.S. markets between 1994 and 1997. Foods were prepared as they would typically be prepared in homes. Results of the analysis showed that seven kinds of produce had relatively high pesticide residues: apples, grapes, green beans, peaches, pears, spinach, and winter squash. The levels detected were considered unsafe for children and others who regularly eat them or are highly sensitive to the toxic effects of pesticides. Foods with much lower pesticide levels included apple juice, bananas, broccoli, canned peaches, milk, orange juice, canned or frozen peas, and sweet corn. The pesticide residues found most often included methyl parathion, dieldrin (banned in 1974 but still persisting in the environment), and aldicarb.

The U.S. Food Quality Protection Act of 1996 focuses on these issues, especially the cumulative risks of pesticide residues on food products. Through this act, the EPA must establish new tolerance levels for every pesticide, considering not only dietary exposure (which was all the EPA had considered up to this point) but also exposure in domestic and residential products such as lawn and garden chemicals. Existing tolerances on more than 9,700 pesticide formulations are to be reassessed by the year 2006.

Regulating pesticides is difficult because many of them are essential to the success of green revolution agriculture. Eliminating or limiting their use could, in some cases, reduce crop yields, cause considerable economic hardship for farmers, and increase the risks of hunger and starvation in some countries.

Beyond the Green Revolution

Integrated Pest Management

When it was realized that chemical pesticides were not a panacea, and were in fact often part of the problem rather than the cure, innovative approaches to agricultural pest control appeared. Best known among these new techniques is **integrated pest management (IPM)**. The principal elements of IPM include:

- Planting the best disease- and pest-resistant crop varieties available.
- Rotating crops from one growing season to the next instead of continuing to grow a single crop variety in the same area year after year. This approach minimizes the buildup of pest species.
- Monitoring crops periodically to assess the actual numbers of pests and to determine the optimal timing of pesticide applications if needed.
- Limiting the use of chemical pesticides to the amount required to control pest infestations while protecting beneficial nontarget organisms.
- Substituting biological pest control strategies for chemical pesticides when possible.

An example of IPM can be found in the Pacific Northwest where parathion was once commonly used to control aphids on snow peas in Washington State's Skagit Valley. The insecticide was sprayed by air on a regular schedule during the growing season. However, this approach is both inefficient and expensive because there are times when very few aphids are actually present. Furthermore, beneficial non-pest species were being killed. In this case, the IPM approach involves monitoring aphids at random sampling sites and assessing the extent of pest damage. If and when the aphid density justifies spraying, the pesticide is applied.

Biological pest control is proving to be more sophisticated and effective than total dependence on chemical pesticides. For example, *Bacillus thuringiensis* (Bt), a common soil bacterium, produces a natural pesticide called delta endotoxin. Bt is effective in controlling many insect pests, particularly those whose development includes a larval or caterpillar stage. Hundreds of different Bt strains exist in nature, each one releasing a somewhat different endotoxin that exhibits selective toxicity against different pests. Various types of Bt can be used to control mosquitoes, black flies, potato beetles, and gypsy moths without harming fish, wildlife, or humans. Although slower acting and costing more than some chemical pesticides, Bt is environmentally much safer.

Some IPM programs have been very successful. Among these is the Indonesian government's effort to control the brown planthopper, a pest that caused great damage to the country's rice agriculture in the 1980s (see Regional Perspectives).

INSIGHT 7.4
Bioengineering

Selecting desired crop characteristics and cross-breeding them to produce new crop varieties is an old practice. When successful, crossbreeding results in certain genes being introduced into new plant varieties—genes that express desired characteristics. Crossbreeding is limited to plants of the same species, but bioengineering is different in that genetic material from a totally different species (plant or animal) can be introduced into a crop variety. For example, genes that express the making of an antifreeze chemical in marine finfish, thus protecting them from freezing, have been inserted into corn, cotton, soybeans, and other plants to protect them from early frosts.

Genetically modified crops are bioengineered from traditional crop varieties by altering the DNA within cell nuclei. Seeds produced by a genetically modified crop contain the modified DNA and, when planted, grow a modified crop having the particular characteristics that were selected and engineered into the plant's genes. In this way, new

crop varieties can be created that, for example, have insect resistance, faster growing rates, higher yields, or perhaps enhanced ability to resist the effects of applied herbicides.

Two of the ways genetic engineering is being carried out are:

1. The bacterium *Agrobacterium tumefaciens* attacks plants by injecting its own DNA into plant cellular material. It can be used as a modified gene carrier by replacing its DNA with genetic material from another organism—material known to express desired characteristics. Cell division provides each new cell with an exact copy of the introduced genes.

2. Gene splicing, a much quicker way of developing genetically modified DNA, can be accomplished using what is called a DNA "gun" that inserts desired genes directly into a DNA target.

Other Approaches to Pest Control

During the summer of 1997, a serious outbreak of the Mediterranean fruit fly, or medfly, occurred near Tampa, Florida, threatening the state's $7-billion-a-year agricultural industry, known best for growing citrus fruit. The medfly is a tiny insect pest whose larvae can destroy several kinds of fruits, nuts, and vegetables. The state developed and activated a plan to release sterilized male medflies from low-flying aircraft to mate with female medflies and produce infertile eggs. The strategy worked, and a marked decline in the medfly population occurred.

Another intriguing pest control strategy involves the use of **pheromones**, natural sex-attractant chemicals released by the females of some insect species. Chemicals similar to natural pheromones can be synthesized in the laboratory and used to lure unsuspecting males of a particular species into specially designed traps. This biological strategy to pest control is highly specific to particular pest species and harmless to other species. But since this approach depends on extensive research to identify and mimic the chemistry of particular pheromones, it is expensive and not a generally available technique.

Bioengineering: Genetically Modified Crops

Molecular genetics research enables scientists to purposefully modify plant and animal DNA by altering or inserting genetic material that will express desired changes in future generations of a plant or animal. For example, in the late 1970s, Monsanto Chemical Company and Calgene Incorporated developed a genetically altered, or **bioengineered** tomato plant, the FlavrSavr tomato, whose fruit does not soften, spoil, or rot during shipment to distant markets. Typically, if tomatoes are to be shipped, they are harvested green and artificially ripened after shipment using ethylene gas, which causes them to soften and turn red. (Tomatoes produce this gas naturally when allowed to ripen on the vine.) The delayed ripening technique allowed them to be transported long distances, although some consumers found FlavrSavr tomatoes pale and tasteless.

The development of FlavrSavr tomatoes was based on the following insights. Tomato plant cell nuclei contain 12 chromosomes, each made up of double-helix strands of **DNA** (deoxyribonucleic acid) that carry coded, hereditary information. A single strand of DNA comprises thousands of genes, which determine a tomato's color, size, shape, taste, rate of ripening, and other characteristics. Scientists discovered that when

a)   b)

Figure 7.11 Maximizer Hybrid Corn, a product of Ciba Seeds. Photo (*a*) shows traditional corn growing in McLean County, Illinois, and under attack by European corn borers. Photo (*b*) shows Maximizer corn growing in an adjacent field. Neither field was treated with insecticides. (*Reprinted courtesy of Ciba Seeds Inc.*)

tomatoes ripen on the vine, they produce the enzyme polygalacturonase—PG-enzyme for short. PG-enzyme initiates a softening process that, if allowed to continue, leads to spoilage. Spoilage was delayed in the FlavrSavr by inserting a specially designed synthetic gene into a strand of tomato DNA, which blocks the formation of PG-enzyme. **Gene insertion** is the basis of bioengineering (see Insight 7.4).

As a result, fully ripened FlavrSavr tomatoes could be harvested and shipped to distant locales with little spoilage. The tomato was approved by the FDA in 1994 and became the first genetically engineered food on the market. Later, however, public uncertainty about the safety of bioengineered tomatoes led producers to withdraw the FlavrSavr tomato from the marketplace.

Transgenic Bioengineered Crops In 1996, Monsanto introduced Roundup Ready soybeans, a transgenic, bioengineered soybean variety. **Transgenic** implies that genetic material from one species is bioengineered into the DNA of another species. In this case, selected genes from the soil bacterium *Agrobacterium* sp., strain cp4, were inserted into soybean DNA to produce a new soybean variety that is tolerant to the weed killer Roundup, one of Monsanto's most important agricultural pesticides. This means that farmers can safely spray Roundup to kill weeds without damaging soybean plants. Roundup Ready soybeans have been approved by the FDA, EPA, and U.S. Department of Agriculture.

Another example of a transgenic crop is Maximizer Corn, an insect-resistant variety developed by Ciba Seeds Inc. Insect resistance in this case was

bioengineered into the crop by inserting modified Bt genes into corn DNA. The genes direct the corn plant to manufacture its own pesticide. The target pest is the European corn borer, an insect that inflicts an estimated $1 billion damage on U.S. corn annually by boring holes into plant leaves and stalks. It is reported that little if any chemical insecticide is needed to control pests where Maximizer Corn is grown (Figure 7.11).

In 1997, a World Bank study of the potential of gene-modified crops concluded that bioengineered crops may prove essential in meeting world food demands. It is believed that bioengineered crops will have improved disease and pest resistance, enhanced ability to withstand adverse weather, and superior nutritional value. Biotechnology may also lead to totally new kinds of foods.

By 1999, 60 genetically engineered crops had been approved in the United States and Canada, including new varieties of corn, potatoes, tomatoes, soybeans, and squash. Seven genetically modified crops were being grown commercially in 12 countries including 35 percent of U.S. corn acreage, 55 percent of U.S. soybean acreage, 60 percent of Canadian canola acreage, and 90 percent of Argentinean soybean acreage. Neither the USDA nor the FDA requires bioengineered foods or products made from them to be labeled to reflect their origin. Furthermore, no safety testing of these foods is required.

Genetically Modified Crops: Safety Issues Many scientists, agriculturists, government leaders, and consumers are concerned that genetically modified crop varieties may turn out to have unexpected negative effects on human health and the environment. This is why Britain and the European Union have thus far banned the import of bioengineered foods and food products as a precautionary move. Others are also taking precautions. For example, Gerber and Heinz, two of the largest makers of baby foods, have decided not to include genetically modified produce in these products. But is there scientific evidence that bioengineered crops and foods made from them may be harmful? For the most part, there is no evidence that genetically modified foods are unsafe for human consumption, but a few recent studies are troublesome. For example:

- In 1997, agricultural researchers at Scotland's University of Cambridge allowed ladybirds to consume aphids that had been feeding on potatoes bioengineered to kill aphids. They found that female ladybirds lived only half as long as those that consumed aphids feeding on normal potatoes.

- In 1999, scientists at Cornell University reported that monarch butterfly larvae feeding on milkweed leaves (their main food source) that had been dusted with pollen from genetically modified Bt corn were killed or seriously harmed. Milkweed dusted with pollen from unmodified corn did not harm larvae. The monarch butterfly is so widely known and appreciated for its beauty and elegance that the Cornell report has been likened to a "canary in the coal mine" bioindicator. These reports and others suggest that genetically modified crops may create environmental problems, but it is too early to know for sure whether this is true.

Terminator Seed Technology   Toward the end of the 1990s, bioengineering led to a new genetic development referred to as **terminator seed technology.** Terminator seeds, which can be developed for many different crops, are genetically modified so that crops grown from them will produce seeds that cannot germinate. The technology is based on introducing a gene that directs a plant to produce a toxin that sterilizes the next generation of seeds. The new technology, patented by the U.S. Department of Agriculture and a Monsanto Chemical Company subsidiary, has been shown to work on cotton and tobacco seeds; it was thought that it could be applied to wheat, rice, and soybeans within a few years.

Farmers who buy the new seeds, would not be able to save seeds from the crops they grow and use them for future planting because those seeds would be infertile. They would have to buy seed for the next growing season. The incentive for seed producers to adopt terminator seed technology is linked to the high cost of developing new crop varieties that are high-yielding, disease-resistant, early-maturing, and frost-resistant. Their investment in creating and producing these miracle seeds would be protected if seeds produced by the new crops were sterilized by the plant itself.

A number of agricultural groups worldwide foresee serious problems associated with terminator seed technology, including the following:

- Poor farmers, especially those in lesser developed countries, would be denied their right to an ancient practice—saving seed to grow the next year's crops. They would be forced to return to the commercial seed market each year to buy seed.

- The new technology would limit genetic diversity in new crops by propagating a smaller number of crop varieties.

- Farmers would be unable to select and breed superior performing plants by themselves, thereby limiting the future of sustainable agriculture.

- Pollen that carries terminator genes and is released to the environment could pose potential risks to plants of all kinds.

- Terminator-seed-modified crops produced for human consumption pose unknown health risks.

In 1998, the Consultative Group on International Agricultural Research (CIGIAR) meeting at the World Bank in Washington, D.C., adopted a strong and unambiguous policy banning terminator seed technology in their crop breeding research, development, and seed dissemination programs worldwide. This action raised serious questions about the wisdom and morality of continuing down that path. As a result, Monsanto has decided not to commercialize terminator seed technology.

Limits to Food Production

Thus far, green revolution agriculture has kept pace with human population growth in many parts of the world. Plant breeding, mechanized farming, commercial fertilizers, crop irrigation, and chemical pesticides have delayed the dire predictions of worldwide hunger and famine forecast by Thomas Malthus (see Chapter 6). Nevertheless, the sustainability of green revolution agriculture is in doubt. In many countries, annual yields and total harvests of basic grains are leveling off, even with increased applications of fertilizers. Expanding crop irrigation is proving to be a problem because of competing demands for limited freshwater supplies.

Also, as we have seen, chemically based agriculture has a down side, especially when pesticide bioaccumulation, biomagnification, pest resistance, and pesticide residues in foods are considered. Integrated pest management is beneficial, but many scientists question our ability to push green revolution agriculture beyond its current capabilities. Two possible future scenarios are gaining world attention:

1. Helping all nations work toward zero population growth in order to bring about a more equitable balance between food production, human needs, and societal expectations.
2. Using bioengineering to develop not only improved crop varieties but totally new kinds of high-yielding crops that will hopefully have more resistance to disease and pests and less dependence on chemical fertilizers.

REGIONAL PERSPECTIVES

Indonesia: Integrated Pest Management

In the 1970s, brown planthoppers invaded rice fields on the island of Java in Indonesia, attacking and decimating rice plants by sucking the juices from their stems. With their rice paddies in jeopardy, farmers switched to a new variety of rice, IR-26, which at first appeared to be immune to planthopper attack. But within three years, IR-26 rice had also been devastated by the same pest. Massive spraying with a number of different pesticides failed to control the brown planthopper. By 1986, the situation looked hopeless, and Indonesia began to import rice.

However, government research showed that several species of beneficial insects were natural predators of planthoppers and were capable of limiting the number of pests to tolerable levels. Unfortunately, those beneficial insect populations were being killed by pesticides. So in 1986, the Indonesian government banned indiscriminate pesticide spraying in rice paddies. Instead, selective spraying was allowed only to control serious planthopper infestations. The limited pesticide strategy was to protect beneficial insects, one of the goals of integrated pest management. Using this approach, pesticide use in Indonesia declined 65 percent during the first year, while rice harvests increased 15 percent and have been increasing ever since. Overall, this approach to pest control achieved the following results:

- The trend toward diminishing rice yields was reversed, and by 1988, rice harvests had increased to levels greater than when pesticides were the main pest control strategy.
- Insecticide subsidies were phased out, saving the government close to $120 million a year.
- Species of fish virtually wiped out by pesticide use are once again prevalent.

Africa: Norman Borlaug's Continuing Work

Although green revolution agriculture has led to spectacular harvests and per capita production of grain crops has increased in many countries despite exponential population growth, food production has not kept pace with population growth in every country. Most noteworthy are agricultural yields in Sub-Saharan Africa, which are among the lowest in the world. Only recently have green revolution strategies been applied there.

Green revolution agriculture was not implemented sooner in Sub-Saharan Africa because of the environmental problems the use of chemical fertilizers and pesticides had caused in other countries. But in 1984, former U.S. President Jimmy Carter and Ryoichi Sasakawa, President of the Japanese Sasakawa Peace Foundation, convinced Norman Borlaug to emerge from retirement and join their efforts to advance agricultural practices in Africa. Borlaug agreed and initiated agricultural research and development in Ethiopia, Sudan, Tanzania, Ghana, and other African countries. He found that Africa's agricultural problems are more complex than those he had encountered in other nations, requiring different approaches to chemical fertilization. Progress has been slow, but on some experimental farms yields of corn have tripled and production of wheat, cassava, sorghum, and cow peas has also increased. The most striking result so far is in Ethiopia where, in 1995–96, the largest crop harvests in the country's history were recorded.

North America: EDB from Berry Fields in Groundwater

Raspberry, strawberry, and other berry crops all over the world are often seriously affected by diseases such as root rot, which weakens and ultimately kills the vines. Root rot is caused by soil nematodes (root worms), and one of the few ways to control them is to fumigate the soil with an effective pesticide prior to planting a new berry field. Until the early 1980s, the most effective pesticide used to control root rot in berry fields was ethylene dibromide (EDB).

However, in the 1970s, traces of EDB were found in groundwater on berry and citrus farms in Hawaii, California, Florida, and Washington State where the pesticide was routinely used. It was discovered that EDB is a *leacher*, one of the relatively few chemicals able to infiltrate the soil and contaminate underlying groundwater. This contamination is of concern because most rural populations and even some urban populations obtain their drinking water from wells drilled into the groundwater.

Then in 1981, the EPA determined that EDB is a potent mutagenic agent and a suspected human carcinogen. A health advisory limit of 0.02 μg/L (parts per billion) was established, an extremely low level that reflects how potentially dangerous this chemical agent is. EDB was subsequently banned in the United States in 1983 and in Canada in 1984. This meant it could no longer be used to control root rot in berry crops and other means of pest control had to be substituted. Berry growers are now using several different approaches to limit pest activity, including the application of a widely used chemical pesticide called methyl bromide. At present it appears that methyl bromide is a much safer pesticide, but unfortunately it is also a powerful stratospheric-ozone-depleting agent whose worldwide agricultural use is believed to be an important contributor to the ozone hole (see Chapter 13).

KEY TERMS

annual yields

aquifer

bioaccumulation

bioengineered

biomagnification

chlorinated hydro-
   carbon pesticides

crossbreeding

cross-pollination

cultivar

DNA (deoxyribonucleic
   acid)

erosion

gene

gene insertion

genetically modified
   crops

highly erodable land
   (HEL)

hybrids

integrated pest
   management (IPM)

landrace

LD$_{50}$

lodging

metabolite

monocultures

natural selection

nontarget organisms

organic

organo-phosphorus
   pesticides

pheromones

self-pollination

subsistence agriculture

target organisms

terminator seed
   technology

total harvests

transgenic

DISCUSSION QUESTIONS

1. Norman Borlaug's words recorded at the start of this chapter reflect his views about using chemicals in modern agriculture. Analyze and comment on his opinion and discuss your own thoughts regarding chemically based, green revolution agriculture.

2. DDT and other pesticides banned in the United States are still manufactured there and sold to developing countries where they are used extensively. Are these practices acceptable? Describe and defend your ideas.

3. What is your opinion of bioengineered crops now that they are part of modern agricultural practice? Should foods manufactured from genetically modified crops be labeled as such? What negative impacts do you see possibly arising from this new technology?

4. Scientific and technological innovation has enabled many societies to increase their food production, but can we expect that science and technology will always generate agricultural breakthroughs? Is it possible that the worldwide famine Thomas Malthus envisioned (see Chapter 6) will never occur? Explain your thinking on these questions.

5. Is science going too far by altering the genetic blueprints of living things? Do you favor or oppose bioengineering to create transgenic crops and animals? Discuss these issues.

INDEPENDENT PROJECT

Contact a local farmer who is familiar with grain crops, beef cattle, or dairy produce. Spend some time on the farm and find out what chemical pesticides are used, what precautions, if any, are taken, and what nonchemical practices have been adopted to enhance farm operations. Write a brief paper on what you have learned.

INTERNET WEBSITES

Visit our website at http://www.mhhe.com/environmentalscience/ for specific resources and Internet links on the following topics:

UN Food and Agriculture Organization

Agricultural electronic database

U.S. Department of Agriculture

Pesticides and food products

SUGGESTED READINGS

Bell, M. A., R. A. Fischer, D. Byerlee, and K. Sayer. 1995. Genetic and agronomic contributions to yield gains: A case study for wheat. *Field Crops Research* 44:55–65.

Brown, Lester R., et al. 1998, 1999. *State of the world: A Worldwatch Institute report on progress toward a sustainable society.* Worldwatch Institute: W. W. Norton & Company.

Brown, Lester R., Hal Kane, and David Malin Roodman. 1997, 1998, 1999. *Vital signs: The trends that are shaping our future.* Worldwatch Institute: W. W. Norton & Company.

Eagles, H. A., and J. E. Lothrop. 1994. Highland maize from Central Mexico—Its origins, characteristics, and use in breeding programs. *Crop Science* 34:11–19.

Easterbrook, Gregg. 1997. Forgotten benefactor of humanity. *The Atlantic Monthly* (January):75–82.

Foods: Tests/Report: Seeds of change. 1999. *Consumer Reports* (September): 41–46.

Georghiou, George P. 1990. Overview of insecticide resistance. Chapter 2 in *Managing Resistance to Agrochemicals*, pp. 18–41. Edited by Maurice B. Green et al., American Chemical Society Symposium Series 421, Washington, D.C.

Glaeser, Bernard, ed. 1987. *The green revolution revisited.* London: Allen and Unwin, Unwin Hyman, Ltd.

Holmes, Bob. 1992. The joy ride is over—Farmers are discovering that pesticides increasingly don't kill pests. *U.S. News and World Report* (September 14): 73–74.

Longman, Phillip J. 1999. The curse of frankenfood: Genetically modified crops stir up controversy at home and abroad. *U.S. News and World Report* (July 26): 38–41.

Losey, John E., Linda S. Rayor, and Maureen E. Carter. 1999. Transgenic pollen harms monarch larvae. *Nature* 399: 214.

Pesticides report: How safe is our produce? 1999. *Consumer Reports* (March): 28–31.

Pimentel, D., et al. 1992. Environmental and economic costs of pesticide use. *BioScience* 42 (10): 750–60.

Soule, Judith D., and Jon K Piper. 1992. *Farming in nature's image—An ecological approach to agriculture.* Washington, D.C.; Covelo, CA: Island Press.

Thayer, Ann M. 1999. Ag biotech food: risky or risk free? *Chemical & Engineering News* (November 1): 11–20.

Wellhausen, Edwin J., L. M. Roberts, and E. Hernandez X. 1957. *Races of maize in Mexico.* Washington, D.C.: National Academy of Sciences, National Research Council.

Wilke, Raymond. 1971. *A Mexican collective ejido.* Stanford, CA: Stanford University Press.

Woodwell, G.M., C. F. Wurster, and P.A. Isaacson. 1967. DDT residues in an East Coast estuary: A case of biological concentration of a persistent insecticide. *Science* 156:821–24.

CHAPTER 8

Sustainable Agriculture

Photo Disc, Inc./Vol. 60

Sustainable development is development that meets the needs of the present without compromising the ability of future generations to meet their own needs.

World Commission on Environment and
Development, 1987

The green revolution has been immensely successful in meeting demands for food in many nations, as we saw in Chapter 7. In fact, most farmers in industrialized countries practice green revolution agriculture today. However, some farmers believe this way of growing crops may be unsustainable in the long run because it depends heavily on costly chemical and energy inputs. Also, they argue, its practice degrades soil ecology, pollutes water resources, adversely affects fish and wildlife, and creates potential health risks for consumers due to pesticide residues.

Thus, many farmers are now seeking alternatives to green revolution agriculture, endeavoring to produce high yields of wheat and other crops without relying on chemicals. They are also learning how to protect soils from excessive erosion, minimize pest damage, limit crop diseases, and protect the environment from the adverse impacts of chemically based agriculture. These alternative practices, referred to as sustainable agriculture, are a more natural, environmentally friendly way to produce grains and livestock while safeguarding soils, surface and groundwaters, fish and wildlife, farm families, and consumers.

One of the places where sustainable agriculture is practiced today is Saskatchewan, Canada, the province where 21 percent of Canada's farms were operating as of 1996, according to Statistics Canada. This chapter explores the differences between green revolution and sustainable agriculture, traces the development of agriculture in Canada's prairie provinces, and provides a case study on organic farming in Saskatchewan as an example of sustainable methods in practice. We will begin by examining the single, most important ingredient in successful farming: soil.

BACKGROUND

The Formation of Soils

Soil is a complex ecosystem of inestimable value. Without it, there would be no arable land and little ability to grow food and forage crops. But unfortunately, soil resources are often undervalued if not completely ignored.

Soil is more than just "dirt"; it is a unique resource derived from rock and mineral matter that contains air and water and is enriched with organic matter and living organisms. Soil formation, called **pedogenesis,** occurs when underlying parent bedrock is broken down physically and chemically into smaller fragments and chemical units through weathering, and then mixed with organic matter.

Precipitation is an important factor in soil formation because it is usually acidic and hastens the rate of weathering. As a result, minerals such as limestone, dolomite, and microcline slowly dissolve, adding inorganic nutrients like calcium, magnesium, and potassium to soils. In addition, rocks are broken down by lichens and mosses, which secrete acids and thereby accelerate weathering. Other factors that influence soil formation and affect its quality include:

- The nature of parent rocks and mineral materials.
- Prevailing climate, particularly average levels of precipitation and temperature extremes.
- The rate at which organic matter decomposes.
- Topography or landform, which controls rates of water runoff.
- Time; soils may be immature, mature, or aged.

In addition to being formed through natural processes, agricultural soils may be transported to a particular site by periodic flooding events. This is often the case along rivers and on river deltas. The historic fertility of the Nile River valley and delta is one of many examples.

Soil Composition

Simple soil analyses indicate that soils are made up chiefly of mineral matter, including sand and clay, along with air and moisture. But in addition, organic matter, although it may comprise as little as 5 percent of soil, turns out to be of great importance. Organic matter consists of decomposed and decomposing plant and animal matter such as animal wastes, leaves, needles, seeds, pollen, bark, and woody debris. Much of the decomposition is carried out by soil microorganisms, including bacteria. This process generates **humus,** dark, chemically stable organic matter that regulates soil pH, moisture content, nutrient exchange capacity, and fertility. It is organic matter that enables soil to retain water during times of drought.

A soil's sand content adds to its ability to drain excess moisture and recharge subsurface aquifers. Clay, another important soil component, is derived from the weathering of rock minerals such as feldspars and other silicates. Clay consists of very small, flat particles less than about 0.002 millimeters in diameter having a preponderance of negative charges on their surfaces. Clay's negative surface charges attract and hold positively charged essential soil nutrients such as calcium, magnesium, potassium, sodium, and ammonium ions, thereby contributing to soil fertility.

Soil ecosystems contain more than just mineral, humus, sand, and clay materials. They also include a wide diversity of plant and animal life, defined here in terms of size:

Microbiota	Bacteria, fungi, algae, viruses, and protozoa.
Mesobiota	Nematodes (eelworms), centipedes, millipedes, turbellarians (unsegmented worms), and mites.
Macrobiota	Earthworms, wood lice, slugs, snails, beetles, ants, and moles.

Most soil organisms are highly beneficial because in one way or another they continuously enrich soil quality. Earthworms, for example, ingest and process plant matter; their castings contain partially digested organic material, and their burrowing allows air and moisture to enter the soil. Fungi and bacteria, which are decomposers, transform vegetative litter into humus. Nitrogen-fixing **cyanobacteria** (blue-green algae) as well as some non-photosynthetic bacteria convert atmospheric nitrogen gas into ammonia and other nitrogen-containing plant nutrients; this is part of the nitrogen cycle, discussed in Insight 8.1.

Mature soils generally display horizontal layers or **horizons,** each having a visibly different composition (Figure 8.1). **Topsoil** (the A horizon) is the most fertile and valuable soil layer; it is richest in organic matter and best able to support vegetation, including crops. Thus, topsoil is the layer that concerns us most in our discussion of agriculture.

Soil Erosion and Soil Formation

Topsoil is continuously being worn away and carried to other locations, most commonly by wind, precipitation, and flooding, in the process called **soil erosion.** It is estimated that since the days of early settlement on America's Great Plains, half of the region's topsoil has been lost to erosion by water or wind. The loss, amounting to about 0.1 centimeter of topsoil yearly, adds up to about 9 metric tons per hectare per year ($\approx$ 4.5 tons per acre per year). As we saw in Chapter 7, erosion of U.S. agricultural soils in 1997 averaged 11.6 metric tons per hectare per year ($\approx$ 5.2 tons/acre/year).

On the Canadian prairies, rates of soil loss due to erosion have often been twice that on the Great Plains, mainly because prairie soils, dry most of the year, are prone to erosion by wind. For example, after 100 years of farming in Saskatchewan, as much as 40 percent of prairie topsoil had been lost by 1990. The soil loss was estimated by comparing the ground level of old settlers' cemeteries with adjacent farmland; the cultivated fields were about 20 centimeters ($\approx$ 8 inches) lower. On the average, about 18 metric tons of soil per hectare ($\approx$ 8 tons per acre) were being lost annually. This amounts to 1 centimeter of soil every 5 years ($\approx$ 1 inch every 12.5 years).

Soil is a renewable resource, but the renewal process is slow. On the average in temperate regions, it takes about 200 years for 1 centimeter of new soil to form ($\approx$ 500 years for 1 inch of new soil), which means it would take about 3,600 years for nature to replace the topsoil lost on Saskatchewan's prairies in just 100 years.

INSIGHT 8.1
The Nitrogen Cycle

Nitrogen gas is the most prevalent substance in the earth's atmosphere, but nitrogen in this form is unavailable to living plants and animals, even though it is a nutrient they commonly require. Most plants, including crops such as wheat, maize, and rice, utilize nitrate or ammonia as their source of nitrogen, not nitrogen gas. In order to enter food webs and meet the nutritional needs of plant and animal life, nitrogen gas must be "fixed" and then cycled from one chemical form to another, a natural process called the nitrogen cycle (Figure 8.A). The illustration shows different pathways by which atmospheric nitrogen is fixed or otherwise enters the biosphere:

Intense heat in lightning strikes causes nitrogen (N_2) to react with atmospheric oxygen (O_2), producing nitrates (NO_3^-), which fall in ordinary precipitation and fertilize soils.

- Mutualistic ***Rhizobium* bacteria** colonize as nodules on the roots of legumes such as peas, clover, and alfalfa, fixing atmospheric nitrogen and forming ammonia (NH_3), some of which is used by their legume host and the rest of which is stored in soils.

- Some of the ammonia produced by *Rhizobium* bacteria is oxidized by other soil bacteria, forming nitrates that are used by plants or stored in soils.

- Blue-green algae, also called cyanobacteria, in soils and aquatic environments fix atmospheric nitrogen, forming ammonia and nitrates.

- Soil bacteria decompose plant and animal debris and organic wastes, releasing ammonia to soils.

- The nitrogen cycle is completed by denitrifying bacteria, which chemically reduce ammonia and nitrates to nitrogen gas.

O Horizon—Surface litter, including leaves, bark, needles, branches, and decaying plants and animals. Litter is slowly transformed into organic matter by bacteria, fungi, and other decomposers. O horizon processes enrich soils by adding organic matter to the underlying A horizon.

A Horizon—Topsoil, the most fertile and valuable soil layer. Its composition is based on varying amounts of clay minerals, silt, and sand. Its fertility is based on nutrient content, organic matter, humus, moisture, air, and soil biota.

B Horizon—Soil layer rich in clay minerals collected from the overlying A horizon. Clay minerals add to soil fertility by bonding with and storing nutrients such as calcium, magnesium, and potassium. The B horizon also contains carbonates, iron, and aluminum compounds.

C Horizon—Subsoil layer where bedrock and parent soil materials undergo active breakdown. Soil fertility here is limited.

R Horizon—Bedrock and unweathered parent soil materials.

Typical Prairie Region Soil Profile

Figure 8.1 A typical prairie region soil profile. Nutrients released by organic litter in the O horizon enrich underlying topsoil in the A horizon. Together, the A and B horizons are called "true soil."

Therefore, soil conservation should be of vital concern to humankind, and today's farmers ought to choose farming practices that maintain topsoils and sustain soil quality. As we will see, this is often not the case.

Comparing Green Revolution and Sustainable Agriculture

One of the key differences between sustainable and green revolution agriculture is the philosophy of **soil husbandry.** Sustainable agriculture aims to continuously sustain and even enhance the quality of agricultural soils (husbandry) in the hope of maximizing crop yields, minimizing pest problems, and protecting against adversities such as drought. While green revolution agriculture also depends on soils, it aims to achieve high crop yields by using chemical inputs such as commercial fertilizers and pesticides rather than purposefully pursuing soil husbandry.

Green Revolution Agriculture

As we saw in Chapter 7, green revolution agriculture is often referred to as chemical agriculture because it relies on chemical fertilizers, herbicides, fungicides, and insecticides. The green revolution also depends on crop monocultures and mechanized infrastructure—

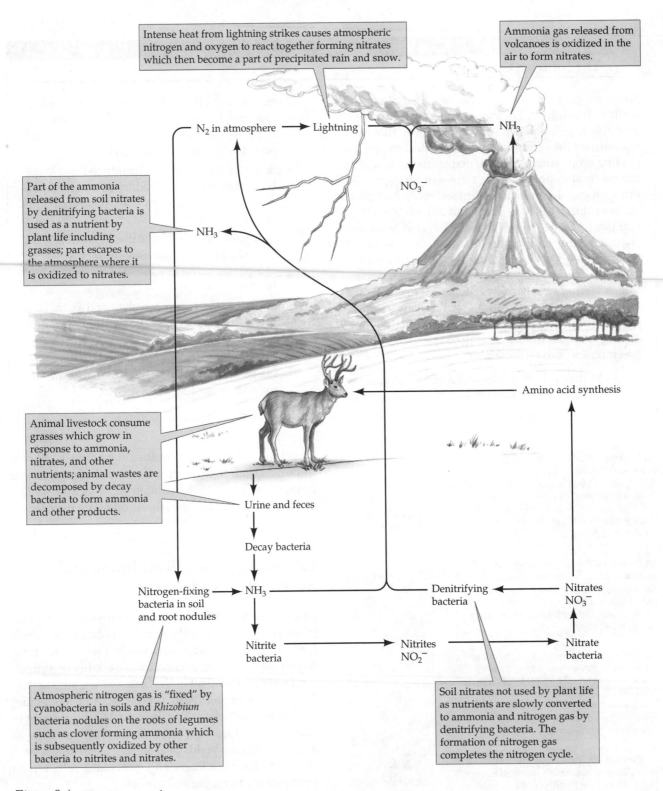

Intense heat from lightning strikes causes atmospheric nitrogen and oxygen to react together forming nitrates which then become a part of precipitated rain and snow.

Ammonia gas released from volcanoes is oxidized in the air to form nitrates.

Part of the ammonia released from soil nitrates by denitrifying bacteria is used as a nutrient by plant life including grasses; part escapes to the atmosphere where it is oxidized to nitrates.

Animal livestock consume grasses which grow in response to ammonia, nitrates, and other nutrients; animal wastes are decomposed by decay bacteria to form ammonia and other products.

Atmospheric nitrogen gas is "fixed" by cyanobacteria in soils and *Rhizobium* bacteria nodules on the roots of legumes such as clover forming ammonia which is subsequently oxidized by other bacteria to nitrites and nitrates.

Soil nitrates not used by plant life as nutrients are slowly converted to ammonia and nitrogen gas by denitrifying bacteria. The formation of nitrogen gas completes the nitrogen cycle.

N_2 in atmosphere → Lightning → NH_3 → NO_3^-

NH_3

Amino acid synthesis

Urine and feces → Decay bacteria

Nitrogen-fixing bacteria in soil and root nodules → NH_3 → Nitrite bacteria → Nitrites NO_2^- → Nitrate bacteria → Nitrates NO_3^- → Denitrifying bacteria

Figure 8.A The nitrogen cycle.

plows, cultivators, planters, irrigation systems, fertilizer applicators, pesticide sprayers, and harvesters —that contribute to agricultural productivity. Unfortunately, this infrastructure incurs economic, environmental, and health costs such as:

- Fertilizer runoff to streams, rivers, lakes, bays, estuaries, and oceans, causing cultural eutrophication.
- Pesticide runoff to aquatic environments, damaging fish and wildlife populations.

Figure 8.2 A no-till drill using discs slices a thin furrow a few inches into the soil, drops seeds into furrows, and presses the ground to close the furrow. No plowing is involved.
(Permission to reprint, courtesy of Case Corporation, Racine, Wisconsin.)

- Adverse impacts of pesticide use on the health of farmers and farm workers.
- Unintended impacts of pesticides on nontarget, beneficial organisms.
- Increasing resistance on the part of agricultural pests to chemical pesticides.
- Risks to consumers of pesticide residues on crops and food products.
- Accelerated soil erosion, depletion of topsoil, and pollution of receiving waters.

Sustainable Agriculture

Sustainable agriculture is emerging as a new agricultural ethic. But it isn't really new. For thousands of years, well before the green revolution, farmers fertilized soils with animal manure and plant residues (green manure)—safe, reliable sources of nutrients and organic matter. They controlled weeds using mechanical devices and curbed insect pests through crop rotation. While they were often not very successful at managing crop diseases, crop yields were usually adequate and sometimes bountiful.

Sustainable agriculture involves agricultural crop and livestock practices that sustain environmental balance in natural systems in the face of intensive land use by humans. The goal of sustainable practices is to meet today's needs without compromising the ability of future generations to meet their needs. Sustainable agriculture allows, but does not advocate, commercial chemical inputs.

First and foremost, sustainable farming implies that the inherent qualities of agricultural soils will be maintained, not impaired or lost. Second, sustainable agriculture seeks to sustain ecological balances essential to all living things including humans, fish, and wildlife. Other important goals of sustainable agricultural practices include the following:

- Agricultural productivity that is capable of meeting the nutritional requirements of the world's human populations and at the same time assuring economic profitability for farmers and others responsible for food production, especially in rural communities and developing nations.
- Production of human food and animal forage that does not compromise the health and well-being of consumers as a result of crop pesticide residues, preservatives in foods, or antibiotics and growth hormones in animal products.
- Soil husbandry that conserves topsoils, retains soil organic matter, and avoids soil amendments that contaminate soils with heavy metals and hazardous wastes.
- Protection of surface and groundwaters from contamination and cultural eutrophication by chemical fertilizers, pesticides, and animal wastes.
- Maintenance of biological diversity among natural plant and animal species.

Sustainable Farming Practices   Today's principal sustainable agricultural practices include:

Conservation tillage	Leaving 30 percent or more of crop residues in and on soils following harvests. The frequency and depth of plowing are limited by using special discs and chisels to cultivate soils. Since moldboard plows, which lift and turn over a foot or more of topsoil, are used less frequently, rates of soil erosion have been reduced, soil organisms protected, and natural soil horizons preserved.
No-till planting	Eliminating plowing by using farm implements that slice narrow furrows in soils, plant seeds, and then close the furrows (Figure 8.2).
Summer fallowing	Allowing soils to remain unplanted for a season. Soil moisture is conserved, and pests are deprived of food sources.

Crop rotation	Planting different crops in rotation over one or more growing seasons, an old agricultural practice that reduces weed and insect problems.
Cover crops	Establishing vegetation between growing seasons, typically during the winter, to lessen soil erosion. Legumes such as clover are often grown as cover crops and later turned over to enrich soils with respect to nitrogen fixed by *Rhizobium* bacteria colonized on their roots (see Insight 8.1).
Shelterbelts	Growing trees around the perimeters of fields to minimize wind-driven soil erosion (Figure 8.3).
Chemical pesticides	Using chemical herbicides and insecticides *only when needed*.

It should be noted that chemical inputs, including fertilizers and pesticides, are sometimes used in the practice of sustainable agriculture, but they are not relied upon nearly as much as in green revolution practices.

CASE STUDY: CANADIAN PRAIRIE AGRICULTURE

About 15,000 years ago, mile high glaciers were slowly retreating from North America, depositing sedimentary materials that were subsequently enriched with organic matter as generation after generation of plants and animals lived and died there. These biophysical actions left parts of North America carpeted with unbelievably biodiverse tall-grass, mid-grass, and short-grass prairie lands whose rich soils had significant agricultural potential.

The prairie biome is one of the planet's most important environments. Prairies exist in Siberia, China, and Argentina, as well as in North America where they stretch from Texas through Oklahoma, Kansas, Nebraska, South Dakota, and North Dakota (America's Great Plains) and into Canada's prairie provinces—Alberta, Saskatchewan, and Manitoba. Today, most prairie grasslands are gone, and in their place are agricultural crop and range lands such as the cornfields of Iowa and the wheat fields of Saskatchewan. However,

Figure 8.3 Shelterbelts planted along perimeters of croplands protect prairie soils and minimize soil erosion.
(Photo courtesy of the Prairie Farm Rehabilitation Administration, Calgary, Alberta.)

Figure 8.4 Mixed species of native grasses—a biodiverse ecosystem that once covered Canada's prairies but is now found only in a few protected places.

remnants of undisturbed grasslands have been set aside on Canada's prairies and America's Great Plains to forever protect them as biological preserves and parks—reminders of what grassland ecology was once like (Figure 8.4).

Settlement of the Palliser Triangle

In the late 1850s, Captain John Palliser, one of the early pioneers to explore Canada's western prairie, traveled to what were soon to become the provinces of Alberta, Manitoba, and Saskatchewan to evaluate the agricultural potential of the region's southern lands. He

INSIGHT 8.2

Homesteading and the Canadian Pacific Railway

As settlement on Canada's prairies was starting, American pioneers to the south were also homesteading on vast buffalo grazing areas that would later become Wyoming and Montana. Cattlemen there were already looking for new grazing areas. They found them to the north in the Palliser Triangle. As a result, concern grew in Canada that if U.S. ranchers were allowed to graze cattle on Canada's prairies, the United States might claim the territory as its own. Therefore, the Canadian government decided to build the Canadian Pacific Railway (CPR) and encourage large numbers of Canadians to move westward to settle.

The route of the CPR was influenced by John Macoun's prediction that southern prairie lands would be ideal for growing wheat. At its completion in 1885, the railway traversed a route connecting prairie settlements that later became Winnipeg, Manitoba; Regina, Saskatchewan; and Calgary, Alberta (see Figure 8.5). Before long, the flow of immigrants to the prairies became a tide of great proportions.

Soon after the railway began operating, the most common of prairie landmarks, grain elevators, began to appear. Many are still standing. They are tall, somewhat narrow, wooden structures usually built alongside railway tracks. Figure 8.B depicts a 1925 scene in Meyronne, Saskatchewan—spring wheat being hauled by horse-drawn wagons to a series of grain elevators where buckets on conveyor belts scooped up the grain, storing it in elevated compartments. Later, the grain was loaded into freight train boxcars and shipped to markets all over the world. Grain elevators are still used on the prairies, but many of these picturesque landmarks have been replaced by large, silo-shaped, poured-concrete structures (Figure 8.C). A typical wooden grain elevator held about 65,000 bushels of wheat; newer concrete elevators hold as much as 375,000 bushels.

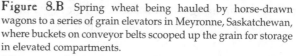

Figure 8.B Spring wheat being hauled by horse-drawn wagons to a series of grain elevators in Meyronne, Saskatchewan, where buckets on conveyor belts scooped up the grain for storage in elevated compartments.
(Photo courtesy of the Prairie Farm Rehabilitation Administration, Calgary, Alberta.)

Figure 8.C A large poured-concrete grain elevator near Davidson, Saskatchewan.

concluded that the southernmost parts of the three provinces, later to be known as the Palliser Triangle (Figure 8.5), were not arable—that is, not suitable for agriculture because the soil was too dry to support most crops. But 20 years later, the naturalist and educator John Macoun surveyed the same area and found an almost ideal agricultural environment with more than adequate rainfall—perfect for growing wheat. Palliser and Macoun had encountered years of quite

different precipitation patterns, a fact that would later have dramatic impacts on those who decided to homestead and farm there.

In 1871, the Canadian Homestead Act opened the door to thousands of settlers eager to tame Canada's prairie frontier. They came from Montana and the Dakotas in the United States and from Canada's eastern provinces (Insight 8.2) as well as from England, Scotland, Wales, Ireland, Ukraine, Russia, Germany,

Poland, and Holland. Some had heard that the region's fertile lands were sure to bring forth bountiful harvests of wheat. Others envisioned new, open territories where herds of livestock could graze. Still others simply wanted to escape one kind of life in hopes of building a new one. They staked their claims to one-quarter sections of land for the cost of a filing fee—160 acres for $10.

Learning to Farm the Prairie

At first, the large and relatively inexpensive land claims on the Canadian prairie appeared to be a bonanza. However, most of the settlers had previously farmed where rainfall was plentiful. They were accustomed to conditions in which enough grass could be grown on one acre to pasture a cow and on another acre to provide forage for a horse. On their new homesteads, they learned, one cow needed 10 acres to graze, and each horse needed another 10 acres.

On the semi-arid Canadian prairies, the most common kind of agriculture is **dryland farming,** which depends solely on natural precipitation, both rain and snow. Crop irrigation has never been a common practice on these lands because surface water is scarce and used mainly to supply drinking water for humans and to water livestock. Furthermore, the groundwater is highly mineralized, even salty. If it were used for crop irrigation, evaporation would deposit mineral salts into soils, a process called **salinization** that greatly decreases soil fertility.

To plant their crops, prairie homesteaders broke grassland sod by turning the earth over with plows pulled behind oxen or horses. They typically used moldboard plows, which have one or more curved metal blades shaped like large spoons to turn the sod over and cut furrows in the ground. Plowing soil to prepare it for crops was done almost every year. Figure 8.6 shows how early plowing was done on the prairies.

Canada's settlers discovered that oats, barley, mustard, lentils, peas, and alfalfa hay could be grown on fertile lands lying mostly to the north of the Palliser Triangle, but the growing season there was too short for the wheat varieties existing at that time. They simply matured too slowly to survive the early frosts that often occurred before harvest. Thus, assuming enough rainfall, wheat could only be cultivated to the south. While wheat is grown successfully in many parts of the world, mainly because it tolerates varying degrees of drought (Insight 8.3), wheat harvests on the Canadian prairies were sometimes dismal owing to years of extreme drought. During years of above average rainfall, wheat crops prospered, but they could not endure the dry years.

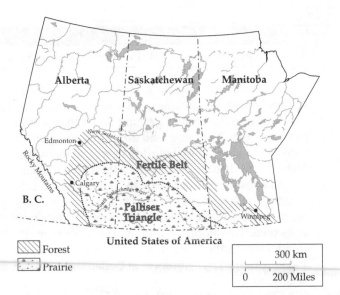

Figure 8.5   The Palliser Triangle.

Figure 8.6   A moldboard plow drawn by a team of horses in the early days of Saskatchewan.
(Photo courtesy of the Prairie Farm Rehabilitation Administration, Calgary, Alberta.)

In addition to unpredictable rainfall and periodic drought, prairie farmers faced other problems: aggressive weeds, crop disease, and insect pests. Common wheat diseases include *rusts,* orange-brown masses on stems and leaves, caused by fungi, which damage and sometimes destroy a farmer's entire wheat crop. Insect pests can also damage wheat; these include sawflies, caterpillars, cutworms, gophers, and grasshoppers, which generally appear during very dry summers. Their onslaught adds to the dilemma of dryland farming, especially when rains fail. Learning how to cope with these problems took a long time, but four developments greatly advanced farming on Canada's prairies: new varieties of wheat, summer fallowing, stubble mulching, and improved farm implements.

INSIGHT 8.3
Wheat

Wheat is one of the world's oldest and most important grains. It was cultivated by early civilizations in the Euphrates Valley 9,000 years ago. Today, three types of wheat are commonly grown around the world: winter wheat, spring wheat, and durum wheat. Winter wheat is sown in the fall of the year, takes root before freezing weather sets in, stands dormant during winter months, resumes its growth in the spring, and is harvested in the summer.

Spring wheat can be planted in any season of the year but requires an uninterrupted growing season of about 100 days. On the prairies, spring wheat is usually sown in early May and harvested in August. Both winter and spring wheat are milled to make flour for bread, rolls, and cake. Durum, a type of spring wheat, is used to make pasta products like spaghetti, macaroni, and ravioli.

New Varieties of Wheat Farmers who settled in the Palliser Triangle brought traditional wheat seed with them to plant on cultivated prairie lands. If adequate spring rains fell and winter frosts were not too early, wheat yields were good. But what they often needed was a wheat variety that could better endure drought and also mature in a shorter growing season. Most prairie farmers did not know that a new wheat variety was already being cultivated by farmers in Ontario. It was discovered by David Fife, who found that some of the wheat grown from seed sent to him from Scotland resisted wheat rust, matured faster, and could be harvested earlier than other varieties. By selecting and crossbreeding these faster-growing varieties, Fife developed a new variety of spring wheat that became known as Red Fife. Once this was known, word spread quickly, and Red Fife became the leading wheat crop on the prairies.

Later, in 1911, Charles Saunders, a scientist at the Canadian Department of Agriculture in Ottawa, developed another new wheat variety by crossing Red Fife with Hard Red Calcutta, a wheat strain from India. The new variety, called Marquis, matured even more quickly than Red Fife and was better able to resist wheat diseases. By 1920, Marquis accounted for 90 percent of the wheat grown on the Canadian prairies. It set a new standard for high-quality wheat for decades. Since that time, several new and improved varieties of wheat have been introduced into Canadian agriculture.

Summer Fallowing Although the new varieties of spring wheat were more drought resistant than earlier ones, they still required close to 460 millimeters (≈ 18 inches) of annual precipitation. Therefore, because the Palliser Triangle averaged only 360 millimeters (≈ 14 inches) of rain each year, the main challenge facing Palliser wheat farmers was how to cope with recurring prairie droughts. The first insight into solving this

problem occurred in 1885 when an uprising took place near Indian Head, Saskatchewan.

Louis Riel, President of a provisional government in western Canada, and Poundmaker, the Plains Indian chief, had incited a rebellion against Canada because they feared losing their property rights as Canada took greater control over its western lands. To quell the rebellion, the government of Canada conscripted horses, including those on Indian Head farms, to carry supplies to soldiers. Since no horses were left on the farms, farmland had to lie **fallow** that summer—that is, no crops were planted even though the land had been plowed. The following year was an exceptionally dry one and most prairie farms were seeded to wheat as usual, but the only good harvests occurred on Indian Head farms where the land had lain fallow the summer before. Soil moisture had been conserved when the plowed land received and stored rainfall, making it available to crops the following year. It was a major discovery to learn that summer fallowing could make wheat farming successful despite repeated years of little rainfall. It not only advanced prairie farming but led the Canadian government to establish experimental farms to initiate continuing programs of agricultural research (Insight 8.4).

Stubble Mulching Summer fallowing succeeded in conserving soil moisture during times of below-average rainfall, but it proved disastrous in very dry years, especially when prairie winds were apt to trigger soil drifting and blow topsoil away. The problem was to figure out how to implement summer fallowing without losing topsoil to wind erosion. Again, quite by accident, a farmer near Sibbald, Alberta, left a field unplowed during the spring of 1918 and allowed the prior year's crop stubble—short stalks of hay left after the last harvest—to remain in the ground. He later noticed that his soil was held in place by crop stubble roots even during

INSIGHT 8.4
Canada's Experimental Research Farms

In 1886, Canada set up the first of several Dominion Experimental Research Farms at Indian Head, Saskatchewan. The research farm program is similar in purpose to the agricultural extension program established in the United States under the 1886 Land Grant Act. The initial purpose of the research farms was to find answers to the appalling and fearful problems prairie farmers were facing in their respective regions. The first project at Indian Head was to study the effectiveness of summer fallowing—plowing the land in the spring to capture spring rainfall and then allowing it to lie unplanted during the summer. The results of the experiments showed that summer fallowing could protect soils against the impact of dry years. The practice of summer fallowing was quickly adopted not only on the Canadian prairies but on the American Great Plains as well.

The scientific study of dryland farming on Canada's prairies began when the Dominion research farm at Swift Current, Saskatchewan, was established in 1922. Research at Swift Current was aimed at improving all aspects of raising crops and

livestock on the Palliser Triangle and focused on mechanized farming, particularly ways the gasoline tractor could be modified to perfect a self-propelled wheat combine. Other mechanical developments at Swift Current helped make dryland farming successful on the prairies.

The Dominion experimental farm at Manyberries, Alberta, was instituted in 1925 to solve problems of rangeland management in the driest parts of southern Alberta and Saskatchewan. Farmers there were facing overgrazed and abandoned lands seemingly no longer capable of supporting grasses for foraging livestock. Continuing studies at Manyberries showed that Russian-created wheatgrass seed mixed with other grasses like alfalfa would take hold and grow well on these lands. Crested wheatgrass not only stabilized these soils but provided forage for cattle and a cash crop for farmers.

These are but a few insights into the many accomplishments of Canada's experimental farms—accomplishments that spelled the difference between agricultural success and failure on the prairies.

---

severe windstorms. It appeared that it might be better *not* to plow prior to summer fallowing.

This event sparked research at the Lethbridge Farm Station in Alberta, and in 1921 a new farming practice called **stubble mulching** was born. Stubble mulching, which involves leaving crop residues and roots in the ground following harvesting and not plowing prior to the next season's fallowing, was found to reduce soil erosion and conserve soil moisture. However, it made weed problems more difficult because spring plowing had been an effective means of weed control. Thus, stubble mulching was not universally adopted.

Improved Farming Implements   Trying to deal with agricultural weeds led to new farm implements designed to uproot them without disturbing soils. One of the most successful new devices was the Noble Blade designed and manufactured by Charles Noble in 1935. It greatly improved mechanical weed control and left the prior year's plant stubble undisturbed. Another important development occurred at the Experimental Farm Station at Swift Current, Saskatchewan, during the late 1930s. Low-pressure rubber tires were designed for use on heavy farm equipment to minimize

soil compaction, in which porous soils are compressed and soil pore spaces collapse, reducing the soil's ability to hold air and water and diminishing agricultural productivity. Today, low-pressure tires are used on most farm equipment all over the world.

Drought, Soil Erosion, and Crop Failure

Because crop irrigation was not an option in Canada's dryland farming areas, spring rainfall to renew soil moisture was critically important. If sufficient spring rains fell, farmers had a good year, but a dry spring augured meager harvests. Adequate rainfall fell on the prairies from 1880 to 1910, and ample wheat harvests were common. But a series of dry years that began in the 1920s caused recurring droughts that ultimately proved disastrous for many farmers.

One of the hard lessons learned on the North American prairies was that native prairie grasses held the soil in place even in times of severe drought. Their roots often reached as deep as six or eight feet. Plowing the prairies removed grass cover, exposing semi-arid soils to erosion by wind and water. In Saskatchewan, where wind is the principal cause of soil erosion, plowing, especially during the dry years of the 1920s and

Figure 8.7 Wind-driven soil erosion frequently darkened the skies during the "dust bowl" era of the 1930s on the American Great Plains and the Canadian prairies.
(Photo courtesy of the Prairie Farm Rehabilitation Administration, Calgary, Alberta.)

1930s, led to drifting soils that at times simply blew off the land. The soil that remained was less fertile.

The summer of 1929 marked the start of the most serious drought, crop disease, insect attacks, and soil erosion ever experienced in North America. These conditions persisted through most of the 1930s on the Canadian prairies and America's Great Plains. Years of summer fallowing without stubble mulching had made prairie soils vulnerable to wind erosion. Those dry years also gave rise to grasshoppers, which appeared seemingly out of nowhere and devoured just about every plant they found. Also, rust diseases occurred more frequently, even attacking fields of Marquis wheat. As the quality of the wheat declined, its price dropped from $1.60 to less than $.40 a bushel.

The 1930s are remembered as the decade of North American dust bowls and worldwide economic depression. Drought prevailed from Texas northward through the American Midwest and into Canada's Palliser Triangle. Soils were lifted from drought-ridden areas by hot summer winds and blown aloft across hundreds if not thousands of kilometers, darkening skies and turning day into night (Figure 8.7). One of the terrible dust bowl incidents, an account of the black blizzard of June 2, 1937, appeared in an Alberta newspaper, *The Medicine Hat News*:

> The worst dust storm in the history of the district struck shortly before 9:00 o'clock. A gigantic black cloud rolling from the northwest plunged the city into darkness as black as night. . . . Furniture, floors and goods in the store windows were covered with heavy gray dust that almost blotted out the original coloring. . . . Grit filtered through the doors and windows, no matter how tightly shut, covering everything with dull gray film. . . . For an hour and a half the storm raged and indistinct lights were all that was visible across a street.

Collapse and Recovery of Prairie Agriculture

North America's dust bowl era, 1931 to 1941, was a major calamity. In Canada, close to 250,000 homesteaders abandoned their farms and the prairies, leaving perhaps 50,000 farm families behind to cope with appalling agricultural problems. Harvests fell to only 10 percent of earlier years, and crop prices dropped as well. For example, 1928 wheat harvests amounted to 321 million bushels worth $218 million, but 1937 harvests of 37 million bushels were worth only $16 million. Families remaining on the prairies did not have enough money to pay their bills, buy food, or feed their livestock. It is estimated that the crisis affected 7.3 million hectares ([a] 18 million acres) of the Palliser Triangle. Many came to believe that John Palliser had been right—agriculture on the grassland prairies was destined to fail.

When the people of eastern Canada learned about the crisis on the prairie farms, relief efforts began. The Canadian Red Cross led an appeal for food and clothing, and the provincial governments of Ontario and Quebec, along with many private citizens, provided assistance. Relief trains made transcontinental missions of mercy, delivering staples such as beans, peas, and corn as well as fodder for horses and hay to feed and shelter cows. Some citizens sent salted and smoked fish. Figure 8.8 poignantly shows prairie farmers lined up at a railway siding during the winter of 1933 waiting for a relief train. What followed was a period of recovery—a long, slow process of learning how to survive and prosper in Canada's unique prairie environment.

Recovery following the dust bowl disaster was indeed slow. There was no quick fix for the desperation prairie farmers and their families faced. What unfolded was a long struggle to learn to cope with unpredictable recurring prairie droughts. Among the most important accomplishments of this time are the following:

- Establishment of Canada's Prairie Farm Rehabilitation Administration (PFRA) in 1935 to help restore the drought-stricken and soil-drifting areas of Manitoba, Saskatchewan, and Alberta; to promote sustainable agricultural practices; and to bring greater economic security to the region and its farm families.

- Initiation of a PFRA program of planting drought-resistant Russian crested wheat grass across the Palliser Triangle to control blowing soils and establish a new cash crop. Crested wheat grass, while developed in the United States, was all but ignored in America. However, its potential was realized by Canadian farmers

Figure 8.8   Prairie farmers waiting at a railway siding for the next relief train bringing needed food and supplies.
(Photo courtesy of the Prairie Farm Rehabilitation Administration, Calgary, Alberta.)

who planted hundreds of thousands of Palliser acres in the late 1930s.

- Granting of PRFA funding to construct individual farm and community prairie dugouts to serve as water supplies for rural farm families and farm animals. Dugouts are artificial surface-water catch basins that conserve rainwater and snowmelt, especially in times of drought.

- Development of regional dams and reservoirs for crop irrigation such as the Adams Lake and Val Marie reservoirs in southwestern Saskatchewan, the St. Mary Dam and Irrigation Project, and the South Saskatchewan River Project, which created Gardiner Dam and Lake Diefenbaker.

Prairie Agriculture in Canada Today

By the late 1940s, Canada's semi-arid, fertile prairies had become one of the world's most productive agricultural regions. Most prairie farmers had adopted green revolution practices, including the use of chemical fertilizers and pesticides, but before long a growing number of them began abandoning those techniques and adopting more sustainable agricultural practices instead. Among them are the Bauml family in Marysburg, Saskatchewan, who decided to become organic farmers.

Organic Wheat Farming in Saskatchewan

In the late summer, from a bluff near Marysburg, wheat fields stretch as far as the eye can see. Here, Clarence and Ray Bauml grow wheat and other crops on their 2,000-acre farm. Wheat is Saskatchewan's

principal agricultural commodity, and for decades the Baumls, like many of their neighbors, practiced green revolution agriculture. They were chemical farmers who started using DDT in 1946 believing it would solve their pest problems. By 1990, however, they were convinced that intensive use of pesticides was not only too costly but was causing health problems. Furthermore, they found that many pests were developing resistance to DDT and similar pesticides. They began *organic farming* that year.

Organic farming is a type of sustainable agriculture except that organic farmers use no commercial chemical inputs whatsoever—no commercial fertilizers, herbicides, pesticides, or food preservatives. Clearly, this is a stricter approach to agriculture than sustainable agriculture requires.

In reality, organic farming is similar to the kind of agriculture that was practiced before the green revolution. Organic farmers aim to cooperate with nature by using natural sources of soil nutrients (animal and vegetative wastes) and minerals, anticipating rainfall cycles or employing irrigation, cultivating the best available crossbred genetic varieties of grains, fruits, and vegetables, and utilizing mechanical and biological means of pest control.

Organic farming often means learning how crops were raised before the green revolution, but it does not mean rejecting all agricultural advances. Organic farmers can and do take advantage of new crop varieties, modern farm implements, and natural pesticides like Bt (see Chapter 7). Plant diseases are all but nonexistent in the many new crop varieties originating from ongoing crossbreeding research in Canada, the United States, and other countries, and it is beneficial for organic farmers to plant them. Thus, the Baumls sow most of their land to two or three different varieties of wheat, including CDC-Teal, a new type released by the University of Saskatchewan's Crop Development Centre in 1991. CDC-Teal is a bread wheat, high in protein, that matures in 98 days. Typically, as many as 40 bushels an acre are harvested on land that lay fallow the prior growing season. By contrast, Saskatchewan's chemical farmers, who typically do continuous cropping (no fallowing), average 28 bushels an acre.

The Baumls use no commercial fertilizers or pesticides. Instead, they rely on legumes such as clover planted as part of a regular crop rotation. In a two-year rotation cycle, half of the land lies fallow while the other half is sown in spring wheat, oats, barley, flax, and other crops. Fallow land is planted in clover or buckwheat and turned over in June to enrich soils with crop residues, so-called "green manure." This

TABLE 8.1   A Comparison of Annual Wheat Yields and their Value on the Bauml Farm and on Conventional Farms in Saskatchewan in 1995

	Bauml Farm	Conventional Farm
Annual yield from 1,000 acres	20,000 bushels[a]	28,000 bushels
Weight in metric tons	545 metric tons	763 metric tons
1995 price of wheat	$210/ton[b]	$150/ton
Value of 1,000 acres of wheat	$114,500	$114,500

SOURCE: Data from personal communication with the Bauml family.

[a]This is one-half of the typical yield of 40,000 bushels, since this land lies fallow every other year.

[b]In 1995, spring wheat grown organically in Saskatchewan was worth about 40 percent more than chemically grown spring wheat.

cropping/fallowing rotation minimizes pests, and shallow cultivation controls weeds.

Certified Organic Foods   Crops grown on the Bauml farm are certified as "organic" by the Organic Crop Improvement Association (OCIA) in Bellefontaine, Ohio, which certifies more than 20,000 growers in 22 countries. Certification agencies enable farmers to assure customers that their grains and produce have been grown without synthetic pesticides and fertilizers. Organic certification generally requires that farmers have not used chemical inputs for at least three years and that their farms are inspected annually.

When farmers go to sell organic grains and produce, they make as much as 75 percent more money than traditional farmers selling nonorganic food products. This is one of the reasons for the rising trend in organic farming. Comparing the economics of organic farming with nonorganic farming is a complex endeavor that perhaps only an accountant should undertake. However, Table 8.1 gives a simplified comparison of the value of wheat grown on the Bauml farm versus that grown conventionally elsewhere in the same province. Because land lies fallow on the Bauml farm during a two-year rotation cycle, average annual yields of organic wheat are less than on conventional farms. But because organic wheat commands a higher price per ton, the value of wheat grown organically turns out to be about the same as that produced more intensively using high-energy chemical fertilizers and pesticides. Also, the costs of chemical inputs are not included in Table 8.1; if they were, net farm income on the organic farm would actually be much greater.

While perhaps only 1 percent of Saskatchewan farms practice organic farming today, the percentage is increasing. Organic farming is growing in importance in other countries as well, especially in the United States.

Current Trends Toward Sustainable Agriculture

Some farmers who are concerned about the future of agriculture believe that sustainable agriculture will ensure that their children and grandchildren will have a stake in the same land they are farming today. One farmer put it this way: "My goal is to leave this soil and this farm in a better condition than when I started." But an important question remains: Can sustainable agriculture, and in particular organic farming, alone satisfy the food requirements of future generations, especially if world population growth continues unabated?

An important step in fostering organic agriculture was taken in the United States with the passage of the 1990 U.S. Farm Bill. It established the first standards and incentives for producing certified organically grown foods. Nonchemical agricultural methods, such as crop rotation, are favored, and organic farmers are entitled to receive higher farm subsidies—dollar levels previously available only to chemical farmers. Since then, increasing quantities of organic foods have been appearing in North American markets from coast to coast. In 1993, the World Resources Institute in Washington, D.C., reported that organic farming has the potential to compete successfully with green revolution agriculture. But to bring this about, federal and state farm policies and regulations must be shifted toward encouraging and supporting sustainable agriculture to the same extent that they have been supporting chemical agriculture. The institute recommended that farm subsidies be slanted less toward farmers committed to chemical fertilizers and pesticides and more toward those willing to adopt more sustainable farming practices.

In 1999, U.S. sales of organic produce totaled $4.2 billion, and the organic food industry was growing about 25 percent annually. There were close to 4,000 certified organic farmers, representing 0.2 percent of all U.S. farmers. In addition, there were an estimated 1,800 noncertified organic growers. The total acreage devoted to organic agriculture added up to about 0.1 percent of all U.S. farmland. Eleven states, together with 33 private agencies, were engaged in certifying organic crops and produce. About 25 percent of U.S. consumers were buying organic products at least once a week, and polls indicated that four out of five consumers preferred organically grown foods to conventional foods and were willing to pay more for them.

A preference for organically grown foods is also beginning to be noticed on a global scale. In Japan, rising demands for organic foods account for the lion's share of U.S. exports of organic produce. Mexico, which has emerged as the world's number one producer and exporter of organic coffee, also exports organically grown beans, bananas, and vegetables. In India, an Institute for Sustainable Agriculture has been established to help farmers shift from chemical to organic methods. Farm areas devoted to organic agriculture in Europe are increasing rapidly. For example, the German government is providing substantial subsidies to farmers who convert to organic practices. In all of these cases, the transition to more sustainable farming is being fueled by consumer demands for safer foods and by the awareness that chemically based agriculture is responsible for many soil and water pollution problems.

REGIONAL PERSPECTIVES

The Pacific Lumber Company: Practicing Sustainable Forestry

It is estimated that there were once two million acres of redwood forests in northwestern coastal California. But by the early 1900s, hundreds of lumber companies and sawmills were harvesting the most valuable trees, including redwoods, which are prized for their natural durability, resistance to insect attack, and pleasing color. Redwood is favored for building homes, patios, decks, deck chairs, picnic tables, and hot tubs. By 1970, high demand had reduced North American stands of old-growth redwood by 97 percent.

The Pacific Lumber Company, the oldest logging and log-sawing firm in California, lies 280 miles north of San Francisco in Humboldt County. Established in 1870 as a family-owned-and-operated business, Pacific Lumber's approximately 200,000 acres of forests include mostly Coast redwood (*Sequoia sempervirens*) and Douglas fir (*Pseudotsuga menziesi*). The trees define an ancient, biodiverse ecosystem, providing habitat for spotted owls, marbled murrelets, red tree voles, fishers, northern goshawks, Olympic salamanders, and tailed frogs. But until 1985, Pacific Lumber's philosophy was different from that of almost every other logging company—it believed in sustainable forestry. From its start, the company adopted two fundamental policies: selective cutting and sustained yield.

Selective cutting requires discriminating between mature and younger trees—that is, harvesting older trees and preserving younger ones. It is different from clear-cutting, the logging industry's standard practice, which involves cutting all trees in a given area. Although clear-cutting is often preferred by loggers because it is efficient and cost-effective, it results in treeless, exposed land that is vulnerable to soil erosion. Erosion on clear-cut slopes depletes the soil of organic matter and nutrients, which impedes the regeneration of redwood trees. Alder, vine maple, and other less valuable trees grow in their place. Soil erosion also threatens salmon-rearing streams and rivers by filling them in with silt, which deprives fish eggs of needed oxygen.

Sustained yield means limiting annual harvests of mature trees, regardless of market demand. This practice guarantees future tree harvesting and protects the forest's natural capacity for new tree growth. It is like choosing to live on the dividends of a financial investment without depleting the investment itself.

These sustainable practices prevailed until 1985 when Pacific Lumber was bought out by Maxxam, Inc., of Houston, Texas, for $800 million in a hostile corporate takeover aimed at acquiring the company's forest resources and other assets. Like other corporate takeovers, this one required the sale of so-called high-yield "junk bonds." The bonds were to be paid off by clear-cutting redwood trees, thus abandoning Pacific Lumber's selective cut and sustained yield policies. Redwoods were cut down at rates two to three times greater than ever before without regard to the future of the forest.

In 1986, Earth First environmental activists focused national attention on the heart of Pacific Lumber's forest lands—the majestic redwoods and unique forest ecology of Headwaters Forest. Many of the old-growth redwoods on the 3,000-acre Headwaters Forest are between 1,000 and 2,000 years old. These trees are up to 250 feet tall, with diameters surpassing 15 feet. In 1999, after years of countless, mean-spirited confrontations with Pacific Lumber and fruitless negotiations, an agreement was finally reached to set aside Headwaters Forest and adjacent forest lands—close to 4,050 hectares (≈ 10,000 acres) of Humboldt County's giant redwoods—as a public preserve. The deal is based on a $250-million U.S. Congressional appropriation and a $230-million State of California contribution. The total, $480 million, is being paid to Pacific Lumber to buy the land.

California: Producing Organic Wine

In California, organic wine is a growth industry. In fact, more acres of wine grapes are certified as organic than any other crop statewide. In 1997, out of a total of 408,000 acres of wine grapes, 2.3 percent, or 9,281 acres, were certified as organic by California Certified Organic Farmers (CCOF), the state's agricultural certification agency. At least 18 of the state's wineries operate as organic or identify their wines as "made from organically grown grapes," including Fetzer-Bonterra Wines in Hopland, Mendocino County.

Fetzer Vineyards, one of California's leading winemakers, sells several organic wines under the label Bonterra, including Chardonnay, Carbernet Sauvignon, and Merlot. Bonterra vineyards are farmed naturally without insecticides, herbicides, fungicides, or commercial fertilizers, and their wines are made from state-certified grapes. Cover crops planted between rows of grapes serve as habitat for insect pests and are part of the vineyard's integrated pest management program.

Although wine contains naturally occurring sulfites, an organic wine may or may not have sulfites added. Sulfites are chemical reducing agents that keep wine fresh by preventing the oxidation of the constituents responsible for its subtle flavor and fragrance. Whether or not sulfites are added to wines can affect the definition of "organic wine"; the definition varies from vineyard to vineyard.

Guatemala: Growing Shade-Tolerant Coffee

Guatemala is one of several Central American countries where farmers are shifting from high-yield, sun-tolerant, hybrid varieties of coffee to shade-tolerant coffee plants that were once commonly grown beneath tropical tree canopies. The move to sun-tolerant hybrids took place in the 1970s when the U.S. Agency for International Development (AID) made funding available to Central American farmers willing to switch to them. The goal was to raise the income of small farmers by encouraging them to grow the new, high-yielding coffee varieties. However, this shift required cutting down trees to gain the benefit of direct sunlight, and led to several problems:

- The new sun-tolerant hybrids require repeated applications of chemical fertilizers and pesticides. The cost of hybrid plants and chemical inputs has resulted in unacceptable financial debts for farm families.
- Soils where hybrid plants are grown have accumulated toxic pesticide residues, which now pose health hazards to farmers and kill fish when transported by soil erosion to nearby streams and rivers.
- The loss of tree canopies in Central American countries where sun-tolerant coffee varieties have been planted deprives migratory birds of needed winter habitat. The National Audubon Society and other conservation groups believe that the trend toward fewer and fewer trees on coffee plantations is contributing to the decline of songbird species such as orioles and warblers. More than 150 bird species are known to seek winter nesting sites in Central America.

In 1996, the Smithsonian Institution's Migratory Bird Center in Washington, D.C., convened a group of environmental scientists, farmers, and gourmet coffee dealers to spearhead a return to native, shade-tolerant coffee. Their strategy was to save the songbirds by saving the trees on Central American coffee plantations. The transition is already beginning, and numerous coffee suppliers and outlets around the world now feature varieties of shade-grown coffee, some of which are certified as organic because they are grown without chemical fertilizers and pesticides.

KEY TERMS

certified organic foods	*Rhizobium* bacteria
conservation tillage	salinization
cover crop	selective cutting
cyanobacteria	soil compaction
dryland farming	soil erosion
fallow	soil husbandry
horizon	stubble mulching
humus	summer fallowing
no-till planting	sustainable agriculture
organic farming	sustained yield
pedogenesis	topsoil

DISCUSSION QUESTIONS

1. Like the Palliser Triangle, perhaps other regions could, with wise stewardship, be transformed into productive agricultural areas. Do you think this might be part of the answer to world food sufficiency? Discuss why this might or might not be possible.

2. Suppose agricultural research results in sophisticated new chemical pesticides that control or kill only target pest species and are harmless to non-target organisms including humans. Would you favor the development and use of chemicals like these? Explain your reasons for or against such new agents.

3. If the practice of sustainable agriculture were to result in more labor-intensive farming requiring a much larger agricultural workforce, would you see this as desirable or undesirable? Explain your answer.

4. Is it important to you whether organically grown food is available in your community? How much more would you be willing to spend for organic foods? Should we wait until more convincing scientific evidence is in hand regarding the risks of pesticide residues in food products before encouraging more sustainable agricultural practices that would lessen the use of chemicals? Explain your answer.

5. Going back to older methods of farming, such as crop rotation, seems like reverting to primitive agricultural methods that might not meet world food demands. Which is a more important goal: sustainable agriculture and conservation of natural resources or exponential growth in food supply to meet the needs of increasing numbers of people? Explain your answer.

INDEPENDENT PROJECT

Find a local food store that sells certified, organically grown products such as tomatoes, potatoes, corn, peas, and rice. Note the price per pound or some other unit price of a few of these food products. Ask the manager how many different organic foods the store sells and whether demand is growing. Are organic foods prominently displayed in this store? Are people willing to pay more for organic food products in this store?

In the same store, or perhaps another store, find the price of similar nonorganic products and determine by what percent organic foods cost more. Talk to the manager of a store that does not stock organic foods and ask why not. Are there plans to add organics to the store's food inventory? Summarize your findings in a report.

INTERNET WEBSITES

Visit our website at http://www.mhhe.com/environmentalscience/ for specific resources and Internet links on the following topics:

Agriculture in Saskatchewan

Biotechnology

Agriculture, pesticides, and organic farming

Transgenic crops

Natural foods

SUGGESTED READINGS

Ainsworth, Susan J. 1996. Changing technologies help herbicide producers compete in mature market. *Chemical and Engineering News* (April 29):35–42.

Brown, Lester R. 1996. *State of the world*. World Watch Institute Report. New York; London: W. W. Norton & Company.

Case, Roslyn A., and Glen M. MacDonald. 1995. A dendroclimatic reconstruction of annual precipitation on the western Canadian prairies since A.D. 1505 from *Pinus flexilis James. Quaternary Research* 44(2):267–75.

Faeth, Paul. 1993. Agriculture policies encourage resource degradation. *Environmental Science and Technology* 27(9):1709.

Gardner, Gary. 1996. Organic farming up sharply. In *Vital Signs,* World Watch Institute. New York; London: W. W. Norton & Company.

Gray, James H. 1996. *Men against the desert*. Saskatoon; Calgary: Fifth House Publishers.

Hassett, John J., and Wayne L. Banwart. 1992. *Soils & their environment*. Englewood Cliffs, NJ: Prentice Hall.

Josephson, Julian. 1993. Delaney Clause. *Environmental Science and Technology* 27(8):1466–67.

Kirschner, Elisabeth M. 1994. Agricultural chemical producers rebound from floods of 1993. *Chemical and Engineering News* (October 3):13–16.

Klinkenborn, Verlyn. 1995. A farming revolution—Sustainable agriculture. *National Geographic* 188(6): 60–89.

Pimentel, David, et al. 1992. Environmental and economic costs of pesticide use. *BioScience* 42(10): 750–60.

Rosegrant, Mark W., and Robert Livernash. 1996. Growing more food, doing less damage. *Environment* 38(7): 7–11; 28–31.

Saskatchewan Agriculture and Food. 1996. *Agricultural statistics fact sheet.* Statistics Branch, Regina Saskatchewan. (June).

Soule, Judith D., and Jon K. Piper. 1992. *Farming in nature's image: An ecological approach to agriculture.* Washington, D.C.; Covelo, CA: Island Press.

Strauss, Stephen. 1995. Telltale trees. *Canadian Geographic* (July/August):14–16.

Zaki, Mahfouz H., Dennis Moran, and David Harris. 1982. Pesticides in groundwater: The Aldicarb story in Suffolk County, N.Y. *American Journal of Public Health* 72(12):1391–95.

PART IV

Energy Production and Consumption

Photo Disc, Inc./Vol. 60

Chapter 9
Petroleum Resources/Fossil Fuels

Chapter 10
Nuclear Power

Chapter 11
Sustainable Energy

Petroleum Resources/Fossil Fuels

Photo Disc, Inc./Vol. 57

The *Exxon Valdez* oil spill injured not only fish and wildlife populations and their

habitats, but also human use of the affected areas. Some people, such as fishers and

recreation guides, could no longer work at their regular occupations. . . . Some peoples'

whole life style changed, especially those who relied on subsistence in the spill area.

Finally, many people who have never been to the spill area, even people who have never

been to Alaska, felt a loss because a pristine area was degraded.

Alaska's Wildlife, January/February 1993

The fossil fuels are natural resources that have shaped the modern world. Petroleum, natural gas, coal, and products derived from them fuel the lifestyle to which most of us have become accustomed, enabling us to drive our cars, heat our homes, and cook our food. Indeed, the use of fossil resources has led to the manufacture of hundreds of contemporary "necessities" as diverse as rubber, plastic, perfume, pharmaceuticals, and ink. However, like all natural resources, supplies are limited; they are also nonrenewable. All of this is especially true for petroleum, whose underground reserves need to be managed judiciously—not only to conserve them but also to keep drilling and refining processes from harming the natural environment.

Ever since the use of petroleum became common, instances of pollution, particularly oil spills, have occurred. The 1989 *Exxon Valdez* oil spill in Prince William Sound off the coast of southern Alaska is especially notable because it was the largest in U.S. history and had devastating effects on local communities and area wildlife. However, there were also positive results because the incident triggered the formulation and enforcement of the first effective state and federal regulations for transporting petroleum products. This chapter traces the discovery and use of petroleum and describes the impact of the historic *Exxon Valdez* oil spill.

BACKGROUND

Origin and Composition of Petroleum

Although petroleum has many modern-day uses, its origins are ancient. Petroleum was formed slowly through geologic eras spanning 60 to 600 million years. The process began when organic matter derived from marine plants and animals settled to the ocean floor and accumulated in sediments. The buildup of successive sediment layers exerted increasing pressure on this organic matter, creating temperatures as high as 100 °C (212 °F). A series of biological, chemical, and physical reactions took place, and ultimately petroleum and natural gas collected in porous sandstone and limestone formations within the earth's crust (Figure 9.1).

The earth's crust consists of a solid layer of continents (**continental crust**) and ocean floor (**oceanic crust**). Continental crust underlies the planet's continents and is about 35 kilometers (≈ 22 miles) thick. Oceanic crust lies beneath the ocean floor and is about 5–7 kilometers (≈ 3–4 miles) thick. Oil and natural gas frequently migrate to porous, permeable rock formations within the earth's crust, displacing water originally trapped there.

Petroleum is a thick, dark oil composed of a complex mixture of organic chemicals, most of which are

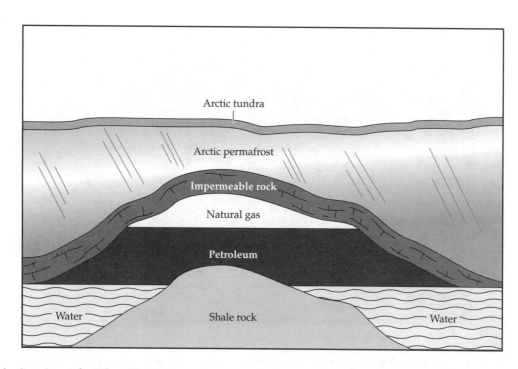

Figure 9.1 Arctic region geologic formations in the earth's crust where petroleum and natural gas have collected.

made up solely of hydrogen and carbon (**hydrocarbons**). For purposes of discussion, the terms *oil*, *crude oil*, and *petroleum* generally mean the same thing and are often used interchangeably. The term *fossil fuels* encompasses petroleum, natural gas, and coal.

Depending on their chemical composition, petroleum derivatives occur as gases, liquids, or solids. The simplest derivative is **natural gas,** which is composed of hydrocarbons containing one to three carbon atoms: methane, ethane, and propane. Natural gas is free of sulfur impurities and burns cleaner than coal or petroleum.

Hydrocarbons with five to about 38 carbon atoms are liquids, such as octane, a component of gasoline. Compounds with 40 or more carbon atoms are solids; an example is tetracontane, which is found in paraffin wax. The names, chemical formulas, and principal uses of a few of the thousands of compounds found in petroleum are shown in Figure 9.2.

Discovering Uses for Petroleum

Petroleum is a relatively new addition to our energy stores. It was not until 1859 that the world's first successful oil well was drilled at Titusville, Pennsylvania. At that time, coal was the world's most important energy resource, fueling the steam engines that powered the Industrial Revolution. Today, coal is still employed to generate electric power and to make coke, a purified

form of carbon used to produce metals from mineral ores. But petroleum quickly surpassed coal as the most highly valued energy resource.

Petroleum originally became important because one of its components, kerosene, could be burned in street lamps and home lanterns instead of whale oil. At first, other petroleum components, such as gasoline, diesel oil, and heating oil, were discarded. However, their importance was soon discovered. Oil was cheaper than coal, burned cleaner, and was easier to transport. In the early 1900s, Henry Ford's automobile created a demand for gasoline, further encouraging the emergence of petroleum exploration, drilling, and refining as new industries.

By 1929, oil supplied one-third of the total U.S. energy demand. Ten years later, almost half of the nation's energy needs were being met by burning fuels derived from petroleum (petrofuels). By the end of World War II, coal was no longer the number one fuel; petroleum and natural gas accounted for two-thirds of all U.S. energy requirements.

During World Wars I and II, the United States, Britain, and their allies depended heavily on petroleum to fuel motorized transport vehicles, merchant ships, naval vessels, aircraft, and military armaments. In 1940, the United States and Russia were the world's largest petroleum producers and consumers, while the rest of the industrial world still depended mainly on coal. But by the end of the second world war, the productivity

Petroleum Hydrocarbon	Empirical Formula	Structural Formula	Principal Product
Methane	CH_4	CH_4	Natural gas
Ethane	C_2H_6	CH_3CH_3	Component of natural gas
Propane	C_3H_8	$CH_3CH_2CH_3$	Liquified petroleum gas (LPG)
n-pentane	C_5H_{12}	$CH_3(CH_2)_3CH_3$	Component of naphtha
n-hexane	C_6H_{14}	$CH_3(CH_2)_4CH_3$	Component of gasoline
Cyclohexane	C_6H_{12}	$\begin{array}{c} CH_2-CH_2 \\ CH_2 \qquad CH_2 \\ CH_2-CH_2 \end{array}$	Component of gasoline
n-octane	C_8H_{18}	$CH_3(CH_2)_6CH_3$	Component of gasoline
Iso-octane	C_8H_{18}	$\begin{array}{c} CH_3 \\ CH(CH_2)_4CH_3 \\ CH_3 \end{array}$	High-octane gasoline additive
n-dodecane	$C_{12}H_{26}$	$CH_3(CH_2)_{10}CH_3$	Component of kerosene
n-eicosane	$C_{20}H_{42}$	$CH_3(CH_2)_{18}CH_3$	Component of fuel oil
n-tetracontane	$C_{40}H_{82}$	$CH_3(CH_2)_{38}CH_3$	Component of paraffin wax

Figure 9.2 Common petroleum hydrocarbons. Smaller molecules, such as methane, ethane, and propane with fewer than five carbon atoms, are normally gases. Larger molecules, such as n-pentane, n-hexane, n-octane, and iso-octane, are normally liquids. Still larger molecules are oils, greases, and waxes. The designation "n," as in n-pentane, means that a molecule's carbon atoms are all connected in a straight chain without any "branching." The designation "iso," as in iso-octane, means that some of a molecule's carbon atoms are "branched" and include the group: $(CH_3)_2CH-$.

and economic prosperity of many nations, including the United States, was linked to petroleum. In 1999, U.S. demand for petroleum averaged close to 19 million barrels a day.

Oil Refining

When the value of petroleum's many components was realized, oil refining became an industry. Fractionating towers were built at refineries to separate oil into these components. When crude oil is heated to about 400 °C (≈ 750 °F) in a fractionating tower, various oil fractions (components) are distilled (boiled off) and collected at different elevations on the tower. Low-boiling fractions collected near the top of the tower consist of relatively small molecules—those containing a small number of carbon atoms; gasoline is one example. Fractions having higher-boiling temperatures, such as kerosene, are collected mid-tower. Still higher-boiling fractions, such as gas oil, are removed near the bottom. Non-boiling residues removed at the bottom of the tower contain materials such as petroleum coke and asphalt (Figure 9.3).

Refining petroleum into useful products marked the beginning of petroleum engineering technology. As the technology matured, new discoveries enabled the industry to convert less valuable crude oil fractions into more valuable products. For example, a process

called **catalytic cracking** was developed whereby hydrocarbon molecules in the gas-oil fraction can be split (cracked) into smaller molecules. The small molecules can be combined and rearranged through **catalytic reforming** processes to make aviation fuel and gasoline. Cracking and reforming operations are employed in other petrochemical operations as well, such as:

- Hydrocarbons in naphtha, a low-boiling, volatile hydrocarbon fraction, are rearranged to form chemicals such as benzene, toluene, and xylenes, which are added to gasoline to improve fuel combustion efficiency.

- Chemicals are made for the manufacture of synthetic rubber, plastics, explosives, refrigerants, pesticides, pharmaceutical drugs, solvents, antifreeze, inks, dyes, perfumes, and hundreds of other important products.

- Methane is converted into hydrogen, which can be reacted with atmospheric nitrogen gas to form ammonia, a chemical fertilizer.

- Methane is reformed to make acetylene, an arc-welding fuel used to create very high flame temperatures. Acetylene is also an industrial raw material used to manufacture synthetic fibers for clothing, paints, adhesives, and vinyl chloride from which polyvinyl chloride (PVC) pipe is made.

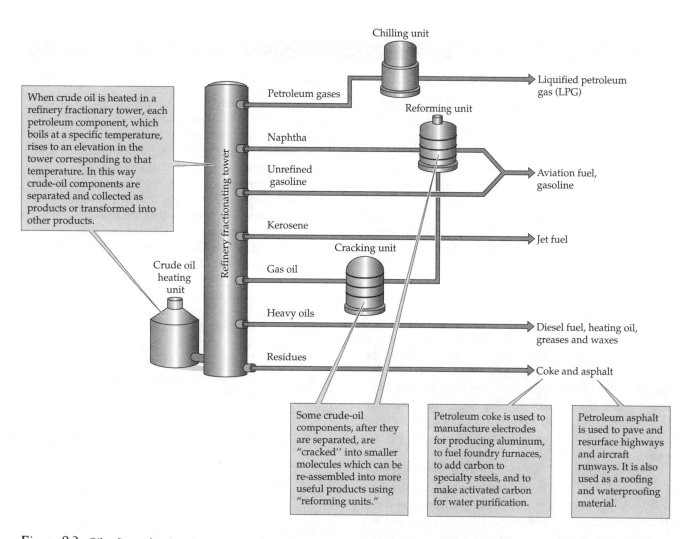

When crude oil is heated in a refinery fractionary tower, each petroleum component, which boils at a specific temperature, rises to an elevation in the tower corresponding to that temperature. In this way crude-oil components are separated and collected as products or transformed into other products.

Crude oil heating unit

Refinery fractionating tower

Chilling unit

Petroleum gases

Liquified petroleum gas (LPG)

Reforming unit

Naphtha

Unrefined gasoline

Aviation fuel, gasoline

Kerosene

Jet fuel

Cracking unit

Gas oil

Heavy oils

Diesel fuel, heating oil, greases and waxes

Residues

Coke and asphalt

Some crude-oil components, after they are separated, are "cracked" into smaller molecules which can be re-assembled into more useful products using "reforming units."

Petroleum coke is used to manufacture electrodes for producing aluminum, to fuel foundry furnaces, to add carbon to specialty steels, and to make activated carbon for water purification.

Petroleum asphalt is used to pave and resurface highways and aircraft runways. It is also used as a roofing and waterproofing material.

Figure 9.3 Oil refinery fractionating tower used to separate crude oil components into various fractions that are then converted into useful products.

Obtaining Petroleum

Following the 1859 Pennsylvania oil strike, worldwide exploration for petroleum began, and oil and natural gas were discovered in Texas, Louisiana, and Oklahoma. In the late 1930s, major oil discoveries were made in the Middle East, where the world's largest reserves exist today. U.S. oil needs were met through domestic production alone until 1947. After that, increasing oil consumption, coupled with cheap overseas sources, led the United States to begin importing petroleum, although reliance on foreign oil has sometimes been complicated by political agendas (Insight 9.1). In 1999, the United States imported about 45 percent of its oil from Saudi Arabia, Venezuela, Nigeria, Kuwait, Algeria, and other countries; the rest was produced domestically. Table 9.1 provides an estimate of worldwide proven oil reserves as of 1999.

Prudhoe Bay Oil Fields: An Important Discovery

A petroleum discovery especially important to the United States was announced in March 1968 by the Atlantic Richfield Company (ARCO) and Humble Oil Company (now Exxon). After many unsuccessful exploratory wells had been drilled in the tundra region of Alaska known as the North Slope, one at Prudhoe Bay (Figure 9.4) finally proved to be a spectacular strike. Almost 9,000 feet deep, the well produced high yields of crude oil and natural gas. Three months later, British Petroleum, a major partner in North Slope oil exploration, announced another significant oil strike in the same area.

By 1969, more than 30 producing oil wells had been drilled on the tundra in areas leased to the oil companies by the state of Alaska. It was estimated that 12 billion barrels of oil and 30 trillion cubic feet of

INSIGHT 9.1
Imports and Politics

By 1970, cheap Middle Eastern oil had flooded world markets, and OPEC, the Organization of Petroleum Exporting Countries, had become the world leader in oil production and pricing. OPEC is a cartel of major oil-producing and oil-exporting nations that includes Saudi Arabia, Iran, Kuwait, Iraq, Libya, Nigeria, Venezuela, and Indonesia

Soon after the start of the 1973 war in the Middle East, OPEC slowed down crude oil production rates and forced higher prices. World crude oil prices escalated from about $3 to $12 a barrel. At the same time, Arab members of OPEC cut off oil ship-ments to the United States, a supporter of Israel. This oil embargo took place at a time when U.S. dependence on Middle East oil accounted for close to 27 percent of total U.S. oil consumption. Soon, long lines of automobiles appeared at gas stations as fuel supplies dwindled. By 1974, the world price of oil had risen even further—to $32 a barrel. However, high oil prices led to expanded petroleum exploration and increased oil production in non-OPEC countries. Today, non-OPEC countries supply about half of the world's crude oil.

natural gas existed at the Prudhoe Bay site, making it the largest and most productive site in North America. While it occupies about 113,000 hectares (≈ 280,000 acres), only 2 percent of the area has been developed so far; the rest is open tundra. When production began there in 1977, close to 2.0 million barrels of oil were produced daily; today production has declined to approximately 1.5 million barrels a day.

It is estimated by ARCO that two-thirds of Prudhoe Bay oil will have been extracted by the year 2000, leaving about 4 billion barrels of recoverable oil. If what is left is pumped out at the rate of 1.5 million barrels a day, the oil will last about 2,700 days (≈ 7.3 years). This is one of the reasons the oil industry is looking for new sources of oil and pressing to open Section 1002 of the Arctic National Wildlife Refuge to oil exploration and development (see Chapter 1).

Drilling for Oil on the North Slope

Drilling rigs, some of which are movable, are used to explore for oil and to drill production wells. On the North Slope, rigs are normally operated on gravel drilling pads to protect the underlying permafrost (See figures 9.1 and 9.5). Some wells are more than two miles deep. The drilling process requires water-based clay formulations called muds to lubricate drilling bits, clean the wellbore, and maintain pressure within the well. Drilling muds are recycled many times before being disposed of. Before 1994, spent muds and other drilling wastes were expelled into reserve pits built as a part of drilling pads. Now, using new technology, these wastes are ground into a slurry and injected several thousand feet into deep wells, thereby eliminating the need for reserve pits.

TABLE 9.1   1999 Worldwide Proven Oil Reserves (Estimated)*

Region of the World	Billion Barrels of Oil
Middle East	673.6
Central and South America	89.5
Africa	75.4
North America	75.3
Eastern Europe and the Former Soviet Union	58.9
Far East and Oceania	43.0
Western Europe	18.9
Total Proven Reserves	1,034.6

SOURCE: Data from *The Oil and Gas Journal*, December 1998.

*The American Petroleum Institute defines proven reserves of petroleum as quantities of crude oil that "geologic and engineering data demonstrate with reasonable certainty to be recoverable from known reservoirs under existing economic and operating conditions."

North Slope petroleum exists in deep geologic formations. It flows out under pressure at about 70 °C (≈ 160 °F) along with natural gas (a part of it dissolved in the crude oil) and a naturally occurring underground water called "produced water." Some natural gas is used to generate electric power to operate pipeline pumping stations. The rest is re-injected into wells to maintain underground pressure, forcing a continuing flow of oil. Produced water is re-injected for the same purpose. Returning natural gas and produced water to wells is an important way to enhance oil recovery.

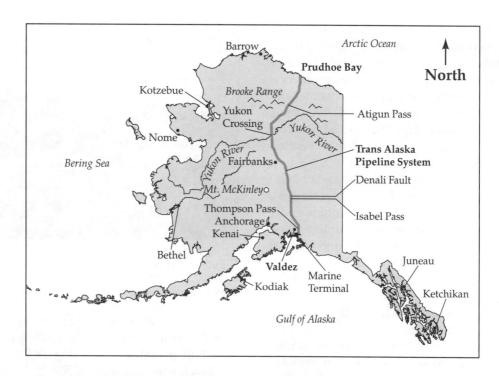

Figure 9.4 Prudhoe Bay, the Trans Alaska Pipeline System (TAPS), and the Port of Valdez.

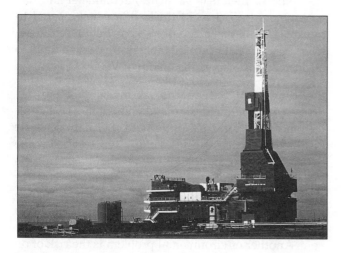

Figure 9.5 North Slope drilling rig.
(Reprinted courtesy of ARCO Alaska, Anchorage, Alaska.)

The Trans Alaska Pipeline Is Proposed

Once oil had been discovered on Alaska's North Slope, a method was needed for transporting it to the lower 48 states. While several plans were considered, the most promising one was to build a pipeline from Prudhoe Bay to the Port of Valdez 1,285 kilometers (≈ 800 miles) to the south on Prince William Sound (see Figure 9.4). Because the Port of Valdez is ice-free year-round, large oil tankers could transport the oil from there to U.S. Pacific coast refineries. This conduit

would come to be known as the Trans Alaska Pipeline System (TAPS).

At first it was feared that oil exploration, drilling operations, and construction of necessary roads, bridges, and airfields would adversely impact caribou herds, waterfowl, shore birds, and the other coastal-plain wildlife that Alaska's North Slope is known for (see Chapter 1). Indigenous people, scientists, and naturalists familiar with the region identified several environmental risks if major oil developments were allowed:

- Disturbance of tundra vegetation, critical habitat for caribou and nesting sites for shore and sea birds.

- Oil spills on the tundra, causing water pollution in streams, wetlands, and nearshore marine environments.

- Obstacles in caribou migration routes, which might force the animals to have more frequent encounters with predators.

Whenever a project is proposed that is federally funded and likely to affect the environment, provisions of the National Environmental Protection Act (NEPA) stipulate that an Environmental Impact Statement (EIS) be prepared. Thus, in March 1970, Walter Hickel, Secretary of the U.S. Department of the Interior, issued an eight-page EIS that declared the Alaska pipeline project would create no environmental problems. This

EIS was regarded by many as a flimsy attempt to rubber-stamp the project. In fact, some regarded it as a violation of NEPA itself. A court suit was joined against the Department of the Interior by The Wilderness Society, Friends of the Earth, and the Environmental Defense Fund. As a result, Federal Judge George Hart ruled that the Department of the Interior had not complied with NEPA and that a more complete and definitive EIS must be developed.

A second EIS was made public in January 1971. Although it was a much longer document (246 pages), it too was judged unacceptable because it did not adequately address important issues that naturalists, conservationists, environmentalists, state and federal officials, and the public at large had raised. Finally, a third EIS was written and published in March 1972, consisting of nine volumes (3,500 pages) and costing $12.7 million. The EIS required that the pipeline's northern half be elevated above the tundra to allow for animal migration beneath it and to protect the underlying permafrost. Based on this EIS, Congress approved construction of the pipeline, and President Richard Nixon signed the pipeline authorization bill on November 16, 1973. Work on the pipeline began in April 1974 and was completed in July 1977.

Constructing the Pipeline

Eight oil companies, including ARCO, Exxon, and British Petroleum (BP), formed the Alyeska Pipeline Service Company to undertake the massive across-Alaska pipeline construction. One daunting challenge was laying pipe over the Brooks Range (Figure 9.6). Another challenge was building the necessary **infrastructure,** the complex of materials, equipment, personnel, and facilities needed to construct, support, and operate the pipeline project. Here is a partial list of the TAPS infrastructure:

- A 579-kilometer (≈ 360-mile) gravel road, called the Haul Road, for building and servicing the northern half of the pipeline.
- 11 airfields and more than 15 access roads.
- 19 construction camps capable of housing 28,000 pipeline workers.
- Bridges across major waterways, including a 700-meter (≈ 2,300-foot) bridge across the Yukon River.
- Hundreds of oil wells in several different oil fields with feeder pipelines connecting production wells with the TAPS pipeline itself.
- A central electric power generating station.
- 10 oil pumping stations.

Figure 9.6 Constructing a section of the Trans Alaska Pipeline across the Brooks Range.
(Reprinted courtesy of ARCO Alaska, Anchorage, Alaska.)

- Oil tanker storage and loading facilities at the Port of Valdez, including 18 oil storage tanks, each with a capacity of 500,000 barrels of oil.

The TAPS pipeline is made of welded sections of steel pipe 1.2 meters (≈ 48 inches) in diameter and 1.59 centimeters (≈ 5/8 inch) thick. It was engineered to transport 2 million barrels of oil a day but presently delivers less. Three or four oil tankers are needed daily to off-load and transport the oil out of Valdez.

Environmental Safeguards A major problem addressed in pipeline design and construction was protection of the tundra's underlying permafrost. Near Prudhoe Bay, permafrost lies only a few feet below the tundra. The permafrost itself is about 600 meters (≈ 2,000 feet) deep. There, as well as at other sensitive sites, the pipeline lies 2.1 meters (≈ 6.9 feet) aboveground to allow caribou to roam freely under it and to protect the permafrost.

Another environmental problem is the risk of oil spills due to pipeline disruption by earthquakes, which are common in Alaska. To deal with this possibility, sections of the aboveground pipeline are built to shift vertically and horizontally on support structures to relieve stress (Figure 9.7). Engineering models predict that TAPS can withstand an earthquake measuring 8.5 on the Richter scale.

The time from the discovery of oil on the North Slope to the completion of TAPS was nine years, and the cost was close to $10 billion. The first tanker to carry North Slope oil, *ARCO Juneau*, left Valdez on August 1, 1977, and delivered its cargo to ARCO's Cherry Point refinery on Puget Sound on August 5. Alaskan oil was shipped only to U.S. markets in the Pacific Northwest, California, and the Gulf Coast until 1995

Figure 9.7 The Trans Alaska Pipeline System is engineered to minimize pipeline disruption and oil spills due to earthquakes. *(Reprinted courtesy of ARCO Alaska, Anchorage, Alaska.)*

when a 22-year-old ban prohibiting the export of North Slope oil to other countries was lifted. In 1999, approximately 20 percent of domestically produced petroleum originated from oil fields on Alaska's North Slope and was transported through the Trans Alaska Pipeline System.

Benefits to Alaska   North Slope oil development has proven to be of great economic value to Alaska and the rest of the United States. Construction of the pipeline and its infrastructure created a large number of well-paid jobs. Oil-drilling leases are yielding billions of dollars to the State of Alaska, funds that in turn can be used to build schools, roads, highways, clinics, and hospitals. In the 1990s, 85 percent or more of Alaska's total state revenues were derived from oil and natural gas royalties and production taxes. The fact that there is no state income tax in Alaska is largely due to the state's oil industry. In addition, the Alaska Permanent Fund, established by the state legislature in 1980, distributes annual dividends to Alaska citizens based on income earned from oil industry lease and tax arrangements. In

1999, every eligible Alaska man, woman, and child received a dividend check of $1,769.84.

However, despite these benefits and the many precautions that were taken to protect the environment, oil-industry-related accidents have occurred; one of the most significant is discussed in the following case study.

CASE STUDY: THE *EXXON VALDEZ* OIL SPILL

It was a typical night departure from the Port of Valdez as the supertanker *Exxon Valdez*, loaded with North Slope crude oil, left port. Three football fields long (≈ 300 meters or 1,000 feet), the tanker was headed for Long Beach, California. As the tanker entered Prince William Sound, icebergs were reported in the shipping lane. Captain Joseph Hazelwood made a change in course, turned on the automatic gyro pilot, and left his command post, placing a Third Officer in charge. Later, an agreed-upon change in direction was made, but not in time. Twenty-five miles from port, the *Exxon Valdez* hit submerged rocks just off Bligh Reef. It was 12:03 A.M. March 24, 1989.

Impacts of the Spill

In the collision, eight of the ship's 11 cargo tanks were ruptured, and about 260,000 barrels (≈ 11 million U.S. gallons) of crude oil spilled into Prince William Sound's pristine waters. Crude oil saturated and spoiled *intertidal* and *subtidal* habitats and damaged submerged eelgrass meadows, which are critical habitat for many species of finfish and shellfish.

> **Intertidal** habitats occupy coastal marine beach areas exposed between high tides and low tides.
>
> **Subtidal** habitats lie below low-tide levels.

The three days following the spill were calm. Massive amounts of oil spread slowly but remained near the tanker. Little was done to recover the spilled oil, largely because there was almost no oil-spill response equipment on hand. Then, on March 27, a major windstorm drove oil slicks southwest toward Smith, Naked, and Knight Islands. Soon the oiled areas included open waters, pristine beaches, and mouths of rivers in the Chugach National Forest, Kenai Fjords National Park, and a number of other protected parks and wildlife refuges. Shorelines as far as 600 miles southwest of Bligh Reef were damaged. In all, close to

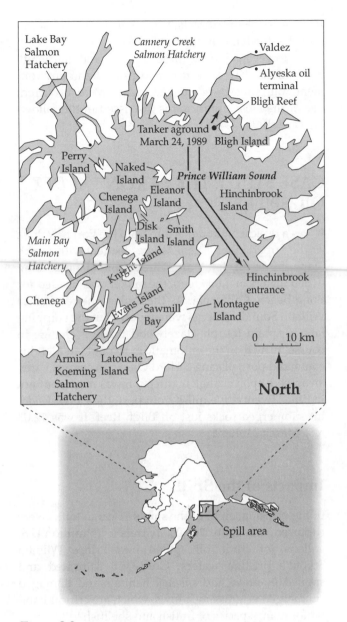

Figure 9.8   Prince William Sound and the site of the grounding of the supertanker *Exxon Valdez* on Bligh Reef.
(Source: Alaska Department of Environmental Conservation.)

1,500 miles of Alaskan shoreline were contaminated (Figure 9.8).

Effects on Wildlife

Unfortunately, the oil spill took place during a critical phase of Prince William Sound's biological year when migration and reproduction were in progress for many species of fish, birds, and mammals. Pacific herring were just beginning to spawn among intertidal and subtidal eelgrasses. Juvenile salmon were starting their migration toward ocean waters. Thousands of sea birds were returning to ancestral nesting sites. There were other concerns as well:

- Prince William Sound pink salmon hatcheries were at risk.
- The livelihood of commercial fishers and their families was in jeopardy.
- Native Alaskans were unsure whether their fish, shellfish, and wildlife harvests were safe to eat.
- The future of Alaskan subsistence cultures and lifestyles was in question.

The oil spill killed hundreds of marine mammals and thousands of birds in Prince William Sound. Some species were affected more than others. Estimates of wildlife killed include: 250,000 sea birds, 2,800 sea otters, 250 bald eagles, 300 harbor seals, 200 harlequin ducks, and possibly 13 orca whales. Bird species most affected included yellow-billed loons, horned grebes, Kittlitz's murrelets, marbled murrelets, and pigeon guillemots. Harlequin ducks suffered nearly complete reproductive failure between 1990 and 1992 as a result of the oil spill. The U.S. Fish and Wildlife Service Research Center in Anchorage estimated that it would take 20 to 70 years for some wildlife populations to recover.

Pink salmon are especially significant to commercial and subsistence fishers in Prince William Sound and are also critical organisms in several food webs. Pacific herring are also important, serving as a food source for sea lions, sea birds, humpback whales, and shorebirds migrating from Central America. Both species were seriously affected by the spill (Insight 9.2).

Many wildlife species escaped injury. Sitka blacktailed deer, though they ate kelp (seaweed) and other plants on oil-contaminated beaches, showed no significant levels of oil in their muscle or liver tissues. Brown bears known to have eaten bird carcasses on oiled beaches in Katmai National Park also did not appear to have suffered at all. Following the spill, native people living near Prince William Sound who depended on fish, shellfish, birds, and deer for their food and livelihood did not know if these food resources were safe to eat. Many abandoned traditional harvesting of fish and wildlife; their lifestyle and means for earning a living had to change.

An Oil Spill Health Task Force was established to test the safety of subsistence foods. Hundreds of samples of finfish, shellfish, and other marine organisms sent to the National Marine Fisheries Service's Northwest Fisheries Center in Seattle were analyzed for **polynuclear aromatic hydrocarbons (PAHs)**—indicators of oil contamination known to be **mutagenic** (capable of altering an organism's DNA). PAH levels were very low in fish because fish rid themselves of oil by breaking it down in their livers, concentrating it in their gall bladders, and excreting it in their bile. The

INSIGHT 9.2
Effects of the Spill on Fish

Pink salmon are consumed by humans, making them an important factor in the commercial fishing industry. They also serve as food for many birds, including gulls, kittiwakes, cormorants, loons, mergansers, murres, auklets, puffins, kingfishers, and dippers, and for such fish as cod, pollock, rockfish, sculpins, porpoises, sea lions, harbor seals, and orcas (killer whales). The main spawning sites for Prince William Sound pink salmon are the mouths of freshwater streams. Unfortunately, many of these sites became oiled following the *Exxon Valdez* incident. Crude oil is not very toxic to adult salmon, but it is toxic to salmon eggs. In the fall of 1989, the year of the spill, salmon egg mortality in oiled streams was 67 percent greater than in non-oiled streams. Mortality rates continued to increase in succeeding

years, and commercial fishing suffered greatly as a result. There is now evidence that high levels of egg mortality among wild pink salmon may be due to mutations in fish that hatched in oiled habitats.

Pacific herring are another important food source in aquatic ecosystems. Alaska's Department of Fish and Game found that the 1989 Pacific herring population, the largest since the early 1970s, produced the fewest offspring ever documented. That is, less than 50 percent of eggs collected from oiled waters hatched successfully. Although herring populations are known to oscillate in size naturally, there is a chance that the observed infertility is due to the oil spill. Further study is needed to learn whether herring will fully recover.

INSIGHT 9.3
Who Cleans Up?

In the United States, oil spill response rests at the federal level with the U.S. Coast Guard and the U.S. Environmental Protection Agency (EPA). The Coast Guard is responsible for spills in U.S. marine waters and the Great Lakes; the EPA has control over spills in inland waters and on land. Individual states may add regulations of their own. In Alaska, the Department of Environmental Conservation (DEC) has

such oversight. According to an oil spill contingency plan worked out by the Alyeska Pipeline Service Company in 1987 and agreed to by state and federal agencies, Alyeska would be responsible for cleaning up spills connected to the pipeline system. However, in the case of the *Exxon Valdez* spill, Exxon corporation assumed full responsibility for the spill.

process can cause injury to fish, but it prevents hydrocarbon buildup to dangerous levels, so it was decided that these fish were safe to eat. However, PAH levels were high in shellfish, and the task force advised community residents not to eat them.

Studies in Alaska showed that petroleum hydrocarbons bioaccumulate in some marine organisms following an oil spill. That is, they concentrate in certain tissues. But biomagnification, the cumulative buildup of a contaminant in a food chain or food web, did not occur.

Cleanup and Restoration

According to earlier agreements, the Alyeska Pipeline Service would be in charge of cleanup in the event of a Trans Alaska Pipeline System oil spill (Insight 9.3).

However, the confusion and shock that followed the *Exxon Valdez* spill met with little or no response by Alyeska. Furthermore, none of the several federal and state agencies that could have assumed responsibility for the cleanup took charge. Given this situation, the Exxon Corporation decided to take control of the cleanup just hours after the spill occurred and to accept responsibility for all cleanup costs.

If an oil slick is successfully dispersed, it will not pollute shallow coastal waters or wash up on beaches. Exxon's initial cleanup strategy was to use **chemical dispersants,** detergent-like substances that enable crude oil components to dissolve in water. But unfortunately, when crude oil components are dispersed, they become toxic to plankton, fish, and wildlife. Many Alaskan residents and fishers were concerned about Exxon's plans to use dispersants, and they were angry that they had not been consulted about handling the spill.

Regardless, very little chemical dispersant was available to Exxon during the days immediately after the spill, and almost no equipment was on hand to apply it. On March 26, Exxon applied 5,000 gallons of dispersant close to the *Exxon Valdez*, but a major windstorm occurred and the oil was driven into open waters of the sound. By the time the storm abated, the opportunity to effectively use dispersants had passed.

Other cleanup maneuvers involved using booms and skimmers. A **boom** is a containment barrier used to corral an oil slick within its perimeter (Figure 9.9). It is deployed using two boats. Local fishing vessel skippers soon became quite skilled at booming since it is a lot like handling fishing nets; booms proved successful in preventing oil from reaching the sound's pink salmon hatcheries. **Skimmers** were used to transfer corralled oil into containment tanks using conveyor belts and vacuum pumps. More than 260 skimmers were used in Prince William Sound during 1989. The oily debris collected by booms and skimmers was eventually landfilled. In early April, crude oil slicks broke up into windrows and patches, and much of the oil began to wash up on island shorelines and coastal beaches.

Except for the successful protection of the pink salmon hatcheries, most of the heroic efforts to save oiled wildlife in Prince William Sound failed. By the third week in April, crude oil was being deposited on many island and coastal beaches in the sound and in the Gulf of Alaska. Recurring tides periodically redistributed these deposits, making shoreline cleanup even more difficult.

After giving up on dispersants, Exxon developed a high-pressure spray technique using hot water at 70 °C (≈ 160 °F). This method was used to clean up 240 kilometers (≈ 150 miles) of oiled shoreline in 1989 (Figure 9.10). But a number of biologists believed that blasting shoreline areas with water that hot would kill microscopic flora and fauna, organisms that would have been capable of restoring oiled environments naturally. The hot-water blasting strategy became very controversial. Two years later, in April 1991, a National Oceanic and Atmospheric Agency (NOAA) study reported that spraying hot water on oiled beaches after the Prince William Sound spill had done more harm than good and had actually slowed the area's recovery.

Bioremediation   In June 1989, the EPA proposed a shoreline and beach cleanup strategy called **bioremediation,** an approach aimed at solving an environmental problem by working with nature. Bioremediation of an oil spill is based on the fact that some naturally occurring bacteria degrade (break down) hydrocarbons.

Figure 9.9   *Exxon Valdez* near Bligh Reef in Prince William Sound 20 days after the oil spill. Oil containment booms are visible, but by this time most of the spilled oil had spread extensively. *(Photo courtesy of the Oil Spill Information Center, Anchorage, Alaska.)*

Figure 9.10   Pressurized hot-water cleaning of Prince William Sound's oiled beaches following the *Exxon Valdez* spill. *(Photo courtesy of Oil Spill Public Information Center, Anchorage, Alaska.)*

They are called "oil-eaters" because they are able to use crude oil as an energy source. Marine studies showed that about 5 percent of the bacteria in Prince William Sound waters were oil-eating organisms. Their occurrence is natural because crude oil has been oozing into these waters from deep ocean "seeps" for millions of years. The bioremediation strategy was simple: Spray fertilizer on oiled beaches to stimulate bacterial growth. As total numbers of bacteria increase, numbers of oil-eating bacteria would also increase and the bacterially mediated cleanup process would be accelerated.

The EPA conducted a 90-day field test of the fertilizer-bioremediation strategy on oiled beaches at Snug Harbor on Knight Island. Within 30 days, tests showed 30 to 100 times more bacteria in treated areas than in untreated areas along with visible disappearance of oil in treated areas. With approval from the U.S. Coast Guard and Alaska's Department of

Environmental Conservation (DEC), Exxon adopted the bioremediation technique and applied fertilizers to a number of oil-contaminated sites.

Bioremediation became Exxon's principal technique for oil spill cleanup in Prince William Sound, even without extensive studies or evidence of its value. In June 1992, after studying the several cleanup techniques used, the EPA's Science Advisory Board reported that applying fertilizers on shoreline beaches had increased the numbers of hydrocarbon degraders. Some board members interpreted the data to show that cleanup occurred three to five times faster at fertilized sites compared to unfertilized sites. Other board members found no statistical difference in cleanup rates between the two types of sites. Thus, the results were inconclusive.

Legal Findings, Fines, and Compensation

The *Exxon Valdez* oil spill forced native Alaskans to suspend their subsistence harvests of fish, shellfish, and wildlife in the Gulf of Alaska because they feared adverse health effects from oil contamination. Commercial fishers also suffered losses. Before the spill, they could get $2.40 a pound for sockeye salmon; afterwards, they got only 70 cents a pound. A commercial fishing license that originally cost $300,000 was worth only $30,000 after the oil spill.

In 1991, after considerable legal wrangling, the U.S. District Court in Anchorage handed down a $1.05 billion penalty against Exxon, the largest penalty in environmental case history. The settlement included criminal penalties of $150 million and civil penalties of $900 million to be paid over a 10-year period. But it did not include claims from private parties—Alaskan natives, native corporations whose properties had been damaged, and commercial fishers whose losses were considerable. Those claims were considered later. The former *Exxon Valdez* captain was fined $5,000, and the following judgments were handed down against Exxon:

- $20 million to compensate native Alaskans for losses suffered in subsistence harvests of fish, shellfish, seals, and kelp.
- $287 million to compensate commercial fishers mostly for the decline in 1989 harvests and their market value.
- $5.2 billion in punitive damages to settle a class action suit by about 14,000 natives, fishers, and property owners.

Even before these penalties were handed down by the court in Anchorage, Exxon had voluntarily spent $2.2 billion in cleanup costs, $1.1 billion in state and federal litigation settlements, and $300,000 on various damage claims. A portion of these payments was credited toward the total penalties for which Exxon was liable. The penalties levied against Exxon were quite large considering that the corporation's annual profits in the 1990s were about $5 billion.

New Laws and Regulations

Prior to the *Exxon Valdez* disaster, efforts to pass effective oil spill legislation had gotten nowhere. For instance, the Oil Pollution Act of 1924 had attempted to deal with oil contamination in U.S. coastal waters, but was largely ineffective. However, oil pollution in Prince William Sound quickly changed state and federal attitudes, forcing the adoption of new, more effective laws and regulations in Alaska, including:

- Comprehensive planning for prevention of and response to oil spills.
- Expansion of the state's oil and hazardous substance emergency response fund from $1 million to $50 million.
- Increased liability and penalties for polluters.
- Authorization of Alaska's DEC to inspect oil tankers.
- Creation of volunteer response teams in coastal communities.

Federal adoption of the Oil Pollution Control Act of 1990 (OPCA) strengthened the national government's hand in dealing with oil spills. Through OPCA, the following policies were established:

1. Oil spills are dealt with through cooperative agreements involving the oil spiller, a state government, and federal agencies. The U.S. Coast Guard may decide to direct all on-site actions if it is found prudent to do so.
2. In the event of large, complex, or nationally significant oil spills, the U.S. Coast Guard will direct all on-site actions.
3. A $1 billion-per-spill pollution response fund will assist in cleaning up accidents.
4. Oil tanker masters, captains, and other vessel-operating license holders must undergo drug and alcohol testing.
5. By 2015, oil tankers operating in U.S. waters must have double hulls to minimize the impacts of shipping accidents. (Although U.S. merchant ships normally have double hulls, up to now most oil tankers have had only single hulls.)

In addition, OPCA increased the liability of oil owners and oil shippers so that oil companies can no longer hire other companies as oil shipping agents to escape liability. Determining when a cleanup operation is complete now rests with the federal government, specifically the U.S. Coast Guard. However, the Coast Guard must consult with an affected state before deciding to end cleanup operations. As with other federal environmental acts, individual states may set more strict regulations than those set forth in OPCA.

Prince William Sound Today

The waters of Prince William Sound appear pristine again, especially to visitors. But those who live and work near the sound say it has not fully recovered. Oil remains hidden beneath seashore rocks, under coastal sediments, and below beach sands. Hard asphalt-like patches can be found embedded along previously oiled seacoast areas. And although many species of wildlife have recovered, some populations still suffer damage. Since 1990, fish, deer, ducks, seals, and sea lions from oiled beaches—subsistence resources in native communities—have all been found safe to eat, but shellfish are still not safe. The decline in the numbers of native pink salmon persists, but hatchery salmon are abundant. Other species, mainly land mammals such as Sitka black-tailed deer and brown bears, remain largely unaffected by the spill.

The impact of the oil spill on many Alaskans has also been significant. Jobs, commercial fishing, subsistence harvesting, tourism, and recreation have still not returned to their levels before the spill. People's lives have been changed, and no one knows how long it will be before Prince William Sound is pristine once more.

The Future of Fossil Fuels

It is estimated that in 2000, about 86 percent of the world's energy depends on fossil fuels—petroleum, coal, and natural gas. Hydroelectric and nuclear power provides about 12 percent more, while alternative sources of energy, mainly solar, wind, geothermal, and biofuels (plant products), contribute close to 2 percent. Figure 9.11 summarizes projected year 2000 U.S. and world energy demands in terms of *quads.*

> 1 **quad** of energy is 10^{15} BTU (a million, billion British Thermal Units). This enormous quantity of energy is equivalent to the energy output of 42 large coal-burning power plants or nuclear reactors operating nonstop for one year.

> 1 **BTU** is the amount of energy needed to raise the temperature of one pound of water by one degree Fahrenheit. It represents a small quantity of energy, enough to power a 100 watt electric lamp for about 10 seconds.

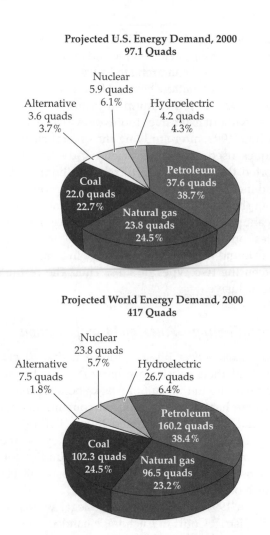

Figure 9.11 Projected Year 2000 U.S. and world energy demands. 1.0 quad = 10^{15} BTUs of energy. The category "alternative energy" includes solar, wind, biofuels, and geothermal energy.
(Source: Data from the Energy Information Agency, U.S. Department of Energy.)

Coal continues to be the world's most abundant fossil fuel and its use is likely to dominate energy production in some countries, especially China and, surprisingly, the United States where cheap Midwest electric power is still generated by burning coal. It is estimated that at current rates of consumption, coal will last for perhaps 400 years.

American Petroleum Institute data indicate that in 1998 approximately 76.5 million barrels of oil were being consumed worldwide daily. The institute projects that, given this demand and assuming the likelihood of future petroleum discoveries, there is a 95 percent probability that global oil supplies will last until 2056. However, two factors influence this calculation: First, world demand for oil is increasing yearly by about 1.8 million barrels per day. Second, proven oil reserves are also rising, making it impossible to predict with certainty when oil will run out.

Today, industrialized and developing nations alike depend on fossil fuels. China, for example, is aggressively mining its vast coal deposits to meet the country's rising energy needs. Worldwide demands for petroleum and natural gas are also rising. In recent decades, petroleum consumption has been increasing 2.2 percent a year, and due to air pollution problems, natural gas has emerged as the fossil fuel of choice in many of the world's cities. But fossil fuels are nonrenewable. Their future is limited to the discovery and extraction of economically recoverable deposits. Predicting when fossil fuels are likely to run out is difficult for two reasons:

1. Sophisticated exploration technologies aimed at discovering petroleum and natural gas are proving highly successful, and as a result, proven reserves are increasing.

2. Advanced extraction technologies make it economically more feasible to recover higher percentages of fossil fuels from proven reserves than were previously possible.

It seems inescapable that within the next several decades industrial nations will have to change their energy policies, particularly as fossil fuels become more costly and environmental problems connected with their recovery, transport, and use worsen. As this occurs, dependence on fossil fuels will decline, energy conservation practices will mature, and a transition to renewable, sustainable sources of energy will take place. Those who study energy policy issues believe that sooner or later solar and wind power will prove to be the most practical, cost-effective, and environmentally friendly energy sources.

REGIONAL PERSPECTIVES

Although the *Exxon Valdez* oil spill was a truly major event, similar occurrences have not been uncommon. In 1989 alone, the U.S. Coast Guard logged about 8,000 spills in U.S. waters—some small and some large. Worldwide, there have been several dramatic and costly crude oil accidents. For example, in 1967, the *Torrey Canyon* released 36 million gallons of oil near the coast of Great Britain. In 1978, the *Amoco Cadiz* spilled 60 million gallons of oil off the coast of France. And in 1993, the tanker *Braer* lost 26 million gallons of oil along the Shetland Islands near Scotland; the effects of that spill are described in the following section.

The Shetland Islands: Oil Spill from the Tanker *Braer*

During the morning of January 6, 1993, the American-owned, Liberian-registered oil tanker *Braer* lost power and ran aground near Britain's Shetland Islands. Hurricane force winds and heavy seas had caused the tanker's engines to fail, and without power, the tanker was driven into submerged rocks at Garth Ness, a peninsula near the southern tip of the islands.

When it struck the rocks, the *Braer* was enroute to Quebec, Canada, to deliver 26 million gallons of Norwegian crude oil. Oil began to spill immediately and continued discharging into the sea for the next eight days as the *Braer*, driven against the rocks by powerful winds and high seas, broke apart. Within hours of the grounding, brown floating oil washed ashore, killing large numbers of fish and marine invertebrates and threatening coastal wildlife as well.

Following the spill, Shags, a local resident sea bird species, and Black Guillemots that migrate to the area from the north suffered marked declines in the numbers of breeding birds. The cause is believed to be related to the demise of local fish populations, their main food source.

The ecological impacts and visible effects of the *Braer* oil spill virtually disappeared within months. Two reasons have been given for such limited impacts compared to those of the *Exxon Valdez* spill:

1. The Norwegian oil spilled by the *Braer*, a light crude oil containing mainly low-molecular-weight hydrocarbons, evaporated quickly in seawater, leaving crude oil components that biodegraded fairly rapidly. This behavior is quite different from that of Alaskan crude oil whose chemical composition favors little evaporation and prolonged, slow biodegradation (Insight 9.4).

2. The *Braer* spill occurred during severe weather— high winds, heavy seas, and great water turbulence. This physical action dispersed the oil within the seawater, diluting its biological impact and diminishing its accumulation on coastal beaches and shorelines.

INSIGHT 9.4
When Crude Oil and Water Mix

The overall fate of crude oil spilled on open waters is generally as follows:

- About one-third of typical crude oil consists of light, low-molecular hydrocarbons like n-pentane, n-hexane, and cyclohexane (see Figure 9.2). These materials evaporate within hours after a spill.

- Another one-third of the oil is made up of higher-molecular-weight materials capable of mixing with seawater to form a dark, frothy mousse. The mousse is an oil-water-air emulsion that coats sea birds, marine mammals, coastal beaches, rocky shorelines, submerged grasses, and seabed sediments.

- The last one-third of the oil consists of even higher-molecular-weight chemicals, which when mixed with seawater form tar balls that sink to the sea floor, covering benthic habitats. These materials break down only very slowly, especially in very cold arctic waters.

The Persian Gulf: An Oil Crisis and a War

The Persian Gulf, known also as the Arabian Gulf, is one of the most biologically productive marine environments in the world. Nutrient-rich waters flowing from the Tigris and Euphrates Rivers into the Gulf support biodiverse salt marshes, coral reefs, seagrasses, mangrove forests, finfish, shellfish, seabirds, and other wildlife.

The Persian Gulf is 917 kilometers (≈ 570 miles) long, 338 kilometers (≈ 210 miles) at its widest point, and has an average depth of about 36 meters (≈ 118 feet), making it a relatively shallow, warm, marine environment. It is semi-enclosed owing to the Strait of Hormuz lying to the south, which restricts water exchange with the Gulf of Oman.

The hostilities of the 1991 Gulf War resulted in the release of an estimated six million barrels (≈ 816 metric tonnes; ≈ 275 million gallons) of crude oil into Persian Gulf waters, making it the world's largest oil spill—25 times greater than the *Exxon Valdez* spill. Some of the oil came from damaged tankers anchored off Kuwait's coast, but most of it was discharged from Kuwait's off-shore Sea Island oil terminal.

In addition, more than 600 Kuwaiti oil wells were set ablaze, igniting more than 500 million barrels (≈ 68,000 metric tonnes; ≈ 22.9 billion gallons) of oil. The resulting pervasive atmospheric pollution of the gulf region consisted of combustion products, oil aerosols, hydrocarbons, heavy-metal particulates, and sulfur dioxide.

Persian Gulf oil pollution following the Gulf War contaminated numerous coastal and intertidal marine environments, particularly those along the Gulf's northwestern perimeter. Adversely affected living resources included clams, scallops, oysters, and several species of fish. Hardest hit of all was Saudi Arabia's multimillion dollar shrimp industry. In addition, oil coated and killed significant numbers of sea birds, including cormorants, grebes, and auks. Less damage occurred along the coastlines of Kuwait, Iraq, and Iran.

By 1993, two years after the war, most of the gulf's coral reefs, seagrasses, and mangrove vegetation had recovered. But the shrimp populations remained well below normal levels. The relatively rapid recovery of the gulf environment may be attributed to the fact that the Gulf is almost continually impacted by oil pollution. Its water quality is influenced by several factors: oil production, refining, and transport; oil from tanker ballast; and natural oil seepage from the Gulf's floor. Between 1978 and 1991, before the Gulf War, five major oil tanker spills occurred in the gulf—each one larger than the *Exxon Valdez* spill. It appears that marine systems such as the Persian Gulf can acclimate ecologically to the presence of crude oil.

Canada: The Sinking of the *Arrow* and the *Irving Whale*

On February 4, 1970, the Liberian tanker *Arrow,* under charter to the Imperial Oil Company and carrying 108,000 barrels (≈ 14.7 metric tonnes; ≈ 5 million gallons) of Bunker C fuel, was enroute to Nova Scotia. A major wind- and rainstorm forced the *Arrow* to run aground in Chedabucto Bay. The tanker's bow hit submerged rocks, and bunker oil began to spill from ruptured tanks. At first, prevailing winds propelled the oil to the north; then oil slicks were driven to the bay's south coast. A few days later, the *Arrow* split in two, releasing more bunker oil, which was transported by then prevailing winds to the north shores of Chedabucto Bay. On February 12, the tanker's stern sank, carrying an estimated one-third of the bunker oil with it to the bottom.

By the time the *Arrow* sank in Canadian waters, considerable environmental damage had occurred. Out of a total of 375 miles of bay shoreline, 190 miles had been oiled to a greater or lesser extent. Fishing, the fish-packing industry, sea birds, and the entire local marine ecosystem had been put at risk. The bay's frigid waters made it all but impossible to clean up Bunker C oil, a substance that is thick, viscous, and intractable at low temperatures.

A second oil spill occurred in Canadian waters that same year. On September 7, the barge *Irving Whale*, carrying 31.3 million barrels (≈ 4,270 metric tons; ≈ 1.5 billion gallons) of Bunker C oil and 6,800 liters of heating oil, sank in the Gulf of Saint Lawrence 60 kilometers (≈ 37 miles) northeast of North Point, Prince Edward Island (PEI). For days, oil spilled from the sunken barge, contaminating as much as 400 square kilometers (≈ 156 square miles) of open sea and oiling parts of PEI and Cape Breton Island.

The sinking of the oil tanker *Arrow* and the barge *Irving Whale* led Canadian authorities to establish the Department of the Environment in 1971. Within a few years, Canada had developed a policy framework to handle environmental emergencies such as oil spills. The basic elements of that framework are prevention, preparedness, response, and recovery. In addition, Canada has made major technical contributions in dealing with oil spills. Among these are techniques and equipment to cope with spills in the arctic region, the development of advanced skimmers and oil-tracking buoys, and the means to employ at-sea oil-burning technology.

KEY TERMS

bioremediation	mutagenic
boom	natural gas
BTU	oceanic crust
catalytic cracking	petroleum
catalytic reforming	polynuclear aromatic
chemical dispersants	hydrocarbons (PAHs)
continental crust	quad
hydrocarbons	skimmers
infrastructure	subtidal
intertidal	

DISCUSSION QUESTIONS

1. The Trans Alaska Pipeline and oil transport via supertankers has proven highly reliable. Except for the one major accident in Prince William Sound, an extraordinary safety record has been established. Should this one accident lead us to question present and future oil industry operations in Alaska?

2. NEPA requires environmental impact assessments for major projects having federal involvement to protect the public interest and the natural environment from serious harm. Comment on this statement.

3. In your opinion, who was really at fault in the *Exxon Valdez* accident? Support your position.

4. Assume that it is known scientifically that the best oil spill cleanup strategy is to not intervene at all. Would the public, including native Alaskans, commercial fishers, and fish and wildlife officials, have allowed Exxon to do nothing to clean up the Prince William Sound oil spill? Discuss your answer.

5. In your view, should we continue extensive, worldwide exploration and mining of fossil fuels before beginning a major transition to renewable energy sources? Explain your answer.

INDEPENDENT PROJECT

There have been many sizable oil tanker spills besides the *Exxon Valdez*. The *Amoco Cadiz* spilled 60 million gallons of oil off the coast of France in 1978. *Torrey Canyon* lost 36 million gallons off the coast of Great Britain in 1967. *Argo Merchant* spilled 7.5 million gallons close to Nantucket, Massachusetts, in 1976. Pick one of these major spills and report briefly on how the accident happened and what the ecological effects were. Are there any residual environmental impacts today?

INTERNET WEBSITES

Visit our website at http://www.mhhe.com/environmentalscience/ for specific resources and Internet links on the following topics:

Prince William Sound—An overview

Oil spill studies in Prince William Sound

Oil spill effects on Pacific herring

Status report on Prince William Sound

International Energy Outlook—U.S. Energy Information Agency

SUGGESTED READINGS

BP Exploration and ARCO Alaska, Inc. 1993. *Arctic oil: Energy for today and tomorrow.* Anchorage, AK.

Brown, Lester R., Hal Kane, and David Malin Roodman. 1994. *Vital signs: The trends that are shaping our future.* New York, London: Worldwatch Institute.

Brown, Tom. 1971. *Oil on ice: Alaskan wilderness at the crossroads.* San Francisco: Sierra Club.

Coates, Peter A. 1991. *The Trans-Alaska Pipeline controversy.* London; Cranbury, NJ: Lehigh University Press.

Kubasek, Nancy K., and Gary S. Silverman. 1994. *Environmental law.* Englewood Cliffs, NJ: Prentice Hall.

Maki, Alan W. 1991. The *Exxon Valdez* oil spill: Initial environmental impact assessment. *Environmental Science and Technology* 25(1):2416.

Oil Spill Public Information Center, Anchorage, Alaska. 1997. *Exxon Valdez* Oil Spill Restoration Web Site. Status Report. http://www. alaska.net/~ospic/

Piper, Ernest. 1993. *The* Exxon Valdez *oil spill: Final report, State of Alaska response.* Alaska Department of Environmental Conservation (June).

Smith, Zachary A. 1992. *The environmental policy paradox.* Englewood Cliffs, NJ: Prentice Hall.

Walker, D. A., et al. 1987. Cumulative impacts of oilfields on northern Alaska landscapes. *Science* 238:757–61.

Wallen, Lynn, ed. 1993. *Exxon Valdez* Oil Spill. *Alaska's Wilderness* 25(1):1–49.

Wolfe, Douglas A., et al. 1994. The fate of the oil spilled from the *Exxon Valdez. Environmental Science and Technology* 28(13):560A–68A.

CHAPTER 10

Nuclear Power

Photo courtesy of the Tennessee Valley Authority

Brothers and Sisters, I want to tell you this. The greatest thing on earth is to have the
love of God in your heart, and the next best thing is to have electricity in your house.

Testimony given in his church by a Tennessee
Valley farmer in the early 1940s after seeing his
house, barn, and smokehouse illuminated
by electric lighting for the first time

People depend on energy from many sources, including the sun, water, and fossil fuels. But at the beginning of the twentieth century, a new form of energy was discovered—the power of the atom. This awesome power was most dramatically displayed in 1945 when the United States dropped atomic bombs on the Japanese cities of Hiroshima and Nagasaki, bringing World War II to an end. Soon after that, the goal of developing atomic power for peaceful purposes emerged, and many countries began programs aimed at building nuclear reactors to generate electric power.

In this chapter, we will examine the nature of nuclear power, the operation of nuclear reactors, and the biological, social, and economic factors affecting nuclear power's future. We will also trace the history of the Tennessee Valley Authority in the United States as an example of commercial nuclear power development. In part, this is a success story that shows how nuclear power plants became major sources of electrical energy. But it is also a story of failure at times—failure to anticipate and manage the unique and potentially serious problems associated with nuclear power.

BACKGROUND

Looking Inside the Atom

The story of nuclear power begins with the structure of the **atom,** the basic building block of all matter,

including chemical elements and compounds. In the early 1900s, many scientists were studying the inner architecture of the atom, particularly its dense central core, or **nucleus.** They found that atomic nuclei consist of elementary particles, mainly **protons** and **neutrons.** Protons are positively charged; neutrons are comparable in mass to protons, but have no charge. The number of protons in an atom's nucleus is the element's **atomic number.** The sum of the protons and neutrons in the nucleus equals the atom's **mass number.**

Surrounding the atomic nucleus in orbitals are one or more **electrons,** each having a negative charge. The mass of an electron is extremely small, about 0.056 percent of the mass of a proton. But the number and geometry of orbital electrons is important because they determine an element's chemical properties.

For example, an oxygen atom has 8 protons in its nucleus, so its atomic number is 8. It also has 8 neutrons, so its mass number is 16. And since its nucleus has a charge of +8, a neutral oxygen atom has 8 orbital electrons (Figure 10.1a).

Isotopes

The situation becomes a bit more complicated when you consider **isotopes,** atoms that are different forms of the same chemical element. Isotopes have the same atomic number (the same number of protons in their nuclei) but different numbers of neutrons and therefore somewhat different physical properties. Chemical

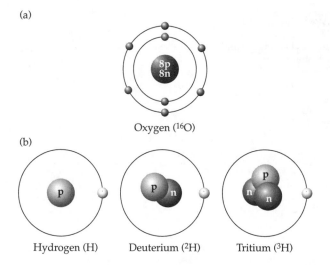

(a)

Oxygen (¹⁶O)

(b)

Hydrogen (H)   Deuterium (²H)   Tritium (³H)

Figure 10.1 Structure of the atom: (a) oxygen; (b) hydrogen and its isotopes.

elements such as hydrogen, oxygen, nitrogen, and uranium are actually families of closely related isotopes. For example, hydrogen, while it is one element, is made up of three different isotopes. The most common isotope, called ordinary hydrogen or hydrogen-1, has one proton in its nucleus. A second isotope, called deuterium or hydrogen-2, has one proton and one neutron in its nucleus, and a third isotope, called tritium or hydrogen-3, has one proton and two neutrons in its nucleus (Figure 10.1b). Each of these three isotopes has one positive charge in its nucleus and one electron in an orbital outside its nucleus.

While an element's various isotopes are virtually the same chemically, they differ in mass numbers due to the different numbers of neutrons in their nuclei. The atomic symbols of hydrogen's three isotopes are shown below. (According to the conventions of nuclear chemistry and physics used in the following expressions, the letter symbol stands for the name of the element (e.g., H = hydrogen), the subscript preceding it refers to its atomic number, and the superscript preceding it refers to its mass number.)

$$_1^1H \qquad _1^2H \qquad _1^3H$$

Ordinary hydrogen   Deuterium   Tritium

Uranium is another element for which isotopes exist. Natural sources of uranium contain close to 99.28 percent of uranium-238 and about 0.72 percent of uranium-235. A third isotope, uranium-234, exists but only to a very minor extent. The atomic symbols of uranium's three isotopes reflect each isotope's atomic number and mass number:

$$_{92}^{238}U \qquad _{92}^{235}U \qquad _{92}^{234}U$$

Uranium-238   Uranium-235   Uranium-234

The differing physical behavior of uranium's isotopes, particularly uranium-238 and uranium-235, is fundamental to the topic of nuclear power, as we will see in the following section.

The Discovery of Atomic Fission

Two German physicists, Otto Hahn and Fritz Strassmann, were studying the atomic structure of the element uranium in 1938 when they found that uranium exposed to neutrons emitted by certain radioactive elements was transformed in part into another chemical element, barium. At first, there appeared to be no explanation for this because scientific theory at that time could not account for one element being transformed into another element. However, the following year, Lise Meitner, a Vienna-born physicist, and her nephew, Otto Frisch, suggested that neutron bombardment of uranium caused one of its isotopes, uranium-235, to split into smaller atoms, a process they called **atomic fission.** This explained how elements such as barium (Ba) and krypton (Kr) were formed (Figure 10.2).

Further studies of atomic fission revealed that splitting atoms of uranium released very large quantities of energy—much more than could be explained by ordinary chemical reactions. It was shown that the collective mass of uranium's fission products, including barium and krypton, was theoretically less than the initial mass of uranium. Therefore, it appeared that physical matter had been transformed into an equivalent quantity of energy. The release of energy obeyed the German-born scientist Albert Einstein's now-famous equation, published in 1905:

$$E = mc^2$$

Einstein's equation connects energy (E) with matter (m) and provides a way to calculate how much energy is released when matter is transformed into energy. The multiplication factor, c^2, is the square of the speed of light. The speed of light, c, is 3.0×10^8 meters per second ($\approx$ 186,000 miles per second). The square of the speed of light is 9×10^{16} meters^2 per second^2—obviously a very large number. For example, using Einstein's equation, if 5 grams of matter (approximately the weight of a nickel) were converted into an equivalent amount of energy, it would supply all the electrical power needs of New York City for more than a year.

The discovery of atomic fission led Leo Szilard, a Hungarian-born physicist, to believe that a self-sustaining atomic **chain reaction** could be initiated

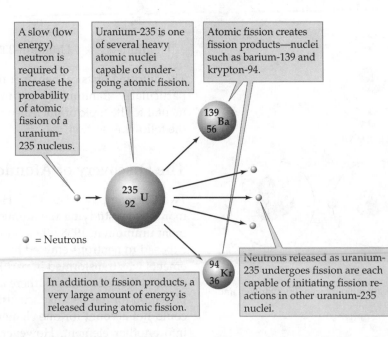

Figure 10.2   An example of atomic fission: A neutron splits a uranium-235 atom, creating two fission products, barium and krypton atoms, and three neutrons. A large amount of energy is released during this process.

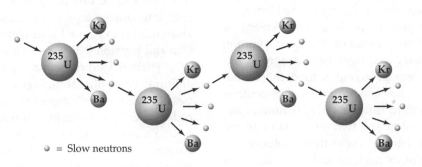

Figure 10.3   A nuclear chain reaction. A uranium-235 nucleus is split (nuclear fission) by a neutron into two smaller nuclei, in this case barium and krypton. Three neutrons are released, which split other uranium-235 nuclei. In an atomic bomb, chain reactions are uncontrolled. In a nuclear reactor, controlled chain reactions release heat energy to generate electric power.

using uranium or perhaps another heavy element. He calculated that relatively slow (low-energy) neutrons would be required (Figure 10.3). Other scientists reasoned that it might even be possible to make an atomic bomb. Coincidentally, this awesome possibility occurred at a time of great political upheaval in Europe. Fascism was on the rise, Adolf Hitler was beginning his reign of terror, and many citizens were fleeing Germany to seek refuge in other countries. Among them were Einstein and Szilard, both of whom came to the United States.

Realizing that Germany had the scientific and technical knowledge to engineer an atomic bomb, Szilard helped his friend Einstein draft a letter to President Franklin Roosevelt urging him to initiate a U.S. program to develop such a bomb before Germany could do so. Thus, in late 1939, Roosevelt authorized and

ordered the start of the highly secret Manhattan Project, which ultimately led to the development of atomic bombs. The first step was to build a reactor to show that a self-sustaining nuclear chain reaction was possible.

The First Nuclear Reactor

A team of scientists led by Enrico Fermi built the world's first nuclear reactor in 1942. It was assembled under the grandstands of the University of Chicago's football stadium, Stagg Field. The reactor consisted of an ordered pile of purified graphite blocks containing natural uranium and uranium oxide, not enriched with respect to any particular isotope. The graphite acted as a **moderator,** a substance that slows down neutrons in order to sustain a chain reaction. On December 2 of that year, a self-sustaining nuclear chain reaction was

INSIGHT 10.1
Nuclear Enrichment

Nuclear reactors can be built using purified natural uranium as a fuel, but nuclear fuels enriched to about 3 percent uranium-235 became common in the United States. Nuclear bombs, particularly those based on uranium-235, require enrichment of 90 percent or more.

Nuclear enrichment can be illustrated by an analogy. Consider a large box containing 10,000 marbles—9,928 yellow ones representing uranium-238 and 72 red ones representing uranium-235. The ratio of 9,928 to 72 closely reflects the naturally occurring abundance of these two isotopes. To enrich the marbles in favor of the red ones (representing uranium-235) to 90 percent, all but 8 of the yellow marbles (representing uranium-238) would have to be removed. This would leave 72 red marbles

(U-235) and 8 yellow marbles (U-238). The percent of red marbles would then be $72/80 \times 100$ percent = 90 percent.

Isotopes of any particular element are essentially identical chemically. A physical means is therefore needed to separate them. One of the enrichment methods used at Oak Ridge during the Manhattan Project was gaseous diffusion. Uranium hexafluoride (UF_6), a gas made from uranium ore, is forced to diffuse through a series of thin porous metal membranes. UF_6 molecules containing uranium-235 diffuse more rapidly than the heavier molecules containing uranium-238. While this is a slow and expensive process, isotopic enrichment can be carried out to any desired level.

---

achieved for the first time in history. The news was flashed to Manhattan Project leaders using the coded message:

> The Italian navigator [Enrico Fermi] has just landed in the new world. The natives are friendly.

Information obtained from Fermi's reactor showed that uranium-235 undergoes fission easily, but uranium-238, the element's most abundant isotope, does not. It was also determined that a uranium-based atomic bomb had to contain at least 90 percent uranium-235. Clearly, this is a much higher percentage than that occurring naturally in uranium ores. Therefore, a means had to be devised to enrich natural uranium in favor of its uranium-235 isotope. A gaseous diffusion, nuclear enrichment process was soon developed for this purpose at Oak Ridge, Tennessee (Insight 10.1).

The First Atomic Bombs

The success of Fermi's reactor led scientists to believe that making an atomic bomb was possible. Two different types of nuclear bombs were planned—one using uranium-235 and one using plutonium-239. Plutonium-239 is made in special reactors by bombarding uranium-238 with neutrons (Figure 10.4).

Uranium-235, prepared at Oak Ridge, was used in the first bomb assembled at Los Alamos, New Mexico. It was designed to explode when two pieces of enriched uranium-235 were driven together in a gun barrel using ordinary explosives thereby forming a

critical mass. Since plutonium-239 was expected to behave differently from uranium-235, a strategy was conceived in which a noncritical shell of plutonium-239 could be exploded into itself (implosion) using externally placed high explosives. The strategy was successfully tested at Trinity Site near Alamogordo, New Mexico, on July 16, 1945. On August 6, a uranium bomb was exploded over Hiroshima, Japan. At least 80,000 people were instantly annihilated. Three days later, a plutonium bomb was dropped on Nagasaki, Japan, destroying half the city and killing an estimated 60,000 people. Japan surrendered to the Allied forces on September 2, ending World War II.

Atoms for Peace

Once the enormous and awesome power of the atom had been demonstrated, there was a move to harness that power for peaceful purposes, particularly for generating electricity. In 1953, U.S. President Dwight Eisenhower delivered his Atoms for Peace speech before the United Nations. In part he said:

> Nuclear reactors will produce electricity so cheaply that it will not be necessary to meter it. The users will pay an annual fee and use as much electricity as they want. Atoms will provide a safe, clean, and dependable source of electricity.

At that time, there was no energy shortage, and energy costs were actually declining. But even though nuclear power plants to generate electricity were not

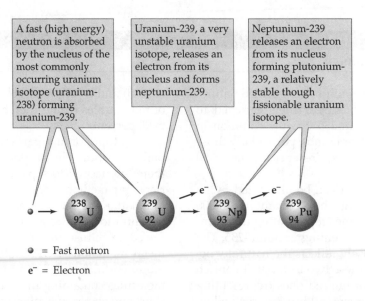

A fast (high energy) neutron is absorbed by the nucleus of the most commonly occurring uranium isotope (uranium-238) forming uranium-239.

Uranium-239, a very unstable uranium isotope, releases an electron from its nucleus and forms neptunium-239.

Neptunium-239 releases an electron from its nucleus forming plutonium-239, a relatively stable though fissionable uranium isotope.

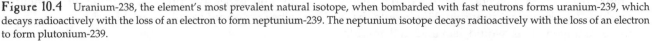

● = Fast neutron

e⁻ = Electron

Figure 10.4 Uranium-238, the element's most prevalent natural isotope, when bombarded with fast neutrons forms uranium-239, which decays radioactively with the loss of an electron to form neptunium-239. The neptunium isotope decays radioactively with the loss of an electron to form plutonium-239.

needed, Eisenhower urged their construction, believing that electric power from atomic energy would be cheap, ushering in unprecedented prosperity. He wanted the United States to be the world leader in developing peaceful uses for nuclear power.

The difficulty of building a nuclear power plant was daunting, and private industries were reluctant to accept the financial risks. But in 1954, a partnership between the U. S. government and Duquesne Light Company in Pittsburgh led to the construction of the first commercial nuclear power plant at Shippingport, Pennsylvania. Part of the plan was that turning to nuclear power would lessen the city's dependence on coal power and thereby diminish its legendary air pollution problems (see Chapter 5). Admiral Hyman Rickover, who had guided the construction of America's first nuclear-powered submarine, the U.S.S. *Nautilus*, was put in charge of the Shippingport project. He decided to employ the same type of reactor used in the submarine, a Westinghouse 60 MW pressurized water reactor (PWR). The Shippingport power plant, completed in 1957 at a cost of $75 million, successfully generated electric power for Pittsburgh until it was decommissioned in 1982. Duquesne Light contributed $20 million to its overall cost.

Types of Nuclear Reactors

Like coal-fired power plants, nuclear reactors are designed to produce heat, make steam to drive turbines, and generate electric power. The advantages and

Figure 10.5 A typical nuclear fuel pellet is about 9.5 millimeters long and 8.0 millimeters in diameter. It is equivalent in energy to 15,700 cubic feet of natural gas, 1,660 pounds of coal, or 157 gallons of gasoline.

disadvantages of both kinds of power plants are listed in Table 10.1. In U.S. nuclear reactors, the fuel is configured as small pellets of uranium dioxide (UO_2), each one about the size of an ordinary pencil eraser (Figure 10.5). The uranium, primarily U-238, is enriched to about 3 percent with uranium-235. Fuel pellets are stacked within **fuel rods** about 4.5 meters ($\approx$ 15 feet) long. The rods are bundled in groups of about 200 to form fuel assemblies. As many as 250 fuel assemblies, made up of approximately 50,000 fuel rods, are loaded into a typical nuclear reactor core. They are immersed in ordinary water, which serves as a neutron moderator and as a reactor coolant.

Two engineering designs dominate the U.S. nuclear power industry—**pressurized water reactors** (PWRs) and **boiling water reactors** (BWRs):

	Coal Power	Nuclear Power
Advantages	• Extensive worldwide supplies of coal exist • No high-level or low-level radioactive wastes result • Few construction delays and cost overruns • Safe, relatively cheap, reliable energy source	• Little or no air pollution • Little or no water pollution • Concentrated energy source • No dependence on fossil fuels at power plant site • No fly ash or bottom ash* disposal problems • Reliable energy source
Disadvantages	• Mining disturbs the land • Acid mine drainage • Coal-mining health issues (e.g., black lung disease) • Pollution: sulfur dioxide, carbon monoxide, and oxides of nitrogen • Atmospheric loading of carbon dioxide, adding to global warming • Fly and bottom ash disposal	• Construction delays and cost overruns • Expensive, redundant backup safety systems • Complex operating systems • Complex procedures to ensure plant safety • On-site storage of high-level radioactive wastes • No acceptable plan for a national high-level nuclear waste repository

TABLE 10.1   A Comparison of Coal Power and Nuclear Power

*Fly ash consists of very small solid particles released in the smoke of burning fuels such as coal. Bottom ash comprises heavier solid particles and fragments that fall to the bottom of a coal-burning furnace.

1. In a PWR, water is heated in a reactor core under pressures as high as 2,000 psi (pounds per square inch) and to temperatures as high as 300 °C (≈ 570 °F). The high pressure prevents the water from boiling. A heat exchanger transfers heat to a secondary loop outside the reactor core where water at atmospheric pressure boils, produces steam, drives a turbine, and generates electricity. Spent steam is condensed and recycled in the secondary water loop (Figure 10.6).

2. In a BWR, water surrounding fuel rods boils, producing steam that drives turbines and generators. Spent steam is condensed, using cooling water or a cooling tower, and then recycled.

The Design of a Nuclear Power Plant

A nuclear power plant may have one or more reactors. Each reactor has a steel reactor core where the fuel rods are positioned. A typical core is about 15 centimeters (≈ 6 inches) thick, 4.0 meters (≈ 13 feet) or more in diameter, and 14 meters (≈ 46 feet) high. A reactor core is surrounded by steel-reinforced concrete typically 1 to 2.5 meters (≈ 3.3 to ≈ 8.2 feet) thick. This is referred to as a **primary containment structure.** In addition, a concrete, dome-shaped **secondary containment structure** as high as 46 meters (≈ 150 feet) encloses each reactor's core, its concrete housing, and its steam

generator. Containment structures such as these are important nuclear power plant safety features.

Nuclear fission in a reactor is regulated by **control rods** positioned above or below a reactor's core. The rods are zirconium stainless steel tubes filled with boron or cadmium compounds capable of trapping neutrons. Control rods inserted into a reactor core trap neutrons, and nuclear fission is slowed down or stopped. Control rods are used to shut down reactors during maintenance or in case of an emergency.

These engineering concepts and safety features are incorporated into most nuclear power plants.

CASE STUDY: NUCLEAR POWER AND THE TENNESSEE VALLEY AUTHORITY

Creation of the TVA

When Franklin Roosevelt began his presidency in 1933, the United States was engulfed in worldwide economic depression. The New York Stock Market had crashed, and close to 14 million Americans were out of work. Many had lost their farms, businesses, and homes. Upon taking office, Roosevelt pledged a New Deal to Americans, giving them renewed hope for

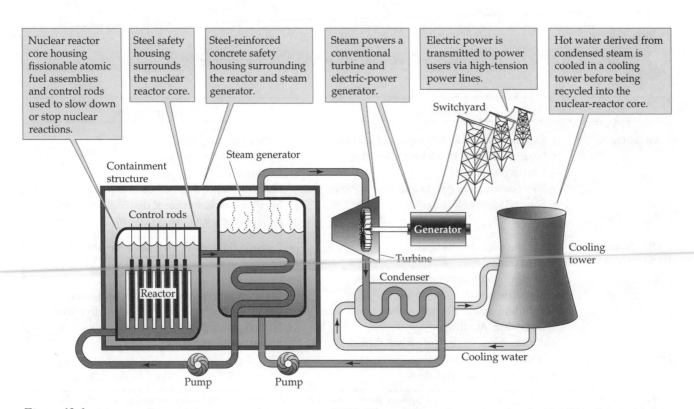

Nuclear reactor core housing fissionable atomic fuel assemblies and control rods used to slow down or stop nuclear reactions.

Steel safety housing surrounds the nuclear reactor core.

Steel-reinforced concrete safety housing surrounding the reactor and steam generator.

Steam powers a conventional turbine and electric-power generator.

Electric power is transmitted to power users via high-tension power lines.

Hot water derived from condensed steam is cooled in a cooling tower before being recycled into the nuclear-reactor core.

Figure 10.6 Schematic diagram of a pressurized water reactor (PWR). This is the type of reactor used in the *Nautilus* submarine and the Shippingport Nuclear Power Plant.

national recovery and prosperity. With the support of Congress, he moved quickly to begin solving the nation's problems.

One region that had been particularly hard hit by the Depression was the Tennessee Valley in the American Southeast, a region including most of Tennessee and parts of six other states: Alabama, Mississippi, Kentucky, Virginia, North Carolina, and Georgia (Figure 10.7). More than half of its three million people lived on farms, and only 3 out of every 100 farms had electric lights. To revitalize the economy of this area, Roosevelt proposed a new federal agency, the **Tennessee Valley Authority (TVA),** which became a reality when Congress passed the TVA Act in 1933. As an independent agency of the federal government, the TVA had several purposes:

- To enhance navigation on the Tennessee River by creating deep-water channels and to regulate floodwaters in the Tennessee and lower Mississippi Valleys. (The Tennessee River stretches 1,050 kilometers (≈ 652 miles) from Knoxville, Tennessee, to Paducah, Kentucky.)

- To foster economic development and improve the quality of life by building hydroelectric power dams, providing cheap electricity, and thus promoting rural and farm electrification.

- To help conserve the Tennessee Valley's natural resources through reforestation and the creation of parks and recreational areas.

The TVA proved to be one of the most important and enduring accomplishments of Roosevelt's New Deal. The dams that were built protected many areas from annual flooding, and the Tennessee River was made navigable from Knoxville to Paducah, Kentucky. But the single most important contribution of the TVA was the production of cheap electric power to electrify farms, homes, towns, and city streets and thus raise the people's standard of living and expectations for a better life. It all started with the building of hydroelectric power dams.

TVA Harnesses Hydroelectric Power

Generating and distributing electric power became the TVA's main purpose. David Lilienthal, one of the agency's first directors, believed that as long as demand for electric power increased, low electric rates would prevail. It turned out that he was right—at least for a while.

The first TVA power project was Norris Dam, a large concrete hydroelectric facility built on the Clinch River in northeastern Tennessee (Figure 10.8).

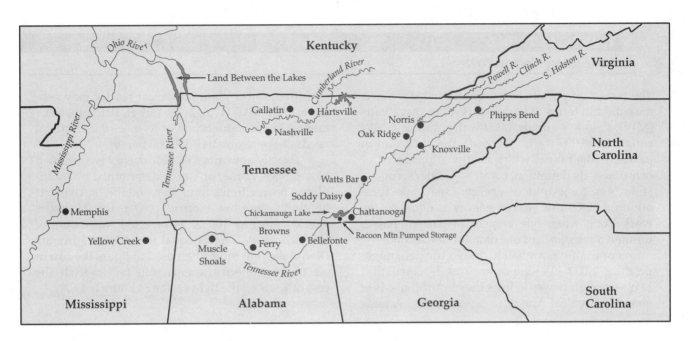

Figure 10.7   The region served by the Tennessee Valley Authority covers 80,000 square miles and includes close to eight million people in seven southeastern States.
(Source: Courtesy of the Tennessee Valley Authority.)

Figure 10.8   Norris Dam on the Clinch River in northeastern Tennessee, the first hydroelectric power dam built by TVA.
(Photo courtesy of the Tennessee Valley Authority.)

INSIGHT 10.2
Measuring Electric Power

The power output of electric generating systems is measured in terms of kilowatts (kW) or megawatts (MW). One kW equals 1,000 watts, and one MW equals 1,000,000 watts. One *watt*, the basic unit of power, is the rate at which energy is produced or consumed. By definition, 1 watt = 1 joule/second; a joule is a basic unit of energy. One joule (pronounced JOOL) of electric energy is equal to the work done when one ampere of current passes through a resistance of one ohm for one second.

For example, a small hydroelectric dam might produce 1,000 kW of power. This is equal to 1 MW—enough power to meet the electrical needs of about 330 typical American homes today. A large coal-fired power plant or nuclear reactor may generate 1,000 MW (1 gigawatt) of electric power or more. This is enough power to meet the electrical needs of approximately 330,000 homes.

Exactly how much electric energy is consumed by a device, home, or city is determined by multiplying power times time. For example, a 100-watt lightbulb drawing electric power for 24 hours would consume 100 × 24 or 2,400 watt-hours of electric energy. This is equal to 2.4 kilowatt-hours (Kwh), the unit most often used to figure the cost of electricity. If electric energy sells for 8¢/Kwh, the cost of keeping the light on for 24 hours is 19.2¢.

Construction began in 1933, but quickly led to controversy when it was realized that the dam would create a 13,800-hectare (≈ 34,000-acre) reservoir (Norris Lake) behind it. This meant that private property would have to be condemned and 2,899 families forced to relocate. Even cemetery grave sites had to be moved. In the end, an entirely new town called Norris was built near the dam with new homes, roads, a school, a post office, parks, a marina, and a regional forest.

Norris Dam was completed in 1936 at a cost of $32.3 million. In terms of year 2000 dollars, that cost would be more than $250 million. The dam still operates today, generating 100,000 kilowatts (kW) of electric power as water impounded behind the dam is channeled through two turbines in the dam's powerhouse. This is equivalent to 100 megawatts (MW) of power—a considerable amount in the 1930s but only 10 percent of what a large coal-fired power plant or nuclear reactor produces today (Insight 10.2).

Producing cheap electric power was the TVA's main mission, and before long electric lighting had replaced candles and oil lamps in many Tennessee Valley barns and homes. Soon farm families acquired radios, refrigerators, washing machines, and electric stoves. Community freezers stored farm produce, vegetables, and meats. Electrified farming equipment appeared—water pumps, incubators, brooders, and grain grinders—fostering increased productivity, greater purchasing power, and economic growth. Norris Dam and several other hydroelectric facilities that were built contributed to the betterment of the area in other ways: Dams store water behind them, creating reservoirs for drinking water, crop irrigation, boating, fishing, water skiing, swimming, and camping. Reservoirs also conserve water and enhance fish and wildlife.

During World War II, TVA hydroelectric power played a number of vital wartime roles. It supplied a considerable amount of electricity to the nearby Aluminum Company of America (ALCOA) for the production of aluminum, a strategic material for constructing military aircraft. Other defense industries in the valley included manufacturers and processors of metals, foods, fibers, timber, and chemicals. Also, though few knew it at the time, TVA hydroelectric power was used to carry out the highly secret Manhattan Project at Oak Ridge, Tennessee. By war's end, the TVA had become the largest public power producer in the United States. It still is.

TVA Turns to Coal Power

Toward the end of the 1940s, the Tennessee Valley's major hydroelectric power sites had been tapped. But the economy continued to expand, and the TVA had to consider new sources of electric power to meet rising demands, both domestic and industrial. Power demands were also escalating due to the Cold War, the confrontation between the United States and the Soviet Union following World War II. The widely differing and incompatible global agendas of the two nations triggered massive production and stockpiling of atomic bombs and intercontinental ballistic missiles, highly energy intensive military operations.

The need for more electricity forced the TVA to turn to coal to produce more power. This was a logical choice because the Tennessee Valley and its

Figure 10.9 The TVA's coal-fired electric power generating steam plant at Gallatin, Tennessee.
(Photo courtesy of the Tennessee Valley Authority.)

surrounding region are rich in coal deposits. The TVA's first coal-powered plant was built at Watts Bar, Tennessee, in 1949 (see Figure 10.7). Within 10 years, eight coal-fired power plants had been built. They generated 70 percent of the TVA's total power supply, while hydroelectric dams provided the rest. By then, the name TVA had become synonymous with electric power. The power plant on the Cumberland River at Gallatin, Tennessee, is typical of TVA's coal-fired facilities (Figure 10.9).

As rising demands for electricity continued, three more coal-fired power plants were constructed. TVA electric power continued to be cheap—about 0.9¢ per kilowatt hour—while the national average at that time was 2.2¢ per kilowatt hour. In the 1960s, more than two million customers were buying TVA electricity, one-third of Tennessee Valley homes were heated electrically, and residential electric use was double that of the rest of the nation. The valley's pride and optimism were reflected in the words of President John F. Kennedy during a TVA visit in May of 1963:

. . . the work of TVA will never be over. There will always be new frontiers for it to conquer. For in the minds of men the world over, the initials TVA stand for progress, and the people of this area are not afraid of progress.

However, by the early 1960s, visible damage and gaping scars could be seen throughout the valley wherever coal was being mined. The TVA had become one of the largest users of coal in North America, and strip mining was ripping the landscape apart. Environmental awareness in America was beginning, and the TVA, like other major utilities, could not escape growing criticism of its massive use of coal and the environmental impacts it was having. Some believed coal mining and power plant smoke were symbols of progress and prosperity, but others demanded new laws and regulations to protect the environment from developments like TVA coal power.

The Clean Air Act of 1963 was the first major air pollution legislation in the United States, and it established strict air quality standards and regulations. The TVA recognized that these regulations would require switching to more expensive, low-sulfur coal or installing costly smoke-abatement equipment. Because its principal goal was to produce cheap electric power, the agency began to consider nuclear power as a more economical way to produce electricity than burning coal.

TVA Goes Nuclear

In 1965, the TVA was producing most of its electricity using 11 coal-fired power plants. More than 22 million tons of coal were burned yearly to operate these plants. Clearly, the availability of cheap coal was critical to the Authority's operations. But Tennessee Valley coal and independent coal producers were being bought out by large oil companies, creating the expectation of higher coal prices. At the same time, demands for electric power were increasing by 7 percent per year, which meant that within 10 years, requirements for electric power in the valley could double.

The TVA started to investigate future power options. In 1966, proposals were requested from manufacturers of nuclear and coal-powered electric generating systems detailing plans and cost estimates to build new power plants. Proposals to build nuclear plants turned out to be cheaper than building coal-burning power plants. Sixteen nuclear plants had already been built in the United States and were proving to be reliable power sources. As a result, the TVA accepted a proposal from the General Electric Company to construct a nuclear power plant at Browns Ferry in northern Alabama.

a)   b)

Figure 10.10 The Browns Ferry Nuclear Power Plant on the banks of the Tennessee River in northern Alabama. (*a*) The large building near the center of the photo is the plant's powerhouse with its three reactors. Reactor cooling is accomplished by low-profile cooling towers in the left foreground. In-plant gases requiring release to the air are charcoal filtered and discharged through the slender 600-feet-high tower to the right of the powerhouse. (*b*) Refueling operations.
(Photos courtesy of the Tennessee Valley Authority.)

Browns Ferry Nuclear Power Plant   The construction of two boiling-water nuclear reactors (BWRs) at Browns Ferry began in 1967. Within a year, it was decided to build a third BWR to meet projected power needs at the Atomic Energy Commission's Oak Ridge facilities. Each of the three reactors was designed to generate 1,070 MW of power. As anticipated, coal prices in the valley escalated from $4/ton to $33/ton, but the cost of nuclear fuel remained low. Consequently, Browns Ferry electric power, which first became available in 1974, was relatively cheap. When all three reactors were on line in 1976, they generated close to 3,200 MW of power, making Browns Ferry the world's largest nuclear power plant at that time (Figure 10.10).

However, by the time Reactor 3 was completed, lengthy construction delays and serious cost overruns had been encountered. The initial estimate for building Browns Ferry had been $900 million; the final cost was triple that amount. Thus, the TVA discovered that constructing nuclear plants could be far costlier than building coal plants.

Besides the escalated construction costs, a number of serious problems occurred at Browns Ferry. In 1975, workmen using a candle to locate air leaks near electric cables under the reactor's control room accidentally set fire to nearby foam insulation. The fire spread, and Reactors 1 and 2 were severely damaged. The power plant was shut down for months, and repairing the electrical controls and other systems cost the TVA $100 million. This incident was only one of many that plagued the TVA's nuclear program. In

1982, the Nuclear Regulatory Commission (NRC) reported that safety devices designed to shut down Browns Ferry during an emergency might not work. In mid-1983, the NRC cited the TVA for 82 operational violations within an 18-month period. The plant received the worst possible rating in five of ten categories. Regulatory problems forced the TVA to shut down all three Browns Ferry reactors in 1985. Only after considerable effort and costly repair work, did two of these reactors later become operational.

Sequoyah Nuclear Power Plant   Construction of the TVA's second nuclear power plant, Sequoyah, began in 1969 at Soddy Daisy, on the banks of Chickamauga Lake, a stretch of the Tennessee River 29 kilometers (≈ 18 miles) north of Chattanooga, Tennessee (Figure 10.11). Sequoyah Reactor 1 became operational in 1981; Reactor 2 followed in 1982. Each is a Westinghouse pressurized water reactor (PWR) capable of developing 1,150 MW of power.

As with the Browns Ferry project, cost overruns plagued the construction of the Sequoyah power plant, with the final costs amounting to three times the initial estimates. Also, the frequency of reactor safety violations at Browns Ferry and Sequoyah was four times that of other U.S. nuclear plants. As more and more problems were identified, the NRC cited the TVA for ineffective management of its entire nuclear program. Ultimately, Sequoyah's two reactors had to be shut down.

The TVA's nuclear program was not the only one to experience regulatory problems; others faced similar

TABLE 10.2   TVA's Electric Power Generating Facilities, 1999

Power Facility	Total Number	Total Power	Percent of Total Power
Hydroelectric dams	29	3,554 MW	10%
Coal-fired plants	11	15,003 MW	60%
Nuclear power reactors	5	5,620 MW	28%
Combustion turbines	48	2,384 MW	1%
Pumped storage	1	1,532 MW	1%
All facilities	94	28,093 MW	100%

SOURCE: Data from the Tennessee Valley Authority.

Figure 10.11 Sequoyah Nuclear Power Plant. The plant's two reactors appear to the right of center, while its two cooling towers are left of center.
(Photo courtesy of the Tennessee Valley Authority.)

difficulties. Following the shutdown of Sequoyah, NRC adopted new power plant construction standards and mandated advanced education and training of nuclear plant operators. Although it was difficult and extremely costly, the TVA complied fully with the NRC. The startup of Sequoyah's Reactors 1 and 2 was approved in 1988, and both are operational today. Browns Ferry Reactor 2 was approved for startup in 1991, and Reactor 3 was approved in 1995. Browns Ferry 1 has been defueled and is in what is called "long-term layup." Its future status has been deferred indefinitely.

Watts Bar Nuclear Power Plant   The TVA's third nuclear plant was built at Watts Bar, which lies halfway between Knoxville and Chattanooga, Tennessee. Work

on two Westinghouse PWRs, each rated at 1,270 MW, began in 1973, but the project faltered when falsified x-ray pipe-welding records were discovered. In 1979, construction at Watts Bar became the object of intense scrutiny due to the Three Mile Island nuclear plant accident earlier that year. These incidents, coupled with less demand for electric power than had been anticipated, led the TVA to complete only Reactor 1 at Watts Bar; its startup occurred in 1996. Reactor 2 has been deferred indefinitely.

TVA Power Facilities Today

The TVA's blueprint to meet future power needs had originally included 17 nuclear reactors at seven power plants. In addition to reactors at Browns Ferry, Sequoyah, and Watts Bar, nuclear facilities were to be built at Bellefonte, Alabama; Hartsville, Tennessee; Phipps Bend, Tennessee; and Yellow Creek, Mississippi. Construction began at some of these sites, but in the end 10 of the planned reactors were deferred or canceled.

The TVA's nuclear program today consists of five reactors at three power plants: Browns Ferry, Sequoyah, and Watts Bar. In 1999, these facilities accounted for 28 percent of the Authority's total power generation. In addition to its hydroelectric, coal-powered, and nuclear plants, the TVA also built 48 natural gas and oil-fired combustion turbines as well as a pumped storage facility. The overall picture of TVA power generation in 1999 is shown in Table 10.2.

The TVA's coal-fired and nuclear power plants meet the area's everyday, minimum power demands, which are known as *base loads*. Its hydroelectric dams, combustion turbines, and pumped storage facility are used for greater-than-normal demands called *peaking loads*. These two types of power demand are described in Insight 10.3.

Demands for electric power vary by time of day, day of week, and season. **Base loads** are minimum power levels needed day and night to meet normally expected urban, suburban, and industrial electrical requirements. Most electric utilities meet base loads using coal-fired power or nuclear power if it is available.

However, periodic demands for more power, called **peaking loads,** also occur. They are due to summer air conditioning, heating during cold weather, and evening-hour cooking. TVA peaking loads are met using hydroelectric power, which can be cycled into operation as needed. Exceptionally high peaking demands are met by firing up the Authority's combustion turbines. The pumped storage system on Raccoon Mountain can also be used to satisfy TVA peaking demands.

Dealing with the Dangers of Nuclear Power

High-energy radiation can produce harmful biological effects in living organisms (Insight 10.4). Intense exposure, even for short periods of time, can cause radiation sickness, which sometimes results in death. But it can take awhile for some effects to show up. For example, by 1996, an estimated 187,000 people had died in Hiroshima due to the explosion, radiation, burns, and subsequent infections. Even moderate exposure can disrupt basic cellular-level functions, trigger tumor growths and cancer, and cause genetic damage to the DNA in cell nuclei. As our knowledge of these effects has grown, so have regulations by the U.S. government regarding not only the operation of nuclear power plants, but also the safe disposal of nuclear wastes.

Regulating Atomic Energy

At the end of the Manhattan Project in 1946, the U.S. government established the Atomic Energy Commission (AEC) to oversee all nuclear activities, including the production and management of military-related nuclear materials. In 1974, the Nuclear Regulatory Commission (NRC) took over AEC responsibilities to assure that civilian uses of nuclear power would be carried out safely to protect public health, national security, and the environment. The NRC controls the licensing of U.S. power plant reactors as well as all commercial, industrial, academic, and medical uses of nuclear materials. The commission also regulates the day-to-day operation of U.S. nuclear power plants as well as the transport, storage, and disposal of nuclear wastes.

Short- and Long-Term Storage of Radioactive Materials

Nuclear wastes generated at military facilities and nuclear power plants contain radioactive materials that release harmful levels of high-energy radiation. Attempting to ensure their safe disposal has created monumental problems. It is easy to see why when you consider that a high-energy radioactive waste material must be isolated from the environment for a time period equal to 20 times its *half-life* (see Insight 10.4). For example, strontium-90 has a half-life of 29 years (that is, it takes 29 years for one-half of its potential radiation to be released). So nuclear wastes containing strontium-90 must be safely stored for at least 580 years (20 × 29 years). It was hoped that a national repository for *high-level radioactive wastes* from nuclear plants would have been built by now, but an acceptable site has not yet been approved (Insight 10.5).

Notable Nuclear Accidents

Despite safety features and government regulations, nuclear accidents can, and have, occurred. Among the most important and defining events in the history of nuclear power were accidents that occurred at Three Mile Island in the United States in 1979 and at Chernobyl in Ukraine in 1986. Although scientists, policymakers, and government officials are divided in their analyses of the causes, these events led to a decline in public confidence in the safety of nuclear-reactor-generated electric power.

Three Mile Island   Three Mile Island (TMI) is a two-reactor nuclear power plant 16 kilometers (≈ 10 miles) southeast of Harrisburg, Pennsylvania. On March 28, 1979, Reactor 2 was operating normally while Reactor 1 was shut down for refueling and maintenance. At 4 A.M., Reactor 2 cooling water unexpectedly shut down, precipitating a loss-of-coolant accident (LOCA). This is a serious accident that can cause fuel rod temperatures to rise as high as 2,700 °C (≈ 4,900 °F) and melt down.

Reactor 2 was fitted with backup water pumps, but a relief valve unexpectedly jammed, allowing

INSIGHT 10.4
The Dangers of Radioactivity

One of the earliest insights into the biologically damaging effects of radioactivity occurred in the 1920s. Radium, a chemical element discovered by Marie Curie in 1898, was used to make a phosphorescent paint that glowed in the dark. Factory workers, mostly women, applied the paint to the hands and numerals on watch and clock dials so they would glow at night. To paint precisely, the workers wet their paintbrush tips with their tongues and then applied the paint, ingesting small amounts of radium as they repeated the procedure. Many of these workers later died of bone tumors and leukemia due to radium exposure. (Radium is no longer used to make objects glow in the dark; other, safer techniques have been discovered.)

Radioactive elements like radium are composed of unstable atoms—atoms that decay (break down) and release high-energy radiation, mainly alpha particles, beta rays, and gamma rays:

- *Alpha particles* are positively charged, high-energy nuclei of helium atoms. They do not penetrate matter deeply, but can cause significant damage to living tissues. This is the principal kind of radiation released by radium.

- *Beta rays* are negatively charged, high-energy electrons. They possess moderate penetrating power, can cause skin burns, and are severely damaging if released by radioactive contaminants within an organism.

- *Gamma rays* consist of high-energy radiation similar to X rays. They can penetrate several inches of lead and are capable of causing considerable biological damage.

TABLE 10.A   Typical Radioactive Fission Products in Spent Nuclear Reactor Fuel Rods

Fission Product	Half-Life	Type of Radiation
Iodine-131	8.0 days	Beta rays
Barium-140	12.8 days	Beta rays
Cesium-134	2.1 years	Beta rays
Cobalt-60	5.3 years	Beta rays
Krypton-85	10.7 years	Beta rays
Strontium-90	29.1 years	Beta rays
Cesium-137	30.2 years	Beta rays
Radium-226	1,599 years	Alpha particles
Plutonium-239	24,110 years	Alpha particles

SOURCE: Data from Canadian Coalition for Nuclear Responsibility, Montréal, QC, Canada; *CRC Handbook of Chemistry and Physics.*

The biological impact of a radioactive substance is determined in part by the kind of radiation it releases and by its **half-life,** the time it takes for one-half of its total potential radiation to be released. Examples of radioactive products resulting from nuclear fission, their half-lives, and the types of radiation they emit are given in Table 10.A.

Iodine-131 illustrates the harm a radioactive substance can inflict on humans. While its half-life is only eight days, it can be quickly concentrated in an exposed person's thyroid gland. Here, the emission of beta particles causes thyroid cancer.

---

water to drain out of the reactor core. A second set of backup water pumps activated automatically, but faulty control room instrument gauges led plant operators to turn the pumps off. Within minutes, what little water remained in the reactor core boiled off, and between one-third and one-half of the core's fuel rods melted. TMI safety systems had failed, partly due to defective instrument gauges and partly due to human error.

During the TMI meltdown, hydrogen gas was detected in the reactor's secondary containment structure, the reinforced concrete dome built over the facility. Hydrogen is not normally produced in nuclear

reactors, and its presence in this case indicated that something unforeseen had occurred. The possibility of a chemical explosion existed, an event that could release significant radioactive material to the environment. The governor of Pennsylvania declared a state of emergency and recommended that pregnant women and children evacuate the area immediately. At least 140,000 did so. No explosion occurred, and no immediate health effects were documented. However, recent research at the University of North Carolina suggests that rates of lung cancer and leukemia in residents downwind of the TMI power plant are two to ten times greater than in those living upwind.

INSIGHT 10.5

Storing High-Level and Low-Level Radioactive Wastes

Radioactive wastes produced at military sites, research facilities, medical centers, and nuclear power plants are classified as either high-level or low-level wastes. **High-level wastes** are materials that release significant quantities of potentially damaging radiation; spent fuel rods from nuclear power reactors are an example. **Low-level wastes** are materials contaminated with small quantities of radioactive residues; this category includes tools from reactor facilities and radioactive trash from hospitals and research laboratories.

High-Level Radioactive Waste Storage

The 1982 Nuclear Waste Policy Act authorized the Department of Energy (DOE) to identify possible sites for the safe disposal and permanent storage of high-level radioactive wastes, mainly spent fuel from commercial nuclear power plants. The act called for waste disposal in deep underground repositories. After five years of little progress, Congress directed the DOE to evaluate a site at Yucca Mountain, Nevada, as a national high-level waste repository. The Yucca Mountain site, about 160 kilometers ($\approx$ 100 miles) northwest of Las Vegas, is an arid, sparsely populated, isolated region whose subsurface is composed of hard volcanic rock called tuff. The plan is for radioactive wastes sealed in canisters to be buried in storage shafts about 300 meters ($\approx$ 1,000 feet) underground. The Yucca Mountain site is currently under study and may not be ready until 2010. Meanwhile, the NRC is developing radiation protection standards under which the facility will be licensed. The DOE would be responsible for repository operations.

Currently, with no national repository in existence to receive and store radioactive wastes from nuclear power plants, each plant's spent fuel rod assemblies are stored on-site in steel-lined concrete pools filled with water. More than 15 million fuel rods are currently stockpiled in this manner, with an additional one million being added each year and power plants reaching capacity. The NRC has the final word on how such nuclear wastes are stored.

Low-Level Radioactive Waste Storage

Low-level radioactive wastes used to be stored at several U.S. federal sites until 1970 when a number of private companies developed commercial waste facilities. By 1980, due to public pressure and unforeseen costs, only three of these facilities remained operational—in Nevada, Washington, and South Carolina. The Low-Level Radioactive Waste Policy Act of 1980 made each state responsible for its own low-level wastes (except for wastes generated at federal facilities). States have therefore been directed to build their own low-level storage sites.

By 1985, no new low-level nuclear waste facilities had been built. Congress established a more strict timetable and encouraged states to enter into compacts to build regional waste sites. For example, Pennsylvania, Maryland, and West Virginia joined to create the Appalachian Compact, with Pennsylvania agreeing to construct a low-level waste site. Currently there are only two low-level radioactive waste facilities in the U.S.—at Richland, Washington, and Barnwell, South Carolina. Richland is open only to states that are members of the Rocky Mountain Compact (Arizona, Colorado, Nevada, New Mexico, Utah, and Wyoming) or the Northwest Compact (Alaska, Hawaii, Idaho, Montana, Oregon, and Washington). Barnwell, which first opened in 1971, accepts wastes from most other states. It has seven to ten years' storage capacity left.

TMI Reactor 1 is still operational, but Reactor 2 is a total loss. Cleanup costs totaled about $1 billion.

Chernobyl   The Chernobyl Power Complex, which lies 130 kilometers ($\approx$ 81 miles) north of Kiev, Ukraine, included four nuclear reactors of the RBMK-1000 design, reflecting first-generation Soviet-designed nuclear power technology (Figure 10.12). On April 25, 1986, Reactor 4, rated at 1,000 MW, was to be shut down for routine maintenance. Prior to shutdown, it was decided to conduct a test to see whether the

reactor, operating at reduced power, could generate enough energy to run nuclear-core water-circulating pumps and related emergency equipment. No advance warning was given to plant operators that a test was going to be carried out. As a result, adequate safety protocols were not in place, and plant personnel were not alerted to potential dangers.

The test was initiated by partly removing some of the reactor's 211 control rods and shutting off the emergency backup cooling system. Power levels were to be reduced to 300 MW, but an operational error

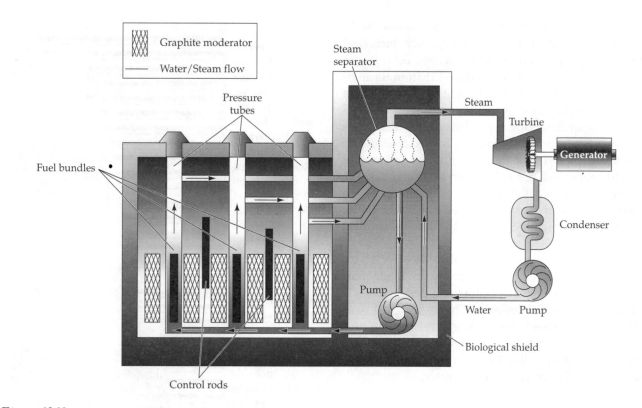

Figure 10.12 Schematic diagram of the Soviet-designed RBMK-1000 boiling-water nuclear reactor, the type of reactor that exploded at Chernobyl. While the reactor core is surrounded by a primary containment structure, there is no protective secondary containment structure. *(Source: OECD Nuclear Energy Agency.)*

Figure 10.13 Aerial view of the Chernobyl Nuclear Power Plant. The concrete sarcophagus built over Reactor 4 appears to the right of the high central tower.
(Photo taken in August 1990, courtesy of Vadim Mouchkin/International Atomic Energy Agency.)

resulted in a power reduction to 30 MW. Within a short time, the reactor became unstable. Plant operators worked to reestablish normal power levels, but in the early morning hours of April 26, a runaway chain reaction occurred, melting fuel rods and flashing cooling water into steam, which exploded the reactor's core and blew away part of the power plant itself.

The blast spewed burning graphite and a massive plume of radioactive material into the atmosphere. Radioactive fallout was carried mainly to the northwest by prevailing winds, contaminating parts of the Soviet Union's former republics of Ukraine and Belarus, as well as Lithuania, Norway, Sweden, and Finland. Portions of Italy, Germany, and France were also contaminated. No announcement of a nuclear accident was made by Soviet officials for several days, but scientists in Sweden, 2,000 kilometers (≈ 1,200 miles) away, detected radioactive fallout and traced its origin to Ukraine.

The World Health Organization estimates that 200 times more radioactivity was released at Chernobyl than was produced by the two atomic bombs exploded over Hiroshima and Nagasaki in 1945. The accident exposed 444 power plant workers to dangerously high levels of radiation, and more than 100,000 people were evacuated from the Chernobyl area. Thirty-one people died. Close to 800,000 cleanup workers (mostly military personnel), called liquidators, were enlisted to build a 300,000-metric-ton concrete and steel sarcophagus to entomb the shattered reactor and prevent further radioactive releases (Figure 10.13).

It is expected that eventually many serious health effects will be shown to have been caused by the

Chernobyl accident. By 1998, however, the only well-documented impact was a significantly increased incidence of childhood thyroid cancer in Belarus and Ukraine due to the fallout of iodine-131 from the accident's radioactive plume (see Insight 10.4). In northern Ukraine, thyroid cancer in children increased by a factor of 200 after the accident, and by 1996 reported cases in former Soviet affected republics totaled 890. Health authorities project that the total number of cases of thyroid cancer will rise to between 4,000 and 8,000. A continuing investigation of the health effects of the Chernobyl accident is being carried out by the Office of International Health Studies at the U.S. Department of Energy.

Many nuclear engineers attribute Reactor 4's explosion and release of radioactivity to inadequate training of power plant operators and design flaws in the reactor itself. For example, Soviet-designed reactors lack steel-reinforced, secondary containment structures (see Figure 10.12), the safety feature at the Three Mile Island facility that prevented the release of significant radiation to the environment.

The Future of Nuclear Power

Worldwide in 1999, 435 nuclear reactors were in operation in 32 countries including 104 in the United States, 58 in France, 53 in Japan, 35 in the United Kingdom, 29 in Russia, 20 in Germany, and 18 in Canada. Another 30 reactors were under construction. Nuclear power provided more than half of the electric power consumed in three of these nations. France, for example, generated 78 percent of its electricity that year. As a leading proponent of nuclear power, France is one of the few Western European nations to have built new reactors following the Chernobyl accident (see Regional Perspectives). Figure 10.14 shows nations that derived 5 percent or more of their electricity from nuclear power in 1999.

In the United States, although 104 commercial nuclear reactors were operating in 1999, no orders for new reactors had been placed since 1973 with the exception of some that were subsequently canceled. Today, American electric power utilities are no longer ordering new nuclear reactors for several reasons:

1. NRC regulations require complex engineering such as redundant safety backup systems, making nuclear plants very expensive. As we have seen, the construction of nuclear reactors often involves lengthy delays and cost overruns.

2. New products, home appliances, homes, and buildings are becoming more energy efficient,

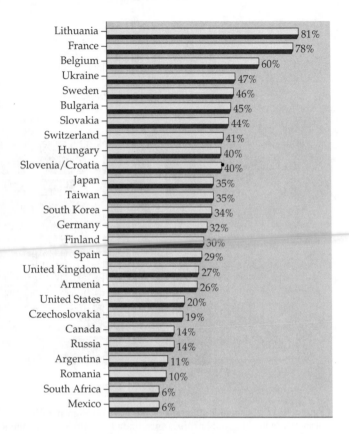

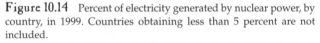

Figure 10.14 Percent of electricity generated by nuclear power, by country, in 1999. Countries obtaining less than 5 percent are not included.

(Source: The Uranium Information Centre, Ltd., Melbourne, Australia.)

thereby reducing the rates at which electric power demands are increasing.

3. The development of alternative sources of energy, including wind, solar-thermal, and photovoltaic energy, is beginning to diminish dependence on both fossil fuel and nuclear power sources (see Chapter 11).

4. The lack of an agreed-upon national plan for storing high-level radioactive wastes has intensified problems of waste storage at nuclear power plants.

5. The accidents at Three Mile Island and Chernobyl have significantly changed people's perception of the safety of nuclear power.

The dream of safe, reliable, cheap electricity from nuclear power has faded. The perception and the reality that nuclear reactors and nuclear wastes pose unacceptable risks to the public have now come into sharp focus. Although it will prove difficult and maybe impossible, the nuclear industry is at work designing new reactors that may be much safer than earlier ones. For example, advanced research programs in the

United States and Japan are investigating the possibility of building **nuclear fusion** reactors in which light elements such as hydrogen are fused (joined together) to form helium, a process similar to that occurring in stars, which releases considerable amounts of energy. Scientists believe that a fusion reactor, unlike a fission reactor, would produce few radioactive wastes and might operate more safely.

However, many energy experts believe that it is time to redirect our technical expertise and resources away from nuclear power toward the development of renewable and sustainable sources of energy, especially wind, solar, and perhaps geothermal power. These options are discussed in Chapter 11.

REGIONAL PERSPECTIVES

Canada: CANDU Nuclear Reactors

In the 1960s, the Canadian government and several private industries designed a nuclear reactor that would not require enriched uranium for fuel. The design is quite different from American nuclear power reactors, which require fuels enriched to about 3 percent uranium-235. Nuclear enrichment is a costly process, and the main reason the United States decided to use enriched fuels was that uranium-235 was being produced to make atomic bombs.

Canada's decision to use nonenriched, natural uranium containing only 0.7 percent uranium-235 led to the use of "heavy water" as a reactor moderator and cooling agent to slow neutrons and make them more effective in sustaining nuclear chain reactions. **Heavy water** is water whose two hydrogen atoms are deuterium, an isotope of hydrogen, instead of ordinary hydrogen (see Figure 10.1b). Deuterium is symbolized as D. Therefore,

Ordinary water is H_2O
Heavy water is D_2O

Using heavy water (deuterium oxide) in Canada's commercial power reactors led to the name CANDU, meaning Canada Deuterium Uranium.

A CANDU nuclear reactor is configured horizontally as a cylindrical vessel surrounded by a shield. Fuel assemblies, each made up of 28 fuel rods 0.5 meters ($\approx$ 1.5 feet) long, are inserted horizontally into reactor tubes around which heavy water circulates under pressure. Heat is transported from the reactor vessel via heavy water to a steam generator that drives an electric-power turbine-generator. Control rods operate vertically within the reactor vessel and are used to regulate power or shut the reactor down.

CANDU reactors incorporate a number of unique design elements:

- For fuel, they use nonenriched, natural uranium configured as uranium dioxide pellets.
- By design, they are capable of being refueled at full power. New fuel assemblies are inserted horizontally on one side of the reactor, while spent assemblies are pushed out the other side.
- The reactor vessel is configured as hundreds of interconnected pressure tubes, each containing a fuel assembly. This design minimizes the potential effect of a loss-of-coolant accident.

A CANDU power plant may have from one to eight reactors. For example, the Pickering facility on Lake Ontario near Toronto has eight reactors whose total electrical output is 4,800 MW. In 1999, there were 18 operating CANDU power reactors in Canada and an additional eight abroad: three in South Korea; one each in Argentina, Romania, and Pakistan; and two in India. Four CANDU reactors are under construction outside of Canada: one in South Korea, one in Romania, and two in China.

New York State: Shoreham Nuclear Power Plant

The Shoreham Nuclear Power Plant was built on the north shore of Long Island, New York, in the hamlet of Shoreham in Suffolk County about 95 kilometers ($\approx$ 60 miles) east of Manhattan Island. It was powered by a single reactor, a General Electric boiling-water unit rated at close to 820 MW. Shoreham was constructed by the Long Island Lighting Company (LILCO) between 1972 and 1984 at a cost of $6 billion—approximately 86 times its projected cost of $70 million. The greatly expanded cost was due to a number of factors that together complicated and delayed construction timetables:

- Shoreham's reactor was redesigned to increase the plant's output from 540 MW to 820 MW.

- LILCO decided to plan the construction of two additional reactors at other Long Island sites.

- Reactor engineering design changes were ordered by the Nuclear Regulatory Commission (NRC) in the wake of the Three Mile Island accident.

Long Island antinuclear sentiment reached a peak following the Three Mile Island incident. In June of 1979, 15,000 people demonstrated against the Shoreham plant; it was the largest public protest in Long Island history. Their concern was the lack of a plan to safely evacuate the island in the event of a nuclear accident. In 1983, the Suffolk County Legislature agreed by a vote of 15 to 1 that the county could not be safely evacuated in case of a nuclear accident.

After LILCO completed the Shoreham plant, the NRC granted permission for low-level testing but withheld a regular license to operate the plant due to unanswered questions about safe evacuation routes. In a test of the Shoreham power plant on July 7, 1985, the reactor was activated and operated for 30 hours at 5 percent of full power, demonstrating its successful completion. However, it was never licensed to operate at full power.

By February 1989, with no evacuation strategy in sight, Governor Mario Cuomo of New York and LILCO's chairman, William Cataconinos, shut down Shoreham and put the cost of its failure on LILCO's electric power customers. That same year, Shoreham, which never produced a single kilowatt of commercial power, was sold to the State of New York for $1. Plans for its decommissioning were approved by the NRC, and the plant was dismantled at an estimated cost of $1 billion.

Having to assume most of the liability of the Shoreham Nuclear Power Plant has placed LILCO among the most expensive electric power utilities in North America. While the national average cost of residential electric power in 1999 was about 8¢/kWh, the cost of LILCO electric power was 16¢/kWh.

Sweden: Reliance on Nuclear Power

In Sweden, with a year 2000 population close to 8.9 million, 51 percent of total electric power is produced by 12 nuclear reactors, which together generate about 10,035 MW—the highest nuclear power production per capita in the world. The balance is supplied through hydroelectric dams (36 percent), biomass (10 percent), and imported energy (3 percent). Finland, with four nuclear reactors, is the only other Nordic country where nuclear power is used to meet domestic and industrial power needs.

The Three Mile Island nuclear accident in the United States caused widespread public concern in Sweden over the safety of nuclear power, concern that led to a national referendum on the future of nuclear power in 1980. A majority of voters preferred no further development of nuclear power and gradual shutdown of all of Sweden's reactors; there were six reactors in operation at that time.

But reactor construction continued, and in 1988 the Chernobyl accident prompted Sweden's Parliament to adopt a new energy policy limiting nuclear power to 12 reactors. Furthermore, all of the country's reactors were to be shut down by the year 2010. However, the Parliament backed down in 1991, repealed the 2010 deadline, and required the shutdown of only one reactor, the Barseback I in south Sweden. This action was subsequently nullified by Sweden's Supreme Court, which suspended the governmental order to shut the plant down. Sweden continues to depend on nuclear power for these reasons:

1. No alternative sources of electric power can be counted on to meet the country's energy requirements.

2. The cost of shutting down and dismantling 12 nuclear reactors would be very high.

3. The loss of safe, reliable, and cheap electricity currently provided by the country's 12 reactors would result in increased prices of goods and services and greater unemployment.

Sweden's nuclear power dilemma is not yet resolved. In the meantime, the country is exploring two key energy strategies: developing renewable, sustainable sources of energy and conserving energy resources. In particular, Sweden's Environment and Energy Ministry is investing 300 million krona ($\approx$ U.S. $ 36.5 million) annually in solar, wind, and biomass energy. While solar power may not be viable in Sweden, wind-powered electric-generating stations are being built, and biomass energy strategies are being investigated.

In addition, Sweden's Public Energy Corporation is studying whether home energy conservation is practical. In 1987, 14 Stockholm homes became part of a carefully monitored energy program called Project 2000. State-of-the-art, energy-saving strategies such as heat pumps, double-paned glass windows, upgraded home insulation, and micro-computers were integrated into each home. Annual energy consumption in these households was cut in half. Sweden's Parliament now grants subsidies to households that choose to implement these energy-saving strategies.

KEY TERMS

atom	moderator
atomic fission	neutrons
atomic number	nuclear enrichment
base loads	nuclear fusion
boiling water reactor (BWR)	nucleus
chain reaction	peaking loads
control rods	pressurized water reactor (PWR)
electrons	primary containment structure
fuel rods	protons
half-life	secondary containment structure
heavy water	Tennessee Valley Authority (TVA)
high-level wastes	
isotopes	
low-level wastes	
mass number	

DISCUSSION QUESTIONS

1. What is your opinion of U.S. federal government agencies (such as the TVA) taking the lead to develop an entire region's power needs as opposed to relying on private business and utilities to meet those needs?

2. In your view, was the Tennessee Valley Authority justified in launching its ambitious nuclear program? Explain your answer.

3. Assume that a nuclear power plant is to be built within 80 kilometers ($\approx$ 50 miles) of your home and you are attending a public meeting to discuss the issue. Present three arguments either pro or con with respect to the proposed project.

4. Assume you have been asked to serve on a distinguished energy-planning panel in your state. It is being argued that renewable energy sources such as solar and wind power are too expensive and too impractical; therefore, nuclear plants should be built. What would you say?

5. Your state government has proposed to set aside a large area as a national repository for high-level nuclear waste materials. This will generate considerable income for schools, hospitals, clinics, and new highways. It will also lower state taxes. Present your view and make your arguments for or against this proposal.

INDEPENDENT PROJECT

Complete one of the following independent projects.

Project A

Imagine that you are invited to serve on your state's Energy Panel. The panel is considering a new kind of nuclear reactor designed so that it cannot overheat or melt down even if cooling water is totally lost. Nuclear experts agree that it will be safe. In fact, a prototype was built, and as a test, the cooling water was purposely allowed to leak out. Although the reactor temperature did rise, it quickly stabilized, and no further problems were experienced. What arguments would you make, pro or con, as to whether your state should allow construction of a series of these new reactors? Write a brief report.

Project B

Organize a debate among your friends or in your class. One debate team will present positive aspects of nuclear power, while the other will argue against nuclear power. You will serve as the moderator, giving each side a chance to present its views. Each side should have a chance to question members of the opposing team on specific points. Finally, give each side a chance to summarize its views.

INTERNET WEBSITES

Visit our website at http://www.mhhe.com/environmentalscience/ for specific resources and Internet links on the following topics:

Nuclear reactors around the world

Nuclear plant information books

Energy production and consumption

Energy and energy statistics

Chernobyl—radiological and health impacts

SUGGESTED READINGS

Beaver, William. 1990. *Nuclear power goes on-line: A history of Shippingport.* New York: Greenwood Press.

Chandler, William U. 1984. *The myth of TVA: Conservation and development in the Tennessee Valley, 1933–1983.* Cambridge, MA: Ballinger Publishing Co.

Creese, Walter L. 1990. *TVA's public planning: The vision, the reality.* Knoxville, TN: The University of Tennessee Press.

Freemantle, Michael. 1996. Ten years after Chernobyl consequences are still emerging. *Chemical and Engineering News* (April 29):18–28.

Groves, Leslie R. 1962. *Now it can be told: The story of the Manhattan Project.* New York: Harper & Brothers.

Hargrove, Erwin C. 1994. *Prisoners of myth: The leadership of the Tennessee Valley Authority, 1933–1990.* Princeton, NJ: Princeton University Press.

Hohenemser, Christoph. 1996. Chernobyl, 10 years later. *Environment* 38(3):3–5.

Newman, Alan. 1991. An international view of nuclear power plants. *Environmental Science & Technology* 25(10):1682–83.

Pilat, Joseph F., Robert E. Pendler, and Charles K. Ebinger, eds. 1985. *Atoms for peace: An analysis after thirty years.* Boulder, CO: Westview Press.

Tennessee Valley Authority, Division of Power Production, Knoxville, Tennessee. 1972. *Introduction to Nuclear Power.*

Tennessee Valley Authority, System Integration, Chattanooga, Tennessee. 1995. *Energy vision 2020: Integrated resource plan and draft environmental impact statement.*

Thurman, Sybil, ed. 1983. *A history of the Tennessee Valley Authority.* Information Office, Tennessee Valley Authority.

Sustainable Energy

Photo Disc, Inc./Vol. 44

I am the proud grandfather of five grandchildren and I sense, maybe even know,

that we must move from conquering nature to living with nature or

there will eventually be no nature to live with.

From a speech in New Mexico by
Carl J. Weinberg, Manager of Research
and Development, Pacific Gas and
Electric Company, February 19, 1992

Throughout history, humans have continuously tapped the earth's natural energy supplies, discovering new sources as well as finding different ways to use the old ones. Fuelwood, crop residues, and animal dung were long used for cooking and heating, and still are in poorer parts of the world. The past 200 years have seen an increasing dependence on fossil fuels; the momentum of the Industrial Revolution was first linked to cheap, readily accessible coal, and later oil and natural gas became the most important energy sources. Today, fossil fuels still supply the lion's share of our energy demands, including the generation of electricity. But limited quantities of fossil fuels, especially oil and natural gas, are left on the earth. Furthermore, they are **nonrenewable** resources, causing concern about what will happen if they run out.

In the years following World War II, some people believed that nuclear power could replace fossil fuels and become the world's most important source of electrical energy (see Chapter 10). Thus, from the 1960s through the 1990s, approximately 430 nuclear reactors were constructed in 30 nations. But serious unforeseen problems developed. Complex engineering designs and redundant backup safety systems led to enormous cost overruns, numerous safety hazards, and frequent start-up delays. In addition, the operation of nuclear reactors led to a number of serious accidents. And finally, no agreement has ever been reached in the United States for establishing a national repository where high-level nuclear fuel wastes can be safely stored.

Most recently, interest is growing in alternatives to fossil fuels and nuclear power—alternative resources that are both renewable and environmentally friendly. The goal of these efforts is to safeguard the future of the planet and the natural resources our children will need to secure a decent and beneficial future. This chapter provides background information about renewable and sustainable energy resources along with a case study citing examples of important ongoing sustainable energy projects.

BACKGROUND

Renewable and Sustainable Energy Sources

Unlike fossil or nuclear fuels, **renewable** sources of energy are not depleted as they are used. The best examples are energy derived from the sun, wind, and water. However, some energy sources, although renewable, can endanger fish and wildlife or disrupt human settlements and cultural traditions. Such effects are called **environmental externalities**—social, cultural, economic, and ecological impacts resulting from large-scale developments. Externalities incur major costs on society that are often unseen and unaccounted for until a project is completed and operational. We will see some examples of externalities later on in this chapter.

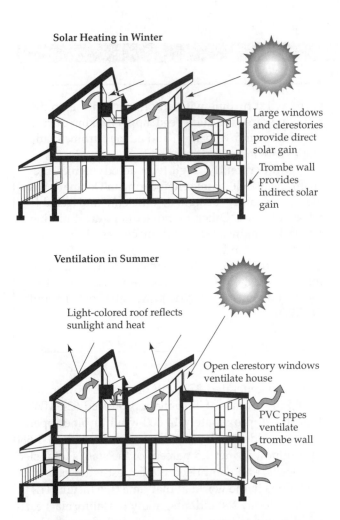

Solar Heating in Winter

Large windows and clerestories provide direct solar gain

Trombe wall provides indirect solar gain

Ventilation in Summer

Light-colored roof reflects sunlight and heat

Open clerestory windows ventilate house

PVC pipes ventilate trombe wall

Figure 11.1 Passive solar heating and cooling residential design concepts.

The terms "renewable" and "sustainable" used to mean the same thing—that is, an energy source that was renewable was understood to be sustainable. But more recently, the term **sustainable** has taken on a new connotation. Applied to agriculture, it means leaving the soil, nearby surface waters, and groundwater in as good or perhaps better condition than when the land was first cultivated. Applied to energy, sustainable means tapping a renewable source of power in ways that create few if any environmental externalities. In other words, sustainable sources of energy are both renewable and environmentally friendly; they do not pollute the atmosphere, contaminate surface or groundwaters, degrade fish and wildlife habitats, or diminish the quality of human life and culture.

Solar Power

Five hundred years ago, South American Incas living in the Andes Mountains constructed rock dwellings

facing the noonday sun. Heat captured by the rocks during the day radiated within the interiors at night. Indian pueblos built in America's Southwest acquired winter warmth from the sun in a similar way. These native people had learned to take advantage of a natural and plentiful energy source that is also sustainable—the sun.

Each year the earth receives 15,000 times more energy from the sun than the entire world uses annually—100 times more energy than all proven reserves of coal, oil, and natural gas could generate. Furthermore, solar energy is more widely distributed than any other source of energy, making it as available to developing nations as it is to the industrialized world. It is well suited not only for cooking and heating, but also for electric power generation. Tapping solar energy does not pollute air or water, create radioactive wastes, require drilling or mining, or add greenhouse gases to the earth's atmosphere—and it cannot be depleted.

Passive and Active Solar Heating and Cooling The heating strategies of the Incas and Indians have evolved into more sophisticated solar heat-trapping techniques called **passive solar heating.** Today, passive solar concepts have matured and are being employed by many architects designing contemporary homes (Figure 11.1). Esperanza del Sol (meaning "hope of the sun"), a housing development in Dallas, Texas, features Spanish-style homes that include passive solar heating concepts. South-facing, double-pane windows collect winter sunlight but are shaded from the summer sun by roof overhangs. Architectural features alone conduct solar-derived winter heat to interior living spaces. Walls and ceilings are well insulated, and deciduous shade trees are planted for summer cooling. Cooling can also be gained through natural ventilation. Passive solar homes can save as much as 50 percent on heating and air conditioning costs compared to similar nonsolar homes.

Active solar heating is closely related to passive solar except that instead of relying solely on architectural design, pumps and fans are used to transport heat (or cooling) to various interior areas. One example of active solar design uses heat pumps to facilitate winter heating and summer air conditioning in some climates. In 1999, there were more than three million passive and active solar homes in the United States.

Wind Power

Like solar power, wind power is also a sustainable energy source. Windmills were used in ancient Persia 15 centuries ago to pump groundwater and irrigate farmland. Later, wind power spread to Europe where it was

INSIGHT 11.1
The Development of Modern Wind Turbines

The evolution of wind turbines from the old wind-mill led first to machines whose rotors (propeller-driven power shafts) were held to a constant speed even at high wind velocities so that a consistent, 60-cycle alternating current would be produced. This assured compatibility with existing power grids to which the turbines were connected. But the limitation of constant-speed rotors prevented turbines from taking advantage of higher wind speeds to generate more power.

In 1991, a consortium of wind power developers, including U.S. Windpower, the Electric Power Research Institute (EPRI), New York's Niagara Mohawk Power Corporation, and California's Pacific Gas and Electric Company, developed a variable

speed wind turbine with automatic controls to detect wind direction and speed, apply braking during very high wind velocities, and adjust turbine blade angles as wind speeds changed.

More recently, high-tech aerodynamic, mechanical, and electronic engineering in Denmark has led to elegantly designed, efficient, and reliable wind turbines. Other manufacturers are building similarly engineered wind machines. For example, Enron Wind Corporation supplies a 750-kW constant-frequency wind machine. Newer turbines installed at California's Tehachapi Pass are of this make and generate electric power at a cost of about 5¢/kWh.

used to water crops and grind grains. Perhaps the best-known examples of early wind power were in Holland where windmills drained polders, lands lying at or below sea level. The heyday of early wind power was the late 1800s when thousands of windmills in Europe and North America partly replaced human and animal power on farms.

When the Industrial Revolution gave birth to coal-fired steam engines and electrically powered equipment, it seemed like the day of windmills was over, but this was not the case. Yesterday's windmills have evolved into today's wind turbines (Insight 11.1), which are making wind power the fastest growing energy source in the world.

Wind Power in the United States

California's Wind Farms In the 1970s, one of the first wind power farms in the United States was built at Altamont Pass, about 30 miles east of San Francisco (Figure 11.2). As many as 40 wind turbines were built each week due to an oil crisis precipitated by OPEC, the Organization of Petroleum Exporting Countries. Wind power at Altamont Pass was spurred by federal and state tax subsidies and long-term, fixed-price contracts with electric power utilities, ultimately creating the world's largest concentration of **wind turbines**—5,041 machines that by 1995 annually generated 544 megawatts (MW) of electric power. (A megawatt is equivalent to one million watts of power.)

Two additional California wind farms, San Gorgonio Pass near Palm Springs and Tehachapi Pass about 60 miles north of Los Angeles were also constructed. In 1995, the 2,898 wind turbines at San

Gorgonio generated 273 MW of electric power, while the 4,714 turbines at Tehachapi produced 624 MW. In 1997, the approximately 14,000 wind turbines operating statewide in California generated close to 1,680 MW of power, met 1.5 percent of the state's total electric needs, and comprised the largest wind power installations in the world at that time (see Insight 11.1).

By today's standards, many of California's early wind turbines were poorly designed. Some had mechanical problems, were unreliable, or didn't work at all; others were simply inefficient and didn't develop their rated power output. On average, they generated electricity at a cost of about 25¢/kWh. In 1998, the Alameda County Board of Supervisors approved a plan to replace old Altamont wind turbines with new, larger, and more efficient ones. The decision to install new wind machines was based partly on their proven reliability and cost-effectiveness and partly on the fact that the slower-rotating blades of the new turbines could better be avoided by golden eagles and red-tailed hawks, many of which had been killed in recent years by the faster-rotating blades of older turbines.

Water Power

Almost 2,000 years ago, water power projects were developed by the Romans, and by the twelfth century, waterwheels were grinding grains, sawing wood, forcing air into iron-making blast furnaces, and hammering metals into useful products. Today, **hydroelectric power** is generated by using the potential energy of water stored behind a dam to drive turbines in powerhouses built at the dam's base. Hydroelectric power is

Figure 11.2   A wind farm at Altamont Pass, about 30 miles east of San Francisco, California.
(Photo courtesy of the photographer, Warren Gretz, and the National Renewable Energy Laboratory.)

renewable because it is linked to water supplies that are continually renewed through the hydrologic cycle.

It is estimated that 8.2 percent of year 2000 energy needs worldwide will be met using renewable energy, mostly hydroelectric power. Although hydroelectric power is the world's leading source of renewable energy, its development and use can generate significant environmental externalities, making it unsustainable in the long run, as will be shown in the following examples.

Washington State's Grand Coulee Dam   In the 1920s, the idea of building a dam on the Columbia River in central Washington State was conceived in part to create a vast agricultural reservoir to irrigate the fertile but arid lands of the Columbia River Basin. In 1931,

plans for a major dam were presented to the U.S. Congress by the Bureau of Reclamation (an agency of the U.S. Department of the Interior) and the Army Corps of Engineers. President Franklin Roosevelt favored these plans and committed $63 million to the first phase of the project. Constructing Grand Coulee Dam began in 1933 and was completed in 1942. It was the largest concrete structure to be built at that time, and it created Lake Roosevelt, stretching 151 miles north to the Canadian border (Figure 11.3).

The construction of Grand Coulee Dam had several purposes: irrigation of arid farmlands, flood control on the Columbia River, hydroelectric power generation, fish and wildlife enhancement, and creation of recreation areas. However, generating hydroelectric power quickly became the highest priority

Figure 11.3 The Grand Coulee Dam on the Columbia River. *(Photo courtesy of the Bonneville Power Administration Library Archives.)*

due to electric power demands by aluminum industries in the Pacific Northwest during World War II. Aluminum, whose production requires large quantities of electricity, was a strategic material because it was needed to make war planes. After the war, added construction enabled the dam to supply enough irrigation water for more than 405,000 hectares (≈ one million acres) of Columbia Basin farmland. Today, the dam's 24 hydroelectric turbines generate 6,500 MW of power, the equivalent of eight large nuclear or coal-burning power plants.

Grand Coulee Dam is one of 11 dams on the main stem of the Columbia River that are part of a complex of Pacific Northwest hydroelectric dams in Canada and the United States. Some are managed by British Columbia Hydro and Power Authority in Canada (BC Hydro), while others are administered by the U.S. Army Corps of Engineers and the Bureau of Reclamation. Power from Grand Coulee Dam is transmitted over high-voltage transmission lines, sold to industries and electric utilities, and in some cases resold to domestic and business customers. It accounts for about 50 percent of the electric power used by the 10 million people living in Washington, Oregon, Idaho, and Montana, and is one of the cheapest sources of electrical energy in the world.

Externalities Created by Columbia River Dams When Meriwether Lewis and William Clark, leaders of President Thomas Jefferson's Corps of Discovery, first saw the Columbia River in the fall of 1805, Clark noted in his journal, "This river is remarkably clear and crowded with salmon in many places." He was astonished to see abounding numbers of the fish swimming upriver to spawn. (Pacific salmon are **anadromous,** meaning that they begin their life cycle in freshwater streams and rivers, migrate downriver to the ocean, spend their adult years in seawaters, and then return to the freshwaters of their birth to spawn.)

Hydroelectric dams on rivers like the Columbia obstruct fish migrations by blocking their natural passageways to upriver spawning sites. In some cases, ladders have been built to allow the fish to reach higher water elevations behind a dam. While as many as 90 percent of adult salmon are able to negotiate fish ladders, young salmon, called smolts, barely survive their downriver migration. It is estimated that 90 percent or more of Columbia River smolts are killed each year as water currents draw them into hydroelectric turbines at dam sites. Columbia River Basin salmon populations, once believed to number 16 million fish or more, are now estimated to total 300,000 or less.

China's Three Gorges Dam Industrialization in China, together with its rapid advancement as a developing nation, is creating exponential increases in the amount of energy needed to build cities, operate factories, construct modern transportation systems, and meet the expectations of the country's almost 1.3 billion people for more affluent lifestyles. Part of this enormous energy demand is being met through coal-fired electric power plants. But the immense potential for hydroelectric power in the Yangtze River led the Chinese government to begin building a major hydroelectric dam at the river's Three Gorges site. When and if it is completed, this dam will be the world's largest hydro-power facility.

Phase One of the Three Gorges project began in 1993 when channels were cut to divert Yangtze River water around the proposed site; this phase was completed in 1997. Phase Two, which is to conclude in 2006, involves the construction of the dam, a flood management system, and 14 hydroelectric turbine generators. Phase Three, to be achieved by 2009, will include a series of ship canal locks and lifts to allow navigation around the dam plus 12 more turbine generators. When it is completed, Three Gorges Dam will be 610 feet high and 1.3 miles wide, with a 410-mile-long reservoir behind it. Its hydroelectric power plant will generate 18.2 gigawatts (= 18,200 MW) of electric power, which is equivalent to the output of 23 large nuclear or coal-fired power plants.

Although the projected power output of Three Gorges Dam is huge in scale, it will satisfy only 10 percent of China's total electric needs in 2009. But it will meet pressing energy demands in central and eastern China, particularly in major cities such as Chongqing and Shanghai. In addition, the dam will accelerate

China's transition from a developing to an industrialized nation.

Externalities Involved in the Three Gorges Dam   Because of its enormous scale, the construction of China's Three Gorges Dam is threatening not only fish and wildlife but humans and their culture as well. The 410-mile-long, 150,000-acre reservoir created by the dam will submerge 13 cities, 140 towns, and 1,352 villages, forcing the resettlement of more than 1.9 million people. Sixteen archaeological sites and 650 factories will also be inundated.

Furthermore, it is expected that the reservoir behind the dam will become an open sewer as untreated sewage and industrial wastes are discharged to the river. The dam will also slow river flow, allowing suspended sediments to settle and build up behind the dam, ultimately reducing its hydroelectric power potential. In addition, damming the Yangtze will impact already endangered species, including the white-fin dolphin, Chinese sturgeon, river sturgeon, and the black finless porpoise. These are some of the reasons the World Bank and the U.S. Export-Import Bank have refused to help finance the project.

CASE STUDY: INNOVATIONS IN SUSTAINABLE ENERGY

We have seen that renewable, sustainable energy options are not really new. Passive solar domestic heating, wind energy to pump water and grind grain, water power to operate factory machinery and, later, to generate electricity, have long been employed in many parts of the world. We have also seen that hydroelectric power, while renewable, is not sustainable. But truly sustainable energy sources have never proved to be practical or capable of meeting large-scale energy demands until now. Today, new knowledge, technical innovation, and advanced engineering materials and concepts are enabling humankind to more efficiently capture and better utilize sustainable sources of energy. As a result, sustainable energy options are becoming cost-effective and reliable.

This case study examines several ongoing, innovative approaches to sustainable power for the future:

- Parabolic-collector solar-electric power
- Central-receiver solar-electric power
- Photovoltaic solar-electric power
- Wind-turbine electric power
- "Green" marketing of electric power
- Hydrogen gas produced from water as a source of fuel

Parabolic-Collector Solar-Electric Power

Arnold Goldman, an American electrical engineer, was one of the first to take large-scale, solar-generated electric power seriously. In 1979, after obtaining financial backing, Goldman built a 14-MW solar power plant in the Mojave Desert near San Bernadino, California. It is comprised of a series of 100-meter-long, silvered-glass, computer-controlled parabolic mirrors that focus sunlight on heat-collecting steel pipes running the length of each collector. Each pipe, 7 centimeters in diameter, is encased in clear, evacuated glass. Under direct sunlight, heat-transfer fluid pumped within the pipes reaches temperatures of about 390 °C ($\approx$ 735 °F) and drives ordinary steam-powered electric turbines. The power plant is called a Solar Electric Generating System (SEGS), and the company Goldman established was called Luz (pronounced "looze") International.

Ultimately, nine SEGS facilities were built. Together, they generate a total of 354 MW of electric power, about 80 percent of the world's solar-generated electricity today. Most of the power is purchased by Southern California Edison, the region's leading electric power utility. It is used to meet peaking load demands such as those on summer afternoons (see Insight 10.3). One of the SEGS facilities, the 30-MW SEGS III Power Plant near Kramer Junction, California, is shown in Figure 11.4, along with a simplified schematic drawing of its operational features. Table 11.1 summarizes basic information about the nine SEGS facilities.

SEGS solar plants utilize electric-generating natural gas turbines as backup power because sunlight is not available on very cloudy days. On an overall annual basis, 75 percent of SEGS electric power is produced from solar energy and 25 percent from natural gas. (To qualify as a renewable energy facility, federal regulations limit fossil fuel backup systems to 25 percent of a facility's total annual power production.) This complementary solar/natural gas design ensures that electric power is supplied reliably and with little air pollution.

The incentive to build Luz International was based largely on energy policies established during Jimmy Carter's administration. The first was the Energy Tax Act of 1978, which provided investment tax credits to encourage renewable energy projects. The second was the Public Utilities Regulatory Policy Act of 1978 (PURPA), which requires energy utilities like Southern California Edison to buy renewable energy. These regulations, along with renewable-energy tax credits guaranteed by the State of California, made Luz International possible.

By the time Luz had become a major power producer, political changes in Washington, D.C., and

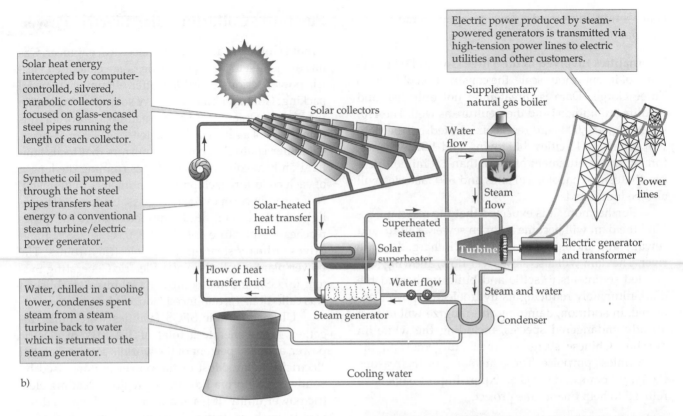

Solar heat energy intercepted by computer-controlled, silvered, parabolic collectors is focused on glass-encased steel pipes running the length of each collector.

Synthetic oil pumped through the hot steel pipes transfers heat energy to a conventional steam turbine/electric power generator.

Water, chilled in a cooling tower, condenses spent steam from a steam turbine back to water which is returned to the steam generator.

Electric power produced by steam-powered generators is transmitted via high-tension power lines to electric utilities and other customers.

Solar collectors

Supplementary natural gas boiler

Water flow

Steam flow

Power lines

Solar-heated heat transfer fluid

Superheated steam

Solar superheater

Turbine

Electric generator and transformer

Flow of heat transfer fluid

Water flow

Steam and water

Steam generator

Condenser

Cooling water

b)

Figure 11.4 (*a*) Aerial view of part of SEGS III, one of the five 30-MW solar thermal power plants operated by Kramer Junction Company in California's Mojave Desert. (*b*) Schematic of a typical SEGS solar-thermal power plant.
(Reprinted with permission of KJR Operating Company, Boron, California.)

California had led to new energy policies that no longer supported renewable energy initiatives. In effect, the rug was pulled out from under solar projects like Luz. By 1991, Luz International was forced to file for bankruptcy. The rise and fall of Luz can be traced to energy policies that at first spurred its development and later caused its decline. Following Luz's bankruptcy, the nine solar power plants were taken over by three private operating companies. The solar plants are still operating today.

Central-Receiver Solar-Electric Power

In 1982, a solar power project called Solar One was built at Barstow, California, in the Mojave Desert about 100 miles northeast of Los Angeles. Solar One was a pilot project to test the central receiver or "power tower" concept of generating solar energy. Using microprocessors and computers, 1,818 sunlight-tracking mirrors, or **heliostats**, were used to reflect sunlight to a central receiver on top of a tower 79 meters (≈ 260 feet) above the ground. (Each heliostat is an array of 122 mirrors;

a)

each mirror has a surface area of 3.25 meters^2 [35 ft^2].) There, water was converted to superheated steam to power a turbine/electric power generator. In the winter, 10 MW of power was produced four hours daily and in the summer almost eight hours daily—enough power to meet the energy needs of 10,000 homes. Solar One operated from 1982 to 1988 and demonstrated that the central receiver concept had commercial promise but lacked the ability to operate on cloudy days or after nightfall.

Plans to convert Solar One to a more sophisticated solar power demonstration project called Solar Two were completed in 1996 (Figure 11.5). Solar Two produced 10 MW of power using 1,926 heliostats and a

TABLE 11.1 Luz Solar-Electric Generating Systems (SEGS Facilities)

SEGS Facility	Capacity (MW)	Power Cost/kWh	California Location	In-Service Date
SEGS I	14	25	Daggett	1984
SEGS II	30	20	Daggett	1985
SEGS III	30	20	Kramer Junction	1986
SEGS IV	30	20	Kramer Junction	1986
SEGS V	30	12	Kramer Junction	1987
SEGS VI	30	12	Kramer Junction	1988
SEGS VII	30	12	Kramer Junction	1988
SEGS VIII	80	10	Harper Lake	1989
SEGS IX	80	8	Harper Lake	1990

SOURCE: Data from Newton D. Becker, "The Demise of Luz: A Case Study" in *Solar Today*, January/February, 1992, pp. 24–26.

Figure 11.5 Solar Two, a 10-MW central-receiver, solar-thermal, electric power generating demonstration project completed at Barstow, California, in 1996.
(Photo courtesy of the National Renewable Energy Laboratory PIX Photo Library and Sandia National Laboratory.)

hot, liquefied (molten) mixture of two salts—sodium nitrate and potassium nitrate—to store and transfer heat energy as needed to its steam-powered turbine-electric generator. This approach resulted in operating temperatures as high as 566 °C (≈ 1,050 °F) as well as the ability to store excess heat in an insulated "hot" salt tank. Molten salts were pumped from this tank for power production after sunset or when clouds obscure the sun.

Solar Two was shut down in 1999 (as planned) after extensive testing and evaluation had been completed. Its molten-salt thermal storage system proved to be highly efficient—capable of storing enough heat to generate electric power around the clock. While the cost of Solar Two electric power is not known, it appears that this type of central-receiver solar-electric power generation is more practical than Solar One and may serve as a model for future solar power plants. The cost of building Solar Two, $48.5 million, was shared by several groups including Arizona Public Service, the Electric Power Research Institute, and the Sacramento Municipal Utility District.

Photovoltaic Solar-Electric Power

Solar cells, also called **photovoltaic** (PV) cells, are semiconductor devices that convert light directly into electricity (Insight 11.2). They are commonly made of silicon and are used to power calculators, watches, shipping lane buoys, emergency telephone call boxes, remote signaling devices, irrigation pumps, and spacecraft. Practical applications of PV devices are increasing as more efficient and cheaper cells are developed. For example, the PV system at Georgetown University's Intercultural Center in Washington, D.C., a showcase solar power system, generates 300 kilowatts of electricity under full sunlight (Figure 11.6).

While a number of PV demonstration projects are in operation today, plans to build a large, utility-scale photovoltaic system were announced by Amoco/Enron Solar together with the Greek Government and the European Union in 1997. The 50-MW solar photovoltaic power plant will be constructed on the Greek island of Crete in the eastern Mediterranean Sea. When

INSIGHT 11.2
Photovoltaic Cells

In the 1950s, scientists at Bell Laboratories in New Jersey discovered that an electric power-generating photovoltaic device could be made using thin layers of silicon semiconductor materials. A simple PV cell consists of a silicon "n" layer and a silicon "p" layer sandwiched together. Each layer is "doped" with a specific chemical additive. A negative charge forms at the "n" layer; a positive charge at the "p" layer. The semiconductors were designed to produce an electric potential upon exposure to light (Figure 11.A). At first, solar cells were only about 6 percent efficient in converting sunlight into electric power, but their efficiency has been gradually increased. Even so, they remained laboratory curiosities due to their high cost. One of the first applications of solar cells was in spacecraft and space satellites.

Photovoltaic cells are being constantly improved. The most efficient ones are made of fairly thick (100–300-micron) layers of single-crystal silicon. Commercially available cells of this type convert about 16 percent of solar energy into electric power but are expensive to make. Polycrystalline silicon cells are less costly and typically about 14 percent efficient. They generate electricity at between 20¢ and 40¢ per kilowatt hour. Thin-film PV solar cells, which are about 7 percent efficient, can generate electricity at 10¢ per kilowatt hour, a cost that is competitive with power needs in remote locations.

Between 1993 and 1999, scientists at the National Renewable Energy Lab in Golden, Colorado, perfected a copper-indium-gallium-diselenide (CIGS) thin-film PV cell with an efficiency of 18.8 percent. This development is considered a major breakthrough and may mark the beginning of cost-effective utility-scale PV power.

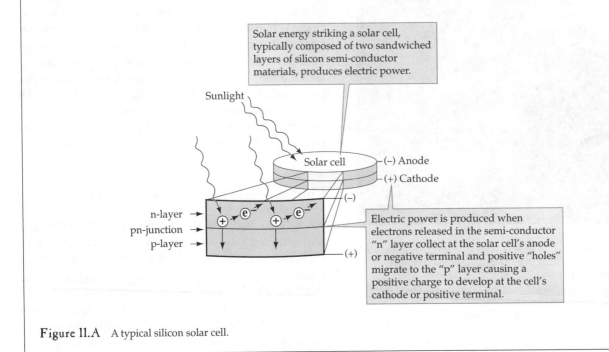

Figure 11.A   A typical silicon solar cell.

completed, it will be the largest PV power plant in the world and will meet the electric needs of 100,000 people, or 12.5 percent of the island's population. The expected cost to consumers will be about 8¢/kWh. The first stage of the Crete project, a 5-MW facility, is currently underway.

Elsewhere, solar cells are effectively supplementing energy needs in individual homes, businesses, and industries. In 1997, Japan launched a major government-subsidized residential rooftop PV program. Thousands of blue PV "shingles" were installed on individual residences that year, and the program is being expanded. Amoco/Enron, a supplier of rooftop solar modules, is working with Misawa Homes Co. Ltd., one of Japan's leading home builders.

Also in 1997, the United States initiated its Million Solar Roofs program, calling for the Department of Energy to help place solar panels on one million public

Figure 11.6 Georgetown University's Intercultural Center in Washington, D.C., is fitted with 4,400 rooftop PV modules that deliver 300 kilowatts of electric power under full sunlight.
(Source: U.S. Department of Energy.)

Figure 11.7 107-MW wind farm at Lake Benton, Minnesota, 240 kilometers (≈ 150 miles) west of Minneapolis, Minnesota, the world's largest single wind energy project.
(Reprinted with permission by the Star Tribune/Minneapolis-St. Paul, 1999.)

and private buildings. It is projected that a program this large will reduce the cost of photovoltaic electric power to 6¢/kWh. New regulations are being written to make federal loans available for financing solar roofs. Proponents of the new program point out that the solar initiative will reduce the production of greenhouse gases, particularly carbon dioxide, thereby helping mitigate climate change. A similar solar photovoltaic roofs program was launched in Germany in 1999, and other photovoltaic energy systems are already in operation in Canada (see Regional Perspectives at the end of this chapter).

Wind-Turbine Electric Power

Minnesota's Wind Energy Program

The development of wind power in Minnesota is an example of what is happening across North America and in other parts of the world today. It was mandated by the state's legislature in 1994 when Minnesota's Northern States Power Company (NSP) was attempting to get permission to store nuclear wastes in its service region. NSP is the largest utility in the upper Midwest, serving the power needs of most of Minnesota and parts of Michigan, Wisconsin, and North and South Dakota. The legislature granted the NSP request but required the utility to integrate wind power into its total energy program. Minnesota's plan is to produce 425 MW of wind power by 2002. The first step (Phase I) was building the Lake Benton wind farm about 240 kilometers (≈ 150 miles) west of Minneapolis.

Enron Wind Corporation completed construction of the 107-MW Lake Benton facility in 1998, making it the world's largest single wind energy project (Figure

11.7). Owned and operated by Enron, the facility delivers electric power to NSP for about 3¢/kWh to be resold to its customers for about 5¢/kWh, demonstrating that wind energy is cost-competitive with other power sources. When completed, the statewide 425-MW project will meet the electrical needs of 43,000 homes.

The Lake Benton wind farm consists of 143 Zond Z-750 (750-kilowatt) wind turbines manufactured by Zond Energy Systems, a subsidiary of Enron. These are the largest wind turbines made in the United States today. Development of a larger, 1,000-kilowatt wind turbine is underway and promises to make wind energy even more competitive with other power sources.

Other states capitalizing on wind power include Hawaii, Iowa, Texas, Wyoming, and Vermont. U.S. states with the greatest untapped wind power potential are North Dakota, South Dakota, Kansas, and Montana. The recent surge in the growth of wind power was due in part to the fact that a federal wind energy tax credit was due to expire in June 1999. Wind power is increasingly favored because construction costs are reasonable, the electric power produced is cheap, and wind power creates few harmful externalities (Insight 11.3).

About 9,600 MW of utility-scale wind turbine facilities were operating worldwide in 1999, the equivalent of 12 large coal-burning or nuclear power plants. Between 1998 and 1999, there was a 35 percent increase in wind-generated electrical capacity worldwide, an all-time, one-year record. Today, wind power meets the electrical needs of 3.5 million homes. But although these numbers are impressive, it is important to keep in mind that 9,600 MW of electric power is still less than 1 percent of the world's electrical needs today.

INSIGHT 11.3
Wind Power and Externalities

One of the principal advantages of wind power is its minimal impact on the environment. A 1990 report published by Pace University Center for Environmental Legal Studies in White Plains, New York, demonstrated that externalities connected with wind power are the lowest for any energy source, including solar power. The Pace report draws attention to these important facts about wind turbines:

1. They add no carbon dioxide to the atmosphere. Therefore, power is produced without exacerbating global warming.

2. Their operation adds no carbon monoxide, sulfur dioxide, oxides of nitrogen, particulates, radioactive wastes, or other pollutants to the environment.

3. They release no stratospheric ozone-depleting substances.

4. Their use reduces dependence on imported foreign oil, thereby strengthening national security and improving the nation's balance of trade.

5. They create no demand for water power such as that required by coal-burning and nuclear power plants for cooling.

Currently, the United States has close to 2,100 MW of wind power, most of it sited in California. The U.S. total includes 235 MW added in 1998, a fairly large one-year increase due mainly to power utilities in 10 states taking advantage of existing wind energy tax credits. The largest U.S. wind power projects under development are in Vermont, Minnesota, Wyoming, and Oregon.

Wind Power in Other Countries

In 1999, Germany, with a total of 2,875 MW of installed wind power, had supplanted California as the world leader in wind energy (Table 11.2). Its commitment to wind energy began shortly after the Chernobyl disaster. Now, in northern Germany, 15 percent of electric power is produced by wind turbines, and the government plans to shut down a nearby nuclear power plant.

Denmark has emerged as a major player in wind power not only in terms of its wind-generating installations but also its commitment to developing what are probably the world's most efficient, reliable, and sophisticated wind turbines. Besides their installation in Denmark, their export to other countries accounts for close to 75 percent of all wind turbines worldwide. In 1998, Denmark's total wind power of 1,350 MW supplied 8 percent of its total electric power needs.

Other countries having significant wind facilities include India and Spain. India has more than 900 MW of wind-generating power plants, and Spain added 395 MW of wind power in 1998, for a total installed wind capacity of 850 MW. It turns out that wind resources are unusually plentiful in Spain, creating the potential

TABLE 11.2   1999 Installed Wind-Power Capacity in Selected Countries

Country	Power (in MW)
Germany	2,875
United States	2,100
Denmark	1,350
India	933
Spain	850
Netherlands	340
United Kingdom	325
China	169
Other countries	658
Total	9,600

SOURCE: Data from *Vital Signs Brief:* 98–7, The Worldwatch Institute, December, 1998.

to meet a high percentage of its electric energy needs. In the Spanish city of Pamplona, wind power currently supplies 23 percent of the city's electricity (see Regional Perspectives).

The aggressive development of wind power in some European nations is driven by new energy policies offering favorable tax incentives. For example, Germany is encouraging independent power producers by offering to buy wind-generated electricity for up to 15¢/kWh. In the Netherlands, the government offers direct subsidies for each wind turbine installed, and

INSIGHT 11.4

Wind Power in Costa Rica

The 20-MW Tejona Wind Power project, with 40 to 100 wind turbines, is under development in Costa Rica's Guanacaste Province. Strong winds blow in that area almost constantly, creating an ideal site for wind-based electric power generation. Constructing the wind farm, which will be operated by a public utility, has eliminated the need to build a fossil fuel power plant.

Costa Rica's wind farm is being financed by the Inter-American Development Bank and the Global Environment Facility, which was set up as a result of the Framework Convention on Climate Change signed by nations party to the 1992 Earth Summit in Rio. Interest in the Tejona wind farm has led to the start-up of five additional wind power projects in Costa Rica.

TABLE 11.3   Typical Electric Power Costs to Consumers in the 1990s Based on Different Energy Sources

Electric Power Energy Source	Cost of Electricity (¢ per kWh)
Old midwest coal-fired power plants	2
Natural gas turbines	5
Current wind-turbine technology	5
Modern coal-fired power plants	6
Most recent solar-thermal power plants	8
Current nuclear power plants	10–20
Most recent photovoltaic power sources	15–25

SOURCES: Data from *Solar Today*, January/February 1992; March/April 1998. *ChemTech*, January 1993.

resources, do not risk air and water pollution, and do not create nuclear and other hazardous wastes.

Green energy markets, such as wind and solar power, are growing. Given the promise of environmental protection, consumers may be willing to pay a premium for electricity that is generated in an environmentally benign way. In other words, some consumers looking toward a sustainable future will pay more to help develop that future.

The cheapest sources of U.S. electric power today are old midwestern coal-fired power plants that are exempt from federal Clean Air Act mandates. While they produce electricity for 2¢/kWh or less, they add immense quantities of greenhouse-gas carbon dioxide to the atmosphere, along with sulfur dioxide and oxides of nitrogen, which create downwind acid rain. Choosing to use an alternative to electric power sources such as these would be a "green" decision.

Table 11.3 identifies some examples of typical electric power costs to consumers in the 1990s based on different energy sources.

Denmark offers grants equal to 30 percent of the cost of buying wind turbines and developing wind farms.

Costa Rica's recent 20-MW wind farm development, while small, is nevertheless an interesting example of what is taking place in several developing countries today (Insight 11.4).

"Green" Marketing of Electric Power

Current federal and state legislation is gradually restructuring the electricity market, making it possible for electric power consumers to choose their source or sources of electricity. The change is already well underway in California, Pennsylvania, and at least 19 other states. For many, the choice will simply be to buy the cheapest power available. But many environmentally aware consumers will likely choose "green" energy sources—sources that do not add greenhouse gases to the earth's atmosphere, do not consume nonrenewable

Hydrogen Gas—Fuel of the Future?

Solar Hydrogen Power

Adding solar and wind power to the mix of worldwide energy sources is a promising development, but neither one can provide uninterrupted power, which limits their practicality for power production. Eventually, it may be possible to store solar and wind-derived energy in batteries, but today's batteries are too inefficient and too costly. Ongoing research in battery technology may solve these problems. Another approach is to use excess power to pump water behind a dam where its potential energy can be tapped as needed.

Meanwhile, one of the most promising energy-storing strategies is to use solar or wind-generated

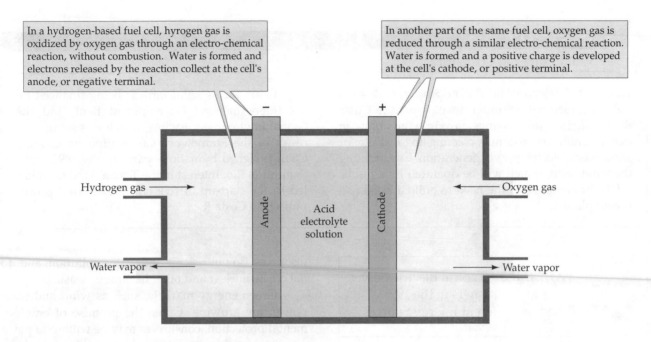

In a hydrogen-based fuel cell, hyrogen gas is oxidized by oxygen gas through an electro-chemical reaction, without combustion. Water is formed and electrons released by the reaction collect at the cell's anode, or negative terminal.

In another part of the same fuel cell, oxygen gas is reduced through a similar electro-chemical reaction. Water is formed and a positive charge is developed at the cell's cathode, or positive terminal.

Figure 11.8   Schematic drawing of a fuel cell.

electric power to produce hydrogen gas from water, a process called water-splitting or *electrolysis.*

Electrolysis is a very simple process in which an electric current is used to "split" water into its basic chemical elements: hydrogen and oxygen.

Hydrogen gas can then be stored or transported to places where it can be used to produce electric power through combustion or oxidation in fuel cells. Thus, hydrogen produced from electric power generated at solar and wind farms could be available for energy production when the sun is not shining or the wind is not blowing. Furthermore, hydrogen could be pumped through existing pipelines presently used to transport natural gas. Linking solar and wind power to hydrogen production makes good sense because hydrogen is an almost perfect fuel, as will be explained in the next section. Thus, fuel cells using solar-energy-derived hydrogen may emerge as practical power sources.

Hydrogen Combustion

Like natural gas, hydrogen burns in the presence of air, releasing energy. It can be used to fuel an automotive combustion engine or an industrial steam engine. The chemical reaction that occurs is:

$$H_2 + 1/2\,O_2 \rightarrow H_2O + \text{Heat}$$

Burning hydrogen as a fuel is not much different from burning gasoline except that it produces no carbon dioxide, carbon monoxide, sulfur dioxide, or particulate matter. However, as in almost all combustion processes, atmospheric nitrogen and oxygen combine to form oxides of nitrogen (NO_x), which contribute to acid rain and generate photochemical smog (see Chapter 5).

NASA space vehicle research and development has led to a better way of releasing hydrogen's energy by using fuel cells instead of burning the gas in a combustion engine. In a fuel cell, hydrogen and oxygen react electrochemically at ordinary temperatures and generate electric power (Figure 11.8). The advantages of fuel cells include all the benefits of using hydrogen in place of fossil fuels plus two more: No oxides of nitrogen are produced, and fuel cells are more energy efficient than internal combustion engines (Insight 11.5). Therefore, using hydrogen as a fuel instead of fossil fuels would greatly reduce air pollution, curtail acid rain, abate the emission of greenhouse gases, diminish dependence on imported oil, and enhance national security.

Hydrogen as an Automotive Fuel

One of the first important demonstrations of automotive solar hydrogen power is DaimlerChrysler's NECAR 4 (meaning "new electric car") announced in 1999 (Figure 11.9). As the first fuel-cell-powered passenger car in the United States, the NECAR 4 travels 450 kilometers ($\approx$ 280 miles) before refueling, reaches top speeds of 145 kilometers ($\approx$ 90 miles) per hour, and

INSIGHT 11.5

Efficiency of Energy Conversions

According to the First Law of Thermodynamics, energy changes from one form to another but is neither created nor destroyed. What appears to be lost energy is due to waste heat, an ineffective power source. Waste heat is related to the efficiency of energy conversion processes. More efficient processes release less waste heat. Here are some typical energy conversion efficiencies:

Device or Process	Efficiency (Percent)
Ordinary incandescent lamp	5
Fluorescent lighting	15
Internal combustion engine	20
Nuclear power plant	40
Coal-fired electric power plant	45
Fuel cell	75
Electric motors and transformers	90

Figure 11.9 DaimlerChrysler's 1999 NECAR 4 concept car fueled with liquid hydrogen and powered by a proton-exchange membrane filter fuel cell.
(Photo courtesy of DaimlerChrysler.)

releases only water vapor in its exhaust. The fuel, liquid hydrogen, is stored in a cryogenic (low-temperature) cylinder in the back of the vehicle. It reacts with oxygen in ordinary air in a proton-exchange membrane filter fuel cell, generating the energy that powers the car's electric motors.

The NECAR 4 fuel cell is small enough to fit in the vehicle's floor, allowing space for five passengers. As a concept vehicle, it is the first practical hydrogen-fueled car. And since its fuel cell releases no pollutants to the environment, it is a **zero-emission vehicle** (Insight 11.6).

The new DaimlerChrysler vehicle is not yet ready for production due to its high cost and some unsolved technical problems. Nevertheless, the development of its fuel cell represents a technical breakthrough that other automotive manufacturers can study and perhaps improve upon.

However, hydrogen gas may not be used in future vehicles, owing to its rather explosive nature. It may instead be employed in solid form, possibly as a metal hydride such as magnesium hydride (MgH_2). Hydrogen in this form is more stable, poses less risk

INSIGHT 11.6
Tomorrow's Zero-Emission Vehicles

In recent years, automobiles have become more fuel efficient and less polluting. Even so, automotive traffic in urban areas such as Los Angeles frequently causes major air pollution. Visibility is often diminished, unsafe high ozone levels are common, and the well-being of living things is compromised. Legislating more stringent air pollution standards, requiring manufacturers to make more efficient cars, and limiting automotive emissions have not solved these problems. Nor have they lessened the buildup of carbon dioxide in the earth's atmosphere—the inevitable consequence of burning fossil fuels such as gasoline and diesel oil to power internal combustion engines (see Chapter 12).

Scientists and government officials recognize that eventually more sustainable energy sources must be created to power cars, trucks, buses, and trains. Because urban areas of California are among the hardest hit by air pollution, the California Air Resources Board (CARB) mandated in 1990 that 2 percent of motor vehicles sold in California in 1998 and 10 percent of those sold in 2003 must be zero-emission vehicles (ZEVs). But in 1996, CARB modified these plans to encourage a market-based approach to ZEVs. Instead of holding to its 1998 mandate, leading U.S. auto manufacturers have agreed to adopt the requirements of a National Low-Emission Vehicle Program beginning in 2001, three years earlier than required under federal law. The year 2003 mandate (10 percent ZEV vehicles) remains in place. Massachusetts and New York have adopted similar regulations, while other states are considering them.

Tomorrow's zero-emission vehicles may well be electric powered if low-cost, lightweight, efficient batteries can be developed. The Clinton administration's Partnership for a New Generation of Vehicles (PNGV) is working with a number of automobile manufacturers to reach this goal. Southern California Edison is already using a fleet of battery-powered Ford Ecostars that have a range of 100 miles, acceleration of 0 to 50 miles per hour in 12 seconds, and top speeds of 75 miles per hour. But what will precede totally electric cars are hybrid electric vehicles such as the Toyota Prius, which is already for sale in Japan and slated to appear in North America in 2000.

The Prius is a four passenger sedan with a 1.5-liter gasoline engine, an electric motor, and a battery pack, all three occupying a space no bigger than an ordinary automatic transmission. When the vehicle's ignition key is switched on, battery power alone quietly provides needed acceleration. At cruising speed, the gasoline engine kicks in, with part of the power used to recharge the batteries and part to run the car. Both power sources are used to develop maximum acceleration (0 to 60 mph in 14 seconds). During coasting or when the brakes are applied, the car's electric motor becomes a generator, and the energy of the rotating wheels recharges the batteries. The Prius gets up to 110 kilometers ($\approx$ 68 miles) per gallon of gasoline and emits about 50 percent less air pollution than a similar-sized automobile but is not a zero-emission vehicle.

---

in case of an accident, and can be released from its hydride form as needed for fuel cell operation. Hydrogen fuel will not be cheap; estimates indicate its cost will be about two and a half times that of gasoline. Nevertheless, operating costs in terms of miles per gallon or kilometers per liter will be about the same as for today's vehicles for these reasons:

1. Fuel cells are 2.5 times more efficient than internal combustion engines, balancing out the higher cost of hydrogen.

2. Electric motors are much lighter than internal combustion engines and automotive transmissions. Their use will improve a vehicle's operat-

ing efficiency and increase its mileage per unit of energy.

3. A vehicle's motors can be designed to engage during coasting and braking so as to generate electric energy that can be stored in a battery.

Fuel cells similar to the one in the NECAR 4 can use fuels other than hydrogen such as methanol (wood alcohol) or even gasoline. Obviously, liquid fuels such as these are easier and safer to handle, but depending on the technology used, they may or may not qualify as true zero-emission vehicles. Furthermore, their use would continue our dependence on fossil fuels.

REGIONAL PERSPECTIVES

Pamplona, Spain: Wind Power

The Spanish city of Pamplona has long been known for its annual running of the bulls. But this mid-sized industrial center, the capital of the state of Navarra in the rugged Pyrenees region, is quickly gaining another distinction: the world's fastest-growing wind energy industry. Starting from scratch in 1995, Navarra was obtaining 23 percent of its electricity from the wind by the end of 1998.

With a population of 180,000, Pamplona's economy is based heavily on manufacturing, including a sizable car industry. But along with much of the rest of Spain, the city has had a relatively stagnant economy and a high rate of unemployment in recent years. In an effort to deal with that problem—and replace the coal and nuclear energy it imports from other parts of Spain with local power—Navarra introduced a set of tax incentives and other inducements for harnessing wind energy using locally manufactured turbines.

These policies paid off well beyond the wildest dreams of the government officials who crafted them. Several wind energy companies were quickly established in Navarra, most of them joint ventures owned in part by the Danish firms that supplied the technology. And much of the investment is coming from Energia Hidroelectrica de Navarra, the regional electric utility. These firms have provided a strong political base for the region's burgeoning wind power industry. Navarra's wind companies are already looking to expand their horizons to even larger potential markets in areas where Spain has strong historic ties, such as North Africa and South America.

The sudden transformation of Navarra's energy mix reflects a much larger trend: During the 1990s, wind power became the world's fastest-growing energy source, and it continues to be propelled by supportive new government policies in many nations around the world.

(This Regional Perspective was adapted from the article "Bull Market in Wind Energy" by Christopher Flavin, which appeared in the March/April, 1999, issue of *WorldWatch* magazine published by the Worldwatch Institute.)

Canada: Photovoltaic Power

In Canada, photovoltaic installed capacity grew at a rate of 30 percent a year between 1992 and 1998, and by the end of that period, about 3.4 MW of PV power was being generated using solar cells. In addition, Canada's investments in PV industrial development, which include PV module suppliers and installers, are also growing and reached $33 million (Canadian) in 1998. Specialized Canadian PV products include systems that can operate reliably in harsh winter climates.

Some of Canada's PV installations are standalone facilities without backup battery storage capabilities, while others employ batteries or are linked to small hybrid auxiliary power systems such as wind or diesel engines to supplement PV power.

Some of the more important PV-operated systems in Canada include remote residential power supplies, water irrigation pumping stations, and telecommunication devices such as those used by the Canadian Coast Guard. It is estimated that close to 7,000 Canadian navigational buoys, beacons, and lighthouses are powered by photovoltaic power.

Germany: Solar Roof Program

In 1999, Germany launched its new "100,000 Roofs Program," which called for PV electric power generating systems to be installed on 100,000 residences by 2005. The solar initiative began when the country's new coalition government, headed by Hermann Scheer, a renewable energy advocate, took power. It was expected that 6,000 PV systems would be in operation by 2000, followed by increasing numbers of annual installations.

Citizens who buy rooftop PV systems will be able to finance their investment over 10 years at zero interest and make no payments until two years after they are installed. Over one-third of the total program cost is being subsidized by the German government. These economic incentives may lead to the world's largest PV rooftop program, which is expected to result in lower solar-cell module costs, a prospect that will propel the future of photovoltaic energy.

California: Geothermal Energy

Seemingly unlimited reservoirs of hot molten rock (magma) lie beneath the earth's surface. To generate power using this heat source, wells would have to be drilled up to 35 kilometers (≈ 22 miles) deep in most places, and current limits to drilling technology make drilling boreholes this deep impossible. However, near-surface magma hot spots exist in a few places. While some of these hot spots are susceptible to depletion, others have been tapped to supply heat or provide power to local users.

One of the best examples of geothermal power generation is The Geysers in northern California, about 130 kilometers (≈ 80 miles) north of San Francisco. There, hot dry steam, hotter than the boiling point of water, is released by wells up to two miles deep and piped to nearby electric power generating plants (Figure 11.10). Some of these are operated by Pacific Gas & Electric Corporation (PG&E); others by independent power producers. The greatest power output, 2,000 MW, was achieved there in 1987. Since then, steam generating power has declined; in 1996, only 1,000 MW of power was produced, reflecting a 50 percent decline. No new power plants have been built at The Geysers since 1989.

Worldwide, close to 8,240 MW of geothermal electric power capacity existed in 1999, and geothermal power has become important in some countries. Table 11.4 identifies the world's top geothermal power-producing countries that year. The United States is the leader with 2,850 MW, most of which is in California, Hawaii, Nevada, and Utah. In Iceland's capital, Reykjavik, hot water pumped from geothermal hot spots below the city is used to heat homes and agricultural greenhouses. Similar projects were developed as early as 1900 in Klamath Falls, Oregon, and Boise, Idaho, to tap shallow geothermal sites for domestic and commercial heating.

Figure 11.10 The Geysers in northern California. Dry steam produced by subsurface hot rocks is tapped by drilled wells and piped to 28 nearby power plants operated by Pacific Gas & Electric Corporation since 1987.

(Photographer: Tony Batchelor, GeoScience Limited.)

TABLE 11.4 1999 Geothermal Electric Power Installed Capacity in Selected Countries

Country	Power (in MW)
United States	2,850
Philippines	1,848
Italy	768
Mexico	743
Indonesia	590
Japan	530
New Zealand	345
Iceland	140
Costa Rica	120
El Salvador	105

SOURCE: International Geothermal Association, http://www.demon.co.uk/geosci/wrtab.html

KEY TERMS

active solar heating	nonrenewable
anadromous	passive solar heating
electrolysis	photovoltaic
environmental externalities	renewable
	sustainable
heliostats	wind turbine
hydroelectric power	zero-emission vehicle

DISCUSSION QUESTIONS

1. Natural gas is a relatively cheap and environmentally clean fossil fuel. Since abundant supplies exist, should worldwide energy needs be met through expanded use of natural gas? Discuss your views.

2. Do you agree that federal and state energy subsidies should be used to encourage if not support certain kinds of energy developments? Explain your views.

3. What single sustainable energy strategy do you believe should be encouraged and subsidized as we enter the twenty-first century? Discuss the reasons for your choice.

4. If battery-powered vehicles are used to meet California's ZEV regulations, electricity generated by

coal-fired power plants in southwestern states such as New Mexico will likely be used to recharge automotive batteries. Discuss California's new air quality regulations in light of this fact.

5. As a consumer of electric power, given the opportunity to buy power from among different providers of electricity, would you choose a "green" source of electricity if it cost 25 percent more than a polluting source? Describe your reasons.

INDEPENDENT PROJECT

Meet with a local architect to discuss solar energy concepts for domestic heating, making hot water, or supplying electric power. Try to find out whether renewable energy options such as solar heating are becoming important in designing houses and commercial buildings. You may find that some architects feel it isn't practical to incorporate these concepts, while others believe the time has come to do so. Write a report on the current thinking in your town, city, region, or state.

INTERNET WEBSITES

Visit our website at *http://www.mhhe.com/environmentalscience/* for specific resources and Internet links on the following topics:

Solar Two power in the United States

U.S. Department of Energy—Renewable fuels

International Energy Outlook, U.S. Department of Energy: 1998 Solar energy resources—Related web sites

Solar cells in spacecraft

Wind energy and climate change

SUGGESTED READINGS

Annin, Peter. 1998. Power on the prairie. *Newsweek* (October 26):6.

Brown, Lester R., et al. 1998 and 1999. *State of the world. A report on progress toward a sustainable society.* New York; London: Worldwatch Institute.

Dunn, Seth. 1998. Green power spreads to California. *WorldWatch* 11(4):7.

Flavin, Christopher. 1998. Wind power sets new record. *Vital Signs*, 58–59.

———. 1999. Bull market in wind energy. *WorldWatch* 12(2):24–27.

——— and Nicholas Lenssen. 1995. Reinventing the automobile. *Solar Today* (January/February): 21–24.

Gray, Tom. 1998. Wind gets competitive in the U.S. *Solar Today* (March/April): 18–21.

Johnson, Jeff. 1998. The new world of solar energy. *Chemical and Engineering News* (March 30):24–28.

Lehman, Peter, and Christine Parra. 1994. Hydrogen fuel from the sun. *Solar Today* (September/October):20–22.

Osborn, Donald E. 1992. Using solar energy at the Sacramento Municipal Utility District. *Solar Today* (July/August): 11–14.

Rackstraw, Kevin. 1998. Wind around the world. *Solar Today* (March/April): 22–25.

Shinnar, Reuel. 1993. The rise and fall of Luz. *Chemtech* (January):50–53.

Smith, Zachary A. 1995. *The environmental policy paradox.* Englewood Cliffs, NJ: Prentice Hall.

Stone, Laurie. 1994. Solar cooking in developing countries. *Solar Today* (November/December):27–29.

Thayer, Burke Miller. 1994. Esperanza del Sol: Sustainable, affordable housing. *Solar Today* (May/June): 20–23.

Thomas, C.E. "Sandy." 1993. Solar hydrogen: A sustainable energy option. *Solar Today* (September/October):11–13.

U.S. Energy Information Agency, Department of Energy. 1998. Hydroelectricity and other renewable resources. Report #DOE/EIA-0484(98).

PART V

Global and Regional Environmental Problems

Digital Stock/Space

Chapter 12
Global Climate Change

Chapter 13
Stratospheric Ozone Depletion

Chapter 14
Managing Hazardous Wastes

246

CHAPTER 12

Global Climate Change

Digital Stock/Space Flight

Background

Factors Influencing the Earth's Climate
  Solar Radiation
  Greenhouse Gases
  Sulfate Aerosols and Global Cooling
  Changes in Ocean Temperature: El Niño and La Niña
  Milankovitch Cycles

Case Study: *Global Warming*

Global Temperature Trends Over Time
Causes of Global Warming
The Future of the Earth's Climate
  Projections Based on Computerized Data
  Projections Based on Cultural Developments
  Current Efforts to Minimize Global Warming

Regional Perspectives

Climate Change and Sea Level Rise
Climate Change and Coral Reefs
Climate Change in the Arctic

The observed increase in global mean temperature over the last century is unlikely to be entirely natural in origin . . . the balance of evidence suggests that there is a discernible human influence on global climate.

United Nations Intergovernmental Panel on
Climate Change (IPCC), 1995

Many people in many countries are concerned about **global warming**, the current worldwide trend toward warmer and warmer temperatures. Global warming has potentially serious consequences for the earth's environment such as rises in sea level affecting coastal marine life, altered precipitation patterns impacting agricultural areas, and higher average temperatures upsetting the balance of nature for both plants and animals. The global warming phenomenon is being studied by leading scientists and discussed in scientific papers, the popular press, and the electronic media. Approximately 2,500 scientists, working as a United Nations panel to study global climate change, issued the statement that appears at the beginning of this chapter. Their consensus is that global warming is occurring and that it is due at least in part to human activities.

Few dispute the fact that over the last 100 years the earth has experienced increasingly higher temperatures, but a good deal of controversy and uncertainty exists as to the cause. Is global warming the result of human activities or simply the outcome of natural planetary cycles? It is important to answer this question because, if humans are responsible, it may be possible to slow the trend. But waiting for scientific proof could take a long time and prevent us from acting soon enough to limit global warming and manage its impacts.

This chapter discusses various factors that affect the earth's climate, traces the development of the global warming trend, and presents possible scenarios for the future.

BACKGROUND

Factors Influencing the Earth's Climate

Any region's climate can be described as a chaotic system—one that is sensitive to small changes among large numbers of variables and behaves erratically. Some of the primary factors that act together to create climate will be discussed in this section, including solar radiation, greenhouse gases, volcanic eruptions, changes in ocean temperatures, and natural variations in the earth's rotation.

Solar Radiation

A key to understanding climate is understanding the role of the sun. As sunlight passes through the earth's atmosphere, about 36 percent of its energy on average is scattered and reflected back into space by atmospheric particles, clouds, and parts of the earth's surface itself, particularly glaciers, ice sheets, and oceans. The remaining 64 percent is absorbed by the lower atmosphere (troposphere) and by vegetation, soils, rocks, and surface waters (Figure 12.1).

Solar radiation comprises a spectrum of electromagnetic wavelengths of three types:

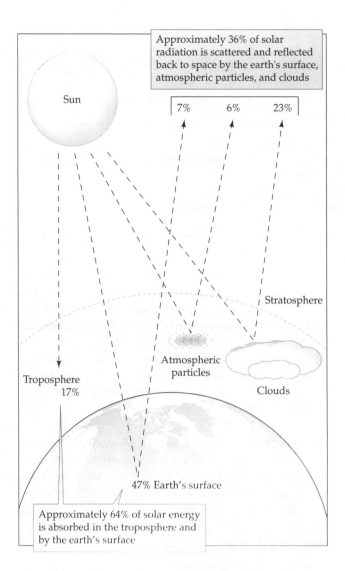

Approximately 36% of solar radiation is scattered and reflected back to space by the earth's surface, atmospheric particles, and clouds

7%   6%   23%

Sun

Stratosphere

Atmospheric particles

Clouds

Troposphere 17%

47% Earth's surface

Approximately 64% of solar energy is absorbed in the troposphere and by the earth's surface

Figure 12.1 Solar radiation is responsible for creating a habitable environment on the earth—the greenhouse effect. Incoming solar energy is partly absorbed by the earth's surface and atmosphere and partly reflected back to space. The planet is warmed by infrared radiation and by light energy converted to infrared energy.

1. *Visible light*, composed of relatively short wavelengths. Most of the solar energy that reaches the earth's surface is received as visible light because the atmosphere is quite transparent to these particular wavelengths. Visible light is the force that energizes photosynthetic primary production.

2. *Ultraviolet (UV) rays* having very short wavelengths. Only a fraction of the sun's ultraviolet radiation reaches the earth's surface because most of it is absorbed in the stratosphere and troposphere. Almost all of the potentially damaging UV-B radiation is blocked by stratospheric ozone, while less harmful UV-A radiation easily penetrates the earth's atmosphere. (The nature of solar electromagnetic radiation was introduced in

Chapter 4: specific impacts of UV radiation are discussed in detail in Chapter 13).

3. *Infrared (IR) rays*, or heat rays, having relatively long wavelengths. Infrared radiation is partially absorbed by carbon dioxide, water vapor, and other gases in the troposphere, which results in warming of the troposphere relative to the much colder stratosphere above it. Infrared radiation that is not absorbed in the troposphere passes through it, warming land areas, surface waters, vegetation, fish, and wildlife. The role of infrared radiation partly explains why planet earth is warm, but a larger warming effect results from the action of visible light, as will be explained in the next section.

Greenhouse Gases

When visible light is scattered and absorbed at the earth's surface, it changes into heat, part of which is trapped in the lower atmosphere by carbon dioxide gas and water vapor and then re-radiated back to the earth's surface. The carbon dioxide and water vapor behave somewhat like the glass in a greenhouse, letting visible light in but holding back some of the heat that would otherwise escape, a process referred to as **radiative forcing.** This is why carbon dioxide, water vapor, and a few other gases in the atmosphere are called **greenhouse gases**. This natural warming process is termed the **greenhouse effect.**

Naturally occurring levels of atmospheric greenhouse gases are highly beneficial to life on earth. In fact, it is estimated that without greenhouse gases, the earth's temperature would average about $-15\ °C\ (\approx 5\ °F)$. The greenhouse effect is therefore a critically important, natural process that makes life on earth possible, at least as we know it. But one thing is sure: Global warming will occur if the levels of greenhouse gases present in the earth's atmosphere increase, particularly levels of carbon dioxide. Atmospheric water vapor, on the other hand, varies considerably from day to day, and month to month, but shows no prevailing trend either to increase or decrease over time.

Carbon Dioxide   Concerns about global warming focus mainly on carbon dioxide because it is the most prevalent atmospheric greenhouse gas and its average annual concentration is rising. Many studies aimed at finding a solution to global warming attempt to identify sources of carbon dioxide, the rates at which they produce it, and the ability of natural processes to consume it. Table 12.1 lists the main sources of the world's carbon dioxide, both natural and anthropogenic (resulting from the actions of humans).

TABLE 12.1   Anthropogenic and Natural Sources of Carbon Dioxide

Anthropogenic Sources	Natural Sources
Large-scale coal-fired electric-power facilities	Volcanic emissions
Industries fueled by coal, oil, and natural gas	Weathering of carbonate minerals and rocks such as limestone
Homes heated by wood, oil, and natural gas	Natural forest fires
Automotive transportation based on fossil fuels	Plant and animal respiration
Land conversion; slash-and-burn agriculture	Decomposition of organic matter
	Release of CO_2 from seawater

Natural processes called **sinks** remove gases like carbon dioxide from the atmosphere. There are two important carbon dioxide sinks:

1. Photosynthesis, whereby vegetation, trees, lichens, and marine algae consume carbon dioxide to make plant matter.

2. Uptake by ocean waters, whereby carbon dioxide dissolves to form carbonates and bicarbonates.

Unfortunately, carbon dioxide is being added to the atmosphere faster than natural sinks are capable of removing it. To illustrate the slowness of ocean uptake, scientists estimate that if 1,000 carbon dioxide molecules had been released to the atmosphere 1,000 years ago, 985 of them would now be dissolved in the ocean and only 15 would be left in the atmosphere. That may sound like good news, but bear in mind that it would take 1,000 years for this to happen. And, unfortunately, the capacity of the oceans to trap carbon dioxide is declining as more of the gas dissolves. Calculations show that for every 1,000 carbon dioxide molecules released to the atmosphere today, 836 will ultimately dissolve in ocean waters, leaving 160 in the atmosphere.

Other Greenhouse Gases   The two most important atmospheric gases, nitrogen and oxygen, are not greenhouse gases—that is, they do not absorb appreciable amounts of infrared radiation. But certain other gases, mainly methane, nitrous oxide, chlorofluorocarbons, and ozone, are greenhouse gases. Like carbon dioxide and water vapor, their molecular structure enables them to absorb infrared energy and thereby trap heat. Currently, the atmospheric levels of all of these gases are increasing.

Methane   Also known as "natural gas," methane is formed when organic matter decomposes in wet, oxygen-deprived environments such as swamps, bogs, wetlands, and wet-rice paddies. The process is facili-

tated by **methanogenic bacteria**. Methane is also released from the gastrointestinal systems of cattle. Methane levels have more than doubled since pre-industrial times due mainly to the worldwide expansion of wet-rice agriculture and cattle ranching. While atmospheric levels of methane are much lower than those of carbon dioxide, its radiative forcing effect is about 20 times more potent per molecule than that of carbon dioxide.

Nitrous Oxide   Nitrous oxide is formed through bacterial decomposition of manure and organic fertilizers, decay of plant and animal matter, and forest fires. Nitrous oxide levels have risen only slightly in recent decades, but the radiative forcing exerted by nitrous oxide is about 200 times more potent than that exerted by carbon dioxide.

Chlorofluorocarbons (CFCs)   These are the chemical cooling agents once widely used in refrigerators, freezers, and air conditioners. They were also used as spray-can propellants and blowing agents in making foam insulation. As greenhouse gases, CFCs are about 10,000 times more potent per molecule than carbon dioxide. Their production in industrial nations was banned by the Montreal Protocol in 1996, and their use in developing countries is being phased out.

Ozone   The fact that ozone is a greenhouse gas is sometimes overlooked. As a naturally occurring gas in the stratosphere, ozone shields life on earth from harmful ultraviolet radiation and plays no role in global warming. But ozone in the lower atmosphere (troposphere), formed for example in photochemical smog, traps infrared radiation and behaves as a greenhouse gas. Ozone levels in the lower atmosphere are too low to affect global warming. The approximate relative heat-trapping contributions of various greenhouse gases is shown in Figure 12.2.

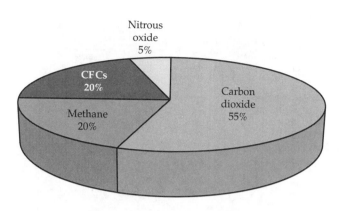

Figure 12.2 The relative contributions of greenhouse gases to global warming as a result of radiative forcing. Water vapor and ozone are not included in this graphic.

Sulfate Aerosols and Global Cooling

The greenhouse effect is not the only natural process influencing global temperatures. Because of their ability to affect the heat balance in the earth's atmosphere, sulfate aerosols must be considered when projecting global climate change. Sulfate **aerosols,** very small solid particulate matter, reflect a fraction of incoming solar energy back into space, causing global cooling. However, the cooling is of short duration because aerosols are particles, not gases. They fall out of the atmosphere or precipitate out in rain and snow.

One major source of sulfate aerosols is volcanic eruptions. The 1991 explosions of Mount Pinatubo in the Philippines and Mount Hudson in Chile ejected about 60 million tons of sulfur dioxide into the earth's atmosphere. The sulfur dioxide reacted with atmospheric oxygen and moisture to form sulfate aerosols that dispersed throughout the troposphere and even parts of the stratosphere. This happened immediately after the two eruptions, and in the next two years, 1992 and 1993, atmospheric cooling was close to 0.3 °C ($\approx$ 0.5 °F)—about 38 percent of the previous 118 years' warming. However, after 1993, global warming resumed, and 1994 became one of the hottest years on record.

Besides volcanic eruptions, sulfate aerosols originate from:

1. Combustion of sulfur-containing fossil fuels by electric utilities.
2. Coal-fired industrial boilers that make steam to operate machinery.
3. Coal smelting of metal ores to produce copper, lead, and nickel.

Since sulfate aerosols cause planetary cooling, they are referred to as **anti-greenhouse gases.** Their importance was once overlooked, but they are now in-

cluded in climate-change computer modeling. Nevertheless, major unknowns still exist concerning the role of aerosols in cloud formation and heat balance in the atmosphere.

Changes in Ocean Temperature: El Niño and La Niña

During normal ocean-circulation patterns in the tropical Pacific Ocean, equatorial trade winds drive warm ocean waters westward into the central Pacific. Seawater there ends up about one-half meter higher and approximately 8 °C ($\approx$ 14 °F) warmer than in the eastern Pacific. At the same time, eastern Pacific surface waters are cooled by the upwelling of colder waters from ocean depths driven by westward-blowing trade winds. This upwelling brings nutrient-rich bottom sediments to the surface along much of Central and South America's west coast, which supports a rich diversity of marine life.

El Niño is a major shift in this normal pattern, particularly along the South American west coast. It can occur every two to ten years, usually near the Christmas season, explaining its name, which is Spanish for "the Christ Child." El Niño interrupts westward-blowing trade winds, allowing warm waters in the central Pacific to flow eastward. This raises the temperature of eastern Pacific waters off northern Chile, Peru, and Central America by as much as 3 °C ($\approx$ 5 °F) above normal. Elevated ocean temperatures during an El Niño adversely affect marine fisheries, driving many species north to cooler waters and sometimes causing increased mortality.

It is thought that the severe 1972–73 El Niño caused the collapse of Peru's renowned anchovy fishery. Typical anchovy harvests of 20 million metric tons a year declined 90 percent in 1973, although overfishing is also believed to have played a role in the collapse. The 1982–83 El Niño elevated offshore water temperatures near Panama and Costa Rica by about 4 °C ($\approx$ 7 °F), inflicting considerable damage on the region's coral reefs. The most serious El Niño event of the past 100 years took place in 1997–98 and led to significant warming of coastal Pacific Ocean waters as far north as central California. Catastrophic impacts included torrential spring rains and mudslides in southern California and horrific rain, wind, and tornadoes in central Florida.

La Niña refers to unusually cold ocean temperatures in the equatorial central and eastern Pacific. It occurs when eastward-blowing trade winds intensify coastal upwelling of deep, cold waters near Peru and Equador. Sea surface temperatures can drop as much as 4 °C ($\approx$ 7 °F). La Niñas are known to affect global weather patterns by causing heavy monsoon rains in India, cool, wet winters in southeastern Africa, above-normal rainfall in eastern Australia, greater-than-normal snowfall west of the Great Lakes, and increased hurricane activity along the mid-Atlantic states.

Milankovitch Cycles

In 1920, Serbian scientist Milutin Milankovitch suggested that periodic variations in the earth's rotation and orbit around the sun could affect climate by triggering recurring climatological events, including ice ages and warm, interglacial periods. His theory is based on three planetary cycles that are known to repeat themselves over long periods of time:

1. Precession of the equinoxes, a gradual shift in the position of earth in its solar orbit when it is closest to the sun (perihelion). Today, perihelion occurs in January, but the date is not fixed. It is becoming later and later in the calendar year and will complete its cycle in about 21,000 years.

2. Variation in the tilt of the earth's axis, a recurring cycle whereby the tilt of the earth's axis with respect to its orbit (currently about 23.5 °) declines and then increases in value, repeating its cycle every 41,000 years.

3. Variation in the geometric shape of the earth's orbit, a very long-term cycle whereby the earth's orbit changes from nearly circular to an ellipse and back again to a circle, requiring about 100,000 years.

Some scientists argue that **Milankovitch cycles,** as they proceed, tend to reinforce natural climatic planetary events that lead to global warming and global cooling.

CASE STUDY: GLOBAL WARMING

Global Temperature Trends Over Time

Geologic research indicates that over the last four million years the earth has experienced at least 30 glacial periods, each lasting 90,000 years or more. They were separated by relatively short interglacial periods, each about 10,000 to 15,000 years long. About 110,000 years ago, an interglacial period was ending and a new ice age was beginning. Planetary ice caps expanded, ocean levels fell, and ice sheets advanced. In North America, this advance, known as the Laurentide glaciation, reached its maximum extent about 18,000 years ago, having drifted as far south as St. Louis, Missouri. In

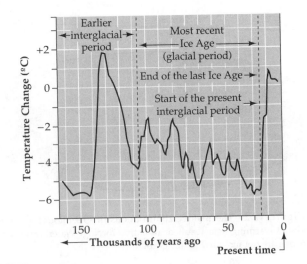

Figure 12.3 Estimated global temperatures during the past 160,000 years based on physical and chemical analyses of the Vostok ice core recovered by Soviet Antarctic explorations in east Antarctica. Changes in temperature in degrees Celsius are reported relative to average global temperatures today.

Europe, a similar ice sheet reached as far south as London, England. Some continental ice sheets became as thick as 3 kilometers ($\approx$ 1.9 miles). Then, about 15,000 years ago, a warming period began, marking the end of the Ice Age and the onset of the present interglacial period. Continental ice sheets started to recede, although they did not retreat totally from North America until about 5,000 years ago.

Temperatures during these ancient times are estimated by analyzing ice cores drilled from polar ice sheets, as described in Insight 12.1. Using this technique, scientists have been able to estimate atmospheric temperatures on earth as far back as 160,000 years ago. The results are plotted in Figure 12.3. They show, for example, that average global temperatures increased by as much as 6 °C ($\approx$ 11 °F) at the start of the current interglacial period.

Figure 12.4 focuses on global temperatures over the last 1,000 years, again based on ice-core analyses. Here we see that a Medieval Warm Period influenced parts of Europe and North America between A.D. 1000 and A.D. 1450. Historical records confirm that a more temperate climate prevailed during those centuries. For instance, agricultural pursuits flourished in Iceland, Greenland, and parts of Europe, though it was too cold to farm in these regions both before and after that time.

Following the Medieval Warm Period, a Little Ice Age descended on Europe and North America. Temperatures fell as glaciers that had retreated at the end of the Laurentide glaciation advanced once again. Colder weather set in, and farmers suffered crop failures due

INSIGHT 12.1

Taking the Earth's Temperature

Ice cores, cylinders of ice drilled out of a glacier or ice sheet, can be dated and used as time capsules for temperature and chemical studies. Several ice cores have been drilled in Greenland and Antarctica, but the most important one is the 3,623-meter-long **Vostok ice core** drilled from the south polar ice sheet in east Antarctica by Soviet scientists beginning in 1970. Analyses reveal that the deepest segments of this core are more than 160,000 years old.

The Vostok core is being used to estimate prevailing atmospheric temperatures on earth during past eras. Segments sliced from the core represent ice of increasing age with depth. The temperature estimates are made by determining the ratio of two oxygen **isotopes**—oxygen-18 and oxygen-16—in ice taken from various core segments. Here is how it works:

Oxygen-16 consists of oxygen atoms that each contain 8 protons and 8 neutrons in their nuclei. The number 16 is the isotope's mass number—the sum of its protons and neutrons. Oxygen-18 consists of oxygen atoms that each contain 8 protons and 10 neutrons in their nuclei. Its mass number is 18.

About 99.8 percent of the oxygen atoms in typical water molecules are oxygen-16, while about 0.2 percent are oxygen-18, the heavier isotope.

When water vapor condenses, the ratio of oxygen-18 to oxygen-16 in the rain, snow, or ice that is precipitated depends on prevailing air temperatures. At higher temperatures, precipitation containing oxygen-18 is favored over oxygen-16. Therefore, the ratio of oxygen-18 to oxygen-16 in an ice-core segment correlates with the prevailing air temperature at the time the ice was formed. During a period when global temperatures decline, a glacial period, the ratio of oxygen-18 to oxygen-16 declines. During planetary warming, an interglacial period, the ratio of oxygen-18 to oxygen-16 increases. Thus, the greater the ratio of oxygen-18 to oxygen-16 in a particular ice segment, the higher global temperatures were when the ice was formed. The deeper the core segment, the older the ice it contains. Ice samples are analyzed physically and chemically and dated using several different techniques, including correlating imbedded volcanic dust layers with historically documented volcanic eruptions.

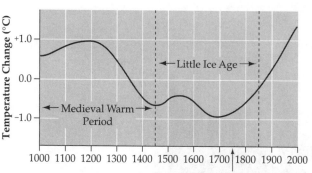

Figure 12.4 Estimated and projected average global temperatures during the past 1,000 years. The horizontal "0.0 °C" line represents average global temperature during the interval 1951–1980.
(Source: Data from United Nations Intergovernmental Panel on Climate Change [IPCC].)

to harsh winters. By the early 1900s, it was evident that the Little Ice Age was ending. Glaciers retreated, ice fields melted, and a warming trend embraced the Northern Hemisphere.

Fortunately, it is now possible to measure actual temperatures instead of relying on ice-core data. For over 100 years, temperatures have been recorded at

more than 2,000 meteorological stations around the world—at urban, suburban, and rural sites as well as on board ocean-going ships. When these data were analyzed by the United Nations Intergovernmental Panel on Climate Change (IPCC) in 1995, the consensus was that worldwide average atmospheric temperatures have increased significantly during the last century.

Average annual global temperatures between 1880 and 1998 are graphed in Figure 12.5. While cultural factors also enter in (Insight 12.2) and there is considerable year-to-year variation, the overall trend is clear: rising temperatures. The actual increase over this 118-year period is about 0.8 °C (≈ 1.4 °F). In addition, it turns out that the 14 warmest years over this time span are clustered within the last 20 years of the twentieth century, with 1998 being the warmest year in recent history.

These data have led most earth scientists to three conclusions:

1. Over very long periods of time, global temperatures have varied greatly, rising as much as 8 °C between glacial and interglacial periods.

2. During the past 118 years, average global temperatures have tended to increase and are continuing to increase.

INSIGHT 12.2

Urban Heat Islands

Although there is little question that global temperatures have increased during the past 100 years, there are some difficulties in interpreting the data. For example, temperatures measured in urban areas tend to overestimate true air temperatures due to the **urban heat island** effect. This effect occurs because urban structures absorb and re-radiate heat better than trees, grass, fields, and forests. Heat absorbed in a city's brick, stone, and concrete during the day is re-radiated to the surrounding air at night, causing higher temperatures to prevail. Increasing urbanization over the last century would be expected to contribute to higher than normal temperatures, but this effect has been carefully studied and can be corrected for.

More than 150 years of weather observations and temperature data have been recorded in Toronto, Canada, a city whose population grew from 13,000 in 1840 to more than four million in 1999. On the average, Toronto is about 2.0 °C (≈ 3.6 °F) warmer today than it was 100 years ago. But in Parry Sound, a rural area 225 kilometers (≈ 140 miles) northwest of Toronto, average temperatures during this time period rose only 0.4 °C (≈ 0.7 °F). Higher temperatures in the city of Toronto are due in large part to the heat island effect, the influence of heat captured each day in urban structures and re-radiated to the atmosphere at night. This effect, added to heat generated by factories and vehicles especially during winter months, creates pronounced localized warming in urban areas.

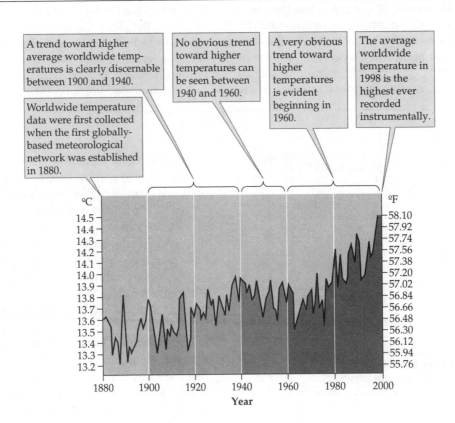

Figure 12.5   Average global temperatures between 1880 and 1998, a 118-year time period.
(Source: Data from James Hansen, NASA Goddard Institute for Space Studies, New York.)

3. On average, the earth's atmosphere is warmer now than at any other time in the last 1,000 years. Furthermore, the rate of atmospheric warming is greater now than at any earlier time during the past 1,000 years.

Causes of Global Warming

The Medieval Warm Period that began roughly 1,000 years ago was marked by temperatures that averaged about 1.0 °C (≈1.8 °F) higher than those recorded

INSIGHT 12.3

Atmospheric Carbon Dioxide

Our knowledge of carbon dioxide levels is based largely on atmospheric data collected at the Mauna Loa Observatory in Hawaii since 1958 (see Figure 12.6). The graph shows a trend toward increasing carbon dioxide levels as well as seasonal changes. Annual peaks (carbon dioxide maxima) recur each winter due to diminished photosynthesis in northern latitudes. Annual troughs (carbon dioxide minima) result from carbon dioxide demand caused by summer vegetative growth.

In Figure 12.6, the Mauna Loa graph is linked to estimates of atmospheric carbon dioxide levels in earlier years determined by analyzing air bubbles trapped in various segments of the Vostok ice core. The overall result is a representation of rising carbon dioxide levels during the last 140 years.

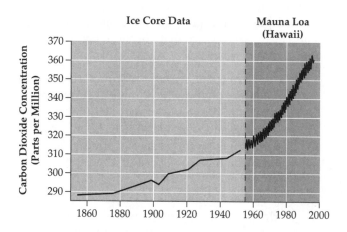

Figure 12.6 Northern Hemisphere carbon dioxide levels measured at the Mauna Loa Observatory in Hawaii and estimated from analyses of Vostok ice-core segments.
(Source: Data from Office of Science and Technology, Office of the President of the United States, 1997.)

1. The current warming trend appears to have started at the time of the Industrial Revolution when the use of coal to fuel industrial growth in England and other nations began to accelerate.

2. Rising levels of atmospheric carbon dioxide have been recorded in laboratory data collected in several countries. Measurements taken at the Mauna Loa Observatory in Hawaii since 1958 are plotted in Figure 12.6, together with estimates of atmospheric carbon dioxide in earlier years based on analyses of air bubbles trapped in segments of the Vostok ice core. The analyses show that preindustrial levels of atmospheric carbon dioxide were about 280 ppmv (parts per million by volume). By 1998, they had risen to 365 ppmv, an increase of 30 percent since the start of the Industrial Revolution. The ice-core studies show that atmospheric carbon dioxide levels today are at their highest level in 160,000 years. Greenhouse warming theory predicts that global temperatures will continue to increase as rising carbon dioxide levels trap more and more heat in the troposphere (Insight 12.3).

during 1951–1980, an arbitrary reference period representing contemporary times. Clearly, human activities were not responsible for this increase. Like other preindustrial temperature variations, it could have been due to natural events such as changes in the intensity of solar radiation reaching the earth or naturally occurring Milankovitch cycles. Given this perspective, some scientists believe that the approximately 0.8 °C (≈ 1.4 °F) rise in average global temperature during the last 118 years is not due to human influence at all.

However, most scientists have little doubt that rising levels of atmospheric carbon dioxide, originating primarily from the combustion of fossil fuels in both industrial and developing nations, are contributing to global warming. Two key observations support this view:

The Future of the Earth's Climate

As we have seen, climate is influenced by a variety of factors, many of which were discussed earlier in this chapter. Scientists are currently using their knowledge of these variables to forecast earth's climate in the future.

Projections Based on Computerized Data

It is possible to deal with atmospheric phenomena mathematically by developing computer-based models

a)   b)

Figure 12.7 (*a*) Typical low-elevation cumulus clouds. (*b*) Typical high-elevation cirrus clouds.

to forecast global climate change. Variables used to generate atmospheric circulation models include:

- Trends in levels of atmospheric greenhouse gases, especially those connected with human activities.
- Geographic landforms, historical temperature data, wind flow patterns, humidity, precipitation, ocean circulation, and atmospheric pressure gradients.
- Small changes in solar irradiance (energy output from the sun).
- The reflectivity (albedo) of clouds, surface waters, polar ice, glaciers, and atmospheric aerosol particulates, including volcanic emissions.
- Rates and patterns of stratospheric ozone depletion and their influence on lower atmospheric processes.
- Extent, types, and trends in forest and vegetative cover.

Here is an example of how these variables can interact: Warmer ocean waters result in increased rates of evaporation, which add more water vapor, a greenhouse gas, to the atmosphere and thereby reinforce a warming trend. This is an example of **positive feedback**, a mechanism that tends to reinforce and amplify an ongoing trend. But this may not happen. If increased atmospheric moisture causes low- and midlevel clouds such as cumulus clouds to form, they will reflect solar radiation away from the planet, creating a cooling effect (Figure 12.7*a*). This would be an example of **negative feedback**, a mechanism that tends to diminish or reverse a particular trend and maintain the status quo.

On the other hand, increased cloud cover does not guarantee cooling. If high-level clouds such as cirrus clouds form, they will behave like a greenhouse gas, reflecting heat back toward the earth (Figure 12.7*b*). At the present time, it is not known whether increased atmospheric moisture will result in more cloud cover. Nor is it known what types of clouds might

form. Thus, the role of clouds is one of the big unknowns in global warming theory, and computer model projections will not be completely useful until this role is understood. Studies of climate change are further complicated by ocean/atmosphere interactions including El Niño and La Niña discussed previously.

Recently, scientists working at the NOAA Geophysical Fluid Dynamics Laboratory at Princeton University developed a 100-year global warming forecast using a computer model they had created (Figure 12.8). The model forecasts a 5 °C (≈ 9 °F) temperature increase for northern mid-latitudes.

Projections Based on Cultural Developments

The United Nations IPCC has projected future levels of greenhouse gases and concurrent climate change scenarios for the year 2100 by considering different rates of world population growth and economic development. The three scenarios that appear in Figure 12.9 are summarized here:

1. The *high-emission scenario* assumes that world population growth will continue at current rates, high rates of economic activity will prevail, and energy demand will continue to be met mainly by fossil fuels. In this scenario, atmospheric carbon dioxide levels will rise to almost 900 ppmv by 2100.

2. The *mid-range emission scenario* assumes that world population growth will slow to about 1 percent per year, economic activity will increase moderately, and energy demand will depend largely on fossil fuels. In this scenario, atmospheric carbon dioxide levels will approach 700 ppmv by 2100.

3. The *low-emission scenario* assumes that world population will stabilize at about six billion, economic activity will not increase significantly, and the use of fossil fuels will decline to 1995 levels. In this scenario, atmospheric carbon dioxide levels will rise to about 470 ppmv by 2100.

If the high-emission scenario prevails, carbon dioxide levels will more than double by the year 2100. In that event, the following climate changes are projected:

- Average global temperature will rise 1.5–4.5 °C (≈ 2.7–8.1 °F).
- Mountain snow cover will diminish, glaciers will retreat, and an unknown but significant fraction of continental ice caps and sea ice will melt.
- Average sea level will rise 10–40 centimeters (≈ 4–16 inches).

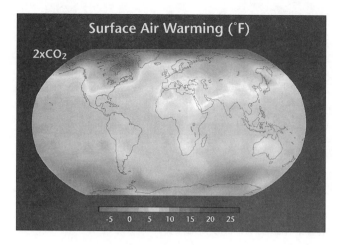

Figure 12.8 Computer-generated world map projecting global warming in the year 2100, assuming doubling of present atmospheric carbon dioxide levels. Northern mid-latitudes reflect average temperature increases of 5 °C (9 °F).
(Source: NOAA Geophysical Fluid Dynamics Laboratory, Princeton, NJ, 1994.)

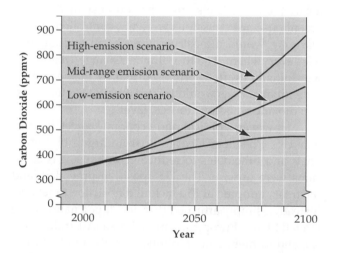

Figure 12.9 Atmospheric carbon dioxide emission scenarios projected to the year 2100.
(Source: D. Scimel, et al., United Nations Intergovernmental Panel on Climate Change [IPCC].)

- Some regions will experience increased precipitation and flooding, while others will face drought and more frequent forest fires.

- Agriculture, fishing, and wildlife will be affected in unpredictable ways. Climate change may also increase rates of species extinction.

- Warmer seas are likely to spawn more frequent, more intense, and more destructive storm events such as tornadoes and hurricanes.

Government leaders and scientists in many countries who take the high-emission scenario seriously are looking for ways to minimize or perhaps avoid these harmful effects on the environment.

Current Efforts to Minimize Global Warming

Limiting Dependence on Fossil Fuels   Today, immense economic and societal interests are connected to the continued use of fossil fuels. Indeed, most of the world's commerce and cultural amenities are energized by coal, petroleum, and natural gas. Also, as the world's population increases, demand for fossil fuels is likely to increase as well because having more people translates into needing more energy to grow crops, raise livestock, manufacture products, transport goods, and build factories, dwellings, bridges, and roads. Also, in both industrial and developing nations, people's expectations for a better lifestyle are rising. They want better housing, modern appliances, shopping malls, new cars, improved roads, and a long list of other amenities. They also want more nutritious food, including protein in meat, fish, and poultry. Almost everything that defines an improved lifestyle requires more energy, which in many countries is derived from fossil fuels.

Two different views prevail regarding the future of fossil fuels. The first is rooted in the knowledge that abundant fossil fuel reserves, mostly coal, still exist in many parts of the world—enough to meet energy needs for hundreds of years. They should therefore be used. The second view is that extracting and burning these fuels will add huge amounts of carbon dioxide to the earth's atmosphere, thereby accelerating climate change.

Those who take the second viewpoint seriously are advocating different ways humans can lessen their dependence on fossil fuels and minimize greenhouse gas emissions. One of the most promising is to use existing energy resources more efficiently—that is, diminish the amount of energy required to make products and provide services. Here are some of the ways energy efficiency is being enhanced:

1. Shifting industries and utilities from coal to natural gas. While both are fossil fuels, natural gas is rich in hydrogen and yields more energy per unit of fuel than coal, thereby lessening carbon dioxide emissions.

2. Converting conventional coal and oil furnaces to natural gas in homes and commercial buildings, thereby increasing energy efficiency from about 50 percent to 90 percent.

3. Improving commercial and residential insulation to conserve heat in cooler seasons and conserve air conditioning in warmer seasons.

4. Minimizing waste and waste products through improved industrial processes. For example, the Minnesota Mining and Manufacturing Company

has developed equipment and methods to recycle the solvents used in making magnetic video tape.

5. Substituting compact fluorescent light fixtures for incandescent lamps. Compact fluorescent lamps produce the same amount of light as incandescent lamps but consume only one-fourth the energy.

Shifting to Alternative Forms of Energy   The most important strategy to control human-promoted global climate change is to begin the transition from fossil fuels to alternative sources of energy, including solar and wind power. The transition has already started, and the examples are many (see Chapter 11). They include:

- Large-scale wind power installations in Germany, India, the United States, Denmark, Spain, and many other countries based on new sophisticated, reliable, cost-effective wind turbines.

- Photovoltaic, solar-powered projects such as the U.S. "million solar roofs" program, a similar large-scale solar roof program in Japan, and Germany's 100,000 solar roofs program.

- Solar-photovoltaic installations in Canada to power Coast Guard telecommunication devices and meet remote residential electricity requirements.

- DaimlerChrysler's hydrogen gas, fuel-cell-powered concept car.

- Designing buildings to (a) capture solar energy passively, (b) generate electric power using photovoltaic (PV) cells, and (c) make hot water using solar panels.

With an estimated two billion people in developing nations still without electric power and the cultural amenities electricity provides, the prospect of perfecting solar, wind, and other alternative energy sources is compelling. As the solar transition takes place, fossil fuels will undoubtedly assume a less commanding role in the overall mix of energy resources. Many believe that sustainable energy will ultimately prevail and global climate change will become less threatening.

The Rio Climate Change Treaty   The United Nations Conference on Environment and Development held June 3–14, 1992, in Rio de Janeiro, Brazil, is referred to as the first Earth Summit. Heads of state, delegates, and diplomats from 178 nations met to discuss environmental issues of global concern. Four treaty documents were developed: the Rio Declaration, the Framework Convention on Climate Change, the Convention on Biological Diversity, and Agenda 21, an environmental action plan for the twenty-first century.

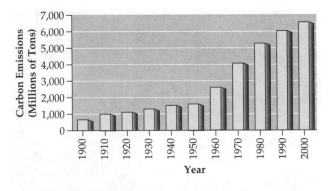

Figure 12.10   Total annual global carbon dioxide emissions in millions of tons, 1900–2000. Year 2000 emissions are estimated. (*Source: Data from State of the World, 1997.*)

At Rio, considerable attention was focused on carbon dioxide and other greenhouse gases, which many who attended the summit felt were already altering the planet's climate. The consensus was that if greenhouse gas emissions were not controlled, global warming would almost certainly harm human societies everywhere. The Rio Climate Change Treaty, as it was initially drafted, proposed that by the year 2000, individual nations would reduce their total annual carbon dioxide emissions to 1990 levels.

However, political and economic forces led former U.S. President George Bush to argue that strict goals and timetables to reduce greenhouse gas emissions were unwise. The U.S. position prevailed, and the treaty adopted at Rio included no action timetable. Instead, it simply affirmed that developed nations (the United States, European Union, Japan, Australia, Russia, and former Soviet republics) would aim to reduce annual greenhouse gas emissions to 1990 levels. However, since Rio, carbon dioxide emissions in most nations have risen rather than declined as many countries had pledged. Worldwide carbon dioxide emission trends over the past century are plotted in Figure 12.10.

In March 1995, the first follow-up meeting of the parties to the Rio Earth Summit was held in Berlin to again consider instituting limits to carbon dioxide emissions and a timetable to enforce them. It was hoped that such an agreement could be reached, but it did not happen. Instead, it was simply agreed that the climate change treaty in its present form was not adequate to protect the earth from global warming. Participants at the Berlin meeting decided that a new treaty should be developed and adopted at their 1997 meeting in Kyoto, Japan.

The Kyoto Protocol   Delegates from more than 160 nations met in Kyoto, Japan, in December 1997 to

TABLE 12.2   Reductions in Carbon Emissions per the Kyoto Protocol

Industrialized North American Economies	Year 1990 Emissions (Million Tons)	Year 2000 Emissions (Million Tons)	Share of Year 2000 Emissions (Percent)	Year 2000 Per Capita Emissions[a]	Kyoto Reduction (Percent)	Year 2010 Kyoto Emissions Target (Million Tons)
United States	1,346	1,577	23.9	5.73	−7	1,252
Canada	126	152	2.3	4.90	−6	118
Mexico	78	99	1.5	0.97	0	—
Western Europe	971	978	14.8	1.51	−8	893
Industrialized Asian Economies						
Japan	274	303	4.6	2.40	−6	258
Australasia[b]	90	107	1.6	4.00	+8	97
Totals for Industrialized Economies	2,885	3,216	48.7	2.98	−7	2,683
Developing Economies						
China	620	978	14.8	0.78	0	—
India	153	281	4.3	0.28	0	—
Other Asia	293	499	7.6	0.52	0	—
Middle East	194	253	3.8	2.10	0	—
Africa	178	219	3.3	0.30	0	—
Brazil	57	85	1.3	0.49	0	—
Other Central and South America	117	165	2.5	0.89	0	—
Totals for Developing Economies	1,612	2,480	37.6	0.56	0	—
Former Soviet Union	991	653	9.9	2.19	0	991
Eastern Europe	299	249	3.8	2.05	−7	278
World Total	5,787	6,598	100%	1.10	—	—

SOURCE: Data from Energy Information Administration, International Energy Annual, 1998, and U.S. Census Bureau.

[a]Year 2000 per capita emissions are given in metric tons per person.

[b]Australasia is defined to include Australia, New Zealand, and the U.S. Territories: American Samoa, Guam, Northern Mariana Islands, Puerto Rico, and Virgin Islands.

draft as strong a global warming treaty as possible, given the powerful political and economic factors at work to maintain the status quo. After 10 days of tough deliberations, it was decided that by 2010, each industrialized country will have reduced its carbon dioxide emissions by an agreed-upon percentage below 1990 levels. Different percentages were established for different countries. For example, European countries agreed to an 8 percent reduction, the United States to a 7 percent reduction, and Japan to a

6 percent reduction. Developing nations are presently exempted from the treaty. If it is to become part of international law, the Kyoto treaty must be ratified by 55 participating nations including the principal CO_2 emitting nations. Table 12.2 sets forth 1990 and projected year 2000 carbon dioxide emissions (in terms of total carbon), the share of total carbon for selected countries, agreed-upon Kyoto percentage reductions, and projected 2010 Kyoto emission target levels.

REGIONAL PERSPECTIVES

Climate Change and Sea Level Rise

The earth's oceans have risen and fallen many times throughout geologic history in response to recurring ice ages and interglacial warm periods. For example, during the last Ice Age, about 18,000 years ago, much of the planet's water was frozen in glaciers covering most of the Northern Hemisphere. With that much water locked up in glaciers, prevailing sea levels averaged about 100 meters (≈ 330 feet) lower than they are today.

The glaciers that existed during the last Ice Age have now melted except for polar ice sheets covering most of Greenland and Antarctica and a large number of smaller mountain glaciers worldwide. It is calculated that if these were to melt, sea level would rise by about 70 meters (≈ 230 feet), causing worldwide calamities. While this is not likely in the very near future, it turns out that western Antarctica's ice sheet is especially vulnerable to melting because it rests on marine waters whose temperature is rising, a trend likely connected to global warming. Geologic evidence suggests that if the western Antarctic ice sheet were to melt, earth's sea level would rise about 7 meters (≈ 23 feet) above present levels.

In addition to the prospect that Greenland and Antarctica's ice sheets will slowly melt as global warming proceeds, a more immediate though less dramatic rise in worldwide sea levels is taking place as a result of heat-driven expansion of ocean waters. Sea water, like most substances, expands as its temperature rises, occupying more space and causing higher sea levels. Studies have shown that average sea levels worldwide have risen approximately 12 centimeters (≈ 5 inches) over the last 100 years due in part to melting glaciers and in part to the expansion of warming oceans.

As we have seen, most scientists believe that today's global warming trend is driven primarily by greenhouse gases in the earth's lower atmosphere. Carbon dioxide is the most important greenhouse gas owing to its prevalence in the atmosphere. The National Academy of Sciences estimates that there is a 75 percent probability that lower atmosphere carbon dioxide levels will double from the year 1900 level of 280 parts per million to 560 parts per million by 2100. While considerable uncertainty exists regarding the consequences of doubling carbon dioxide levels, pro-

jected additions to sea level rise over the next century include:

- Expansion of ocean waters, adding about 50 centimeters (≈ 20 inches).
- Melting of mountain glaciers, adding about 20 centimeters (≈ 4 inches).
- Melting of parts of Greenland's ice sheet, adding about 12 centimeters (≈ 5 inches).
- Melting and breaking off of parts of Antarctica's ice sheet, adding possibly 24 centimeters (≈ 9 inches).

A conservative analysis of these scenarios has led some scientists to conclude that a 1-meter (100-centimeter) rise in sea level is a reasonable expectation over the next 100 years. If this occurs, worldwide coastal flooding and storm surges will be experienced more frequently. In addition, a number of coastal ecosystems and their living resources will be put at risk, including saltwater marshes, coral reefs, river deltas, mangrove forests, and seagrass meadows.

A 1-meter rise in sea level would seriously impact 6 million people in Egypt, 13 million in Bangladesh, and 72 million in China. It would incur an estimated $6.4 billion of damage in the United States alone. Coastal areas in south Florida, including a large fraction of the Everglades, would be severely impacted by a 1-meter rise in sea level.

Climate Change and Coral Reefs

Coral reefs are biologically diverse ecosystems that have existed in equatorial marine waters for close to 200 million years. They are made up of innumerable soft-bodied transparent sea animals called **coral polyps** that live symbiotically with marine algae. The polyps provide habitat and allow sunlight to reach the algae; in return, the algae, through photosynthesis, provide nutrients and energy to the polyps. Symbiotic algae also lend spectacular colors to coral polyps and coral reefs.

Coral polyps and symbiotic algae slowly secrete limestone layers that build up over hundreds of years, forming rocklike coral-reef support structures. The structural complexity of coral reefs provides habitat, protection, feeding, spawning, and nursery grounds for a great variety of fish that school nearby or occupy parts of the extensive network of reef fissures and crevices. The physical and biological complexity of coral reefs explains why they harbor such a rich biodiversity of marine life.

Coral reefs are important in ecotourism because many people who have never seen reefs are eager to snorkel, dive, and sometimes fish near them. Coral reefs are favorite tourism destinations, and a number of Pacific, Caribbean, and Indian Ocean islands depend on income from ecotourism. Reefs are also important because they protect many equatorial islands and coastal areas from storm damage, erosion, and flooding. In addition, some plant and animal biota associated with coral reefs are recognized as potential sources of new medicines and pharmaceutical agents.

Recent research suggests that coral reefs are sensitive to long-term climate change, especially higher-than-normal sea surface temperatures. When exposed to prolonged elevated temperatures, coral polyps lose their symbiotic algae and the diverse colors associated with them. As they lose their color, white limestone support structures become visible through the transparent polyps, a phenomenon known as **coral reef bleaching.** Although corals can recover from short-term bleaching, longer exposure to elevated temperatures kills polyps, causing permanent damage to the reef.

Large-scale coral reef bleaching was first observed in the early 1980s, and by the middle of that decade, reef bleaching was being reported every year at one or more ocean sites worldwide. Severe global-scale coral bleaching occurred in 1987, 1990, and 1998. Studies show that El Niño events by themselves do not explain this increase. Rather, the bleaching appears to be due mainly to global climate change, which is gradually causing rising sea surface temperatures.

Climate Change in the Arctic

The Arctic's floating mass of sea ice, which in winter months typically covers as much as 14 million square kilometers ($\approx$ 5.4 million square miles), is melting. Studies beginning in the early 1950s indicate that this melting trend accounts for a 3.6-percent-per-decade loss in ice cover and is accelerating. While several factors may be responsible, a key factor appears to be a warming trend of close to 0.5 °C per decade over recent decades affecting parts of Siberia, Alaska, and western Canada. In the early 1980s, relatively warm weather in the Arctic lasted for about 55 days. By the late 1990s, the warm season prevailed for about 75 days.

In 1997, scientists working in the Beaufort Sea found that sea ice known from earlier studies to average between 2 and 3 meters thick was typically only 1.5 meters thick. These observations were consistent with the finding that sea surface water temperatures were higher than normal and salinity was lower than normal. Salinity is a measure of dissolved salts in marine waters, and low salinity can be explained by the melting of very large areas of sea ice.

Unprecedented warming in the Arctic began in the mid-1800s. Evidence for the trend is based on lake and marine sediments, glacier ice cores, and tree rings measured at 29 different sites. They indicate that the highest arctic temperatures of the last 400 years occurred during this time. This warming trend coincides with observations of retreating glaciers, thawing permafrost, disappearing sea ice, and ecological shifts in terrestrial and aquatic ecosystems. Warming in the Arctic also parallels rising temperatures in Siberia where it is warmer now than at any other time in the past 1,000 years.

Most arctic marine species depend in one way or another on sea ice, and melting sea ice results in a loss of habitat for a number of arctic species. Even primary production is affected because marine algae utilize submerged sea ice surfaces during the winter and then bloom in the spring, providing an important food source for zooplankton, crustaceans, other invertebrates, and fish. It is believed that the loss of sea ice in the Arctic will threaten marine organisms dependent on ice floes for habitat and food resources. For example, polar bears hunt prey such as ringed seals from the edge of ice floes. The loss of sea ice may limit polar bear hunting grounds and threaten their survival.

KEY TERMS

aerosols	La Niña
anti-greenhouse gas	methanogenic bacteria
coral polyps	Milankovitch cycles
coral reef bleaching	negative feedback
El Niño	positive feedback
global warming	radiative forcing
greenhouse effect	sinks
greenhouse gases	urban heat island
isotopes	Vostok ice core

DISCUSSION QUESTIONS

1. Global temperatures since 1880 are plotted in Figure 12.5. Make a copy of this graph and project a trend line to the year 2020. Given this trend, what average global temperature would you forecast for 2020? By itself, does Figure 12.5 support the

view that global warming is caused by human influences? Discuss your answer.

2. Figure 12.4 shows that the Little Ice Age ended in about 1850. In your view, is it likely that the warming trend observed during the last century is simply due to the ending of this period? Discuss your view.

3. Deforestation in Brazil is due to urbanization, land conversion, and expanding slash-and-burn agriculture. It is argued that deforestation adds large quantities of carbon dioxide to the atmosphere. Identify the biological reasons this is probably true and discuss them in as much detail as you can.

4. In Germany, gasoline taxes are as high as $3 U.S. per gallon. Is this a fair and reasonable way to limit the use of a fossil fuel? Discuss your answer.

5. The abundance of coal in China appears essential to the country's industrial and economic progress. And like other developing nations, China regards the burning of fossil fuels as their inalienable right. Should developed nations such as the United States employ political, economic, or other sanctions to make China use coal more efficiently or shift to renewable/sustainable energy options? Discuss your answer.

INDEPENDENT PROJECT

It may be possible to obtain temperature data for your state or region going back 25 or 50 years. A local public or university library or a state or regional meteorological office or airport may be able to provide you with this information.

Create a graph of your data by plotting each year's average temperature (or average temperature for a particular month) for each of several years. Your y-axis will be temperature, and your x-axis will be particular years for which you have data.

Analyze your graph and decide for yourself whether a warming trend is evident or not. Write a brief report on this project, citing the sources of your data and your own findings.

INTERNET WEBSITES

Visit our website at *http://www.mhhe.com/environmentalscience/* for specific resources and Internet links on the following topics:

U.S.E.P.A. homepage on global warming

World Meteorological Organization

NASA and NOAA global temperature data

Graphic presentations of global warming trends

Environment Canada: Possible impacts of global warming

SUGGESTED READINGS

Brown, Lester R., et al. 1998 and 1999. *State of the world. A Worldwatch Institute report on progress toward a sustainable society.* New York; London: Worldwatch Institute.

———, Michael Renner, and Christopher Flavin. 1998. *Vital signs: The environmental trends that are shaping our future.* Worldwatch Institute. New York; London: W. W. Norton & Co.

———, Michael Renner, and Brian Halweil. 1999. *Vital signs: The environmental trends that are shaping our future.* Worldwatch Institute. New York; London: W. W. Norton & Co.

Climate change—State of knowledge. 1997. Office of Science and Technology Policy (OSTP) (October).

Flavin, Christopher. 1998. Last tango in Buenos Aires. *Worldwatch* 11(6):11–27.

Forsyth, Tim. 1998. Technology transfer and the climate change debate. *Environment* 40(9):16–20; 39–43.

Hansen, J., R. Ruedy, J. Glascoe, and M. Sato. 1999. *Giss analysis of surface temperature change.* New York: NASA Goddard Institute for Space Studies.

Kattenberg, A., et al. 1996. Climate models—Projections of future climate. P. 307 in *Climate Change 1950.* Cambridge, England: Cambridge University Press.

Lanchbery, John. 1998. Climate talks in Buenos Aires. *Environment* 40(8):6–20; 42–45.

McDonald, Alan. 1999. Combating acid deposition and climate change: Priorities for Asia. *Environment* 41(3):4–11; 34–41.

Sarmiento, Jorge L. 1993. Ocean carbon cycle. *Chemical and Engineering News* (May 31):30–43.

Schimel, D., et al. 1995. Radiative forcing of climate change. In *Climate Change 1995: The Science of Climate Change.* United Nations Environmental Programme, Intergovernmental Panel on Climate Change.

Schnoor, Jerald. 1993. The Rio Earth Summit: What does it mean? *Environmental Science and Technology* 27(1):18–22.

U.S. Energy Information Agency, Department of Energy. *International Energy Outlook, 1998, with Projection Through 2020.*

Victor, David G., and Julian E. Salt. 1994. Climate change. *Environment* 36(10):7–15; 25–27.

Zurer, Pamela S. 1995. Expectations sink for first meeting of parties to climate treaty. *Chemical and Engineering News* (March 13):27–30.

CHAPTER 13

Stratospheric Ozone Depletion

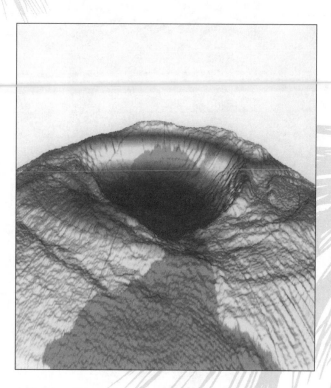

Goddard Space Flight Center, MD/ Russell Woolridge

We have been watching you. It has been at times fascinating, at times confusing, at

times horrifying. . . . We have had to keep reminding each other that what is actually

being debated here is the future of the ozone layer. . . . Remember that we will inherit

the consequences of your decisions. We will not sign the Montreal Protocol–you will.

You will not bear the brunt of ozone depletion–we will.

We demand that you think in the long term.

Australian teenager Susannah Begg, age 17,
speaking at the London ozone meeting,
June 27, 1990

Since the early 1980s, scientists studying the earth's atmosphere have recorded progressively decreasing levels of ozone in the stratosphere. Ozone is an atmospheric gas that surrounds the planet and is concentrated in a protective layer that shields all living things from damaging solar radiation. Normally, relatively constant ozone levels prevail from year to year, although they are known to increase and decrease seasonally especially above the earth's South Pole. However, certain chemicals in the atmosphere, stemming mainly from human activities, are affecting this normal pattern by causing episodes of significant ozone depletion that could threaten life on earth if the problem is not solved.

Abnormal stratosphere ozone depletion was first noted at the South Pole, a phenomenon now known as the "ozone hole." The computer-generated graphic appearing on page 264 is a digital representation of ozone levels above the South Pole during an ozone-hole event. Since the trend was discovered, there has been a concerted global effort to limit the use of ozone-depleting substances. In 1990, delegates met in London to draft regulations aimed at protecting the earth's ozone layer. Young people were present at the proceedings to see for themselves whether the delegates could agree to phase out the production of harmful substances; a quotation from one of the young attendees appears at the beginning of this chapter.

This chapter explains the chemistry behind natural ozone formation and depletion. It then provides a case study tracing the discovery of stratospheric ozone depletion and its causes—a discovery that serves as an example not only of scientists in action but also of nations working together to solve an environmental problem.

BACKGROUND

Stratospheric ozone depletion is a complex physical and chemical process that results from several interacting factors. In order to understand how and why it happens, we must first examine some of those factors, particularly the interplay of the earth's atmospheric layers and solar radiation.

The Role of the Troposphere and Stratosphere

You may recall from Chapter 5 that the earth's atmosphere is divided into four layers, each of them distinctly different from the others in elevation, temperature profile, and chemistry. Studies of ozone depletion focus mainly on the two lower layers, the troposphere and the stratosphere.

The **troposphere** extends from the earth's surface to an elevation of about 12 kilometers ($\approx$ 7.4 miles). Its chemistry is familiar to almost everyone, being composed of approximately 78 percent nitrogen gas, 21 percent oxygen gas, and 1 percent other gases. The troposphere also contains water vapor and aerosol particulate matter.

Lying directly above the troposphere, the **stratosphere** rises to a height of 52 kilometers ($\approx$ 32.4 miles). Because temperatures decline with increasing elevation in the troposphere, the boundary between the troposphere and the stratosphere has a temperature close to $-40\,°C$ ($-40\,°F$), cold enough to freeze many natural and synthetic chemicals.

Temperature is a critical factor because some substances that would otherwise "leak" from the troposphere into the stratosphere freeze at this boundary and go no further. For example, water vapor freezes before reaching the troposphere/stratosphere boundary and becomes ice crystals that in some cases form high cirrus clouds. However, other substances do not freeze at these temperatures and can therefore migrate freely between the troposphere and stratosphere. These include oxygen, nitrogen, and several man-made chemicals that will be described later.

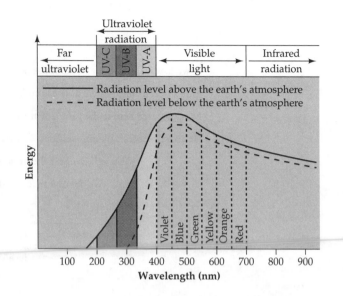

Figure 13.1 Solar energy and planet earth. Radiation from the sun spanning a broad band of different wavelengths reaches the earth's outer atmosphere. However, a significant fraction of energy-intensive, short-wavelength radiation is blocked by the planet's atmosphere.

Solar Radiation

The earth's atmosphere is exposed to radiation from the sun. The energy associated with radiation is a function of particular wavelengths, as diagrammed in Figure 13.1. Shorter wavelength radiation is seen as a color spectrum of **visible light** ranging from red to violet. But the radiation that most concerns us in this discussion is that of even shorter wavelengths (between 200 and 400 nm)—**ultraviolet radiation**. Ultraviolet (UV) rays possess greater energy than visible light and are capable of damaging living organisms. Three kinds of UV radiation are of special interest:

UV-A radiation UV-A rays, with wavelengths of 320–400 nm, have energy levels similar to those of visible light. About half of incoming solar UV-A radiation is absorbed by the earth's atmosphere, while the other half reaches the earth's surface. UV-A is not considered very harmful to plant and animal life, but it can cause sunburns in humans, depending on the sun's intensity and the length of exposure.

UV-B radiation UV-B rays, having wavelengths of 280–320 nm, are more energetic than UV-A rays and can be very harmful to living things. UV-B rays are somewhat like X rays because they can penetrate living matter and damage biological materials such as DNA, the molecule that carries coded hereditary information. Only a small percentage of solar UV-B radiation normally reaches the planet's surface, but susceptible individuals—especially those with blue eyes and fair skin—can suffer serious skin damage that may even trigger skin cancer. Life on earth as we know it would be impossible if a large fraction of solar UV-B radiation reached the planet's surface. But fortunately, UV-B radiation is blocked to a very great extent by the ozone in the earth's stratosphere.

UV-C radiation UV-C rays, having wavelengths of 200–280 nm, possess very high energy levels—enough to split chemical bonds. Although they are potentially very harmful to living things, solar UV-C is totally absorbed in the earth's stratosphere where it reacts with atmospheric gases before reaching the troposphere. But although no UV-C rays reach the earth's surface, they play a critical role in ozone formation, as we will see shortly.

Ozone Formation and Depletion

Natural Formation of Ozone

Relatively low, harmless levels of ozone occur naturally in the troposphere, largely as a result of lightning strikes. Periodically elevated ozone levels occur in urban areas as a result of photochemical smog (see Chapter 5). But higher ozone levels exist in the stratosphere, concentrated in a layer above the earth's surface that blocks harmful solar rays.

Ozone is a naturally occurring form of oxygen whose molecules consist of three atoms of oxygen, instead of the two atoms in ordinary oxygen. Ozone is formed when the high energy associated with short-wavelength UV-C rays splits oxygen molecules (O_2) in the stratosphere into oxygen atoms. Individual oxygen atoms produced this way react with oxygen molecules to form ozone (O_3). UV-C radiation is therefore responsible for producing stratospheric ozone, and this reaction takes place only in sunlit parts of the stratosphere. Here are the two chemical reactions that produce ozone:

$$O_2 + \text{UV-C} \rightarrow O + O$$

$$O + O_2 \rightarrow O_3$$

How Depletion Occurs

Depletion of ozone also occurs naturally in the earth's stratosphere, as molecules of ozone absorb UV-B radiation and are transformed into molecules of oxygen:

$$2\,O_3 + \text{UV-B} \rightarrow 3\,O_2$$

The critical role of stratospheric ozone in protecting life on earth is reflected in this equation. Ozone absorbs (traps) more than 90 percent of incoming solar short-wavelength UV-B radiation, most of which would otherwise reach the earth's surface. In the process, ozone is converted into oxygen.

Under normal conditions, the balance between ozone-forming reactions and ozone-depleting reactions results in a steady-state equilibrium, and stratospheric ozone levels remain fairly constant—about 10 parts per million in a layer approximately 25 kilometers ($\approx$ 15 miles) above the earth's surface. However, this equilibrium can become unbalanced if certain ozone-depleting chemicals, some natural and some man-made enter the stratosphere. This happens when certain chemicals do not react with atmospheric substances such as oxygen, persist long enough to reach the outer limits of the troposphere, and enter the stratosphere. Whether or not this happens depends on the stability of a particular chemical and how it behaves at the very cold troposphere/stratosphere boundary. Most chemicals do not survive long enough in the troposphere to reach the stratosphere.

In order for a chemical to initiate stratospheric ozone depletion, three conditions are required:

1. A chemical containing chlorine or bromine must rise through the earth's lower atmosphere and enter the stratosphere. It is the chlorine and bromine atoms split from these substances that initiate ozone depletion.

2. Intense stratospheric UV radiation must be present to split chlorine or bromine atoms from an ozone-depleting substance. South polar UV radiation becomes most intense at the start of the antarctic springtime in September each year. North polar solar radiation becomes most intense in March of each year.

3. A polar vortex—stratospheric winds circling the polar region—is required to sustain temperatures as low as $-90\,°C$ ($\approx -130\,°F$). These conditions favor the formation of stratospheric ice crystals whose surfaces act as catalysts, triggering the release of chlorine or bromine atoms, which then initiate ozone-depleting chain reactions.

Common Ozone-Depleting Substances

Currently, atmospheric scientists believe that the most important ozone-depleting chemicals are:

1. Chlorofluorocarbons (CFCs), common refrigerants and cooling agents.

2. Halons, a group of fire-extinguishing chemicals.

3. Carbon tetrachloride, a dry-cleaning agent and industrial solvent.

4. Methyl chloroform, an industrial solvent.

5. Methyl chloride, a naturally occurring substance.

6. Methyl bromide, both a natural substance and a synthetic chemical pesticide.

Insight 13.1 describes how some of these chemicals were used before their potentially harmful effect on the ozone layer was realized. They are relatively stable when released to the air, and their stability allows them to persist long enough to enter the stratosphere where their chlorine or bromine atoms trigger ozone depletion. Figure 13.2 shows past and estimated future levels of these chemicals in the stratosphere, expressed in terms of *stratospheric chlorine levels*.

INSIGHT 13.1
Ozone-Depleting Substances

Chemicals that persist in the troposphere, leak into the stratosphere, and react with ozone under the influence of intense solar radiation are referred to as **ozone-depleting substances.** Because they freeze well below the temperatures occurring at the troposphere/stratosphere boundary, they easily rise into the stratosphere. But, though these substances are now known to have harmful effects on the earth's ozone, they originally served useful purposes, and it has sometimes been difficult to find substitutes for them. In the United States, chemicals like the CFCs, halons, methyl bromide, and a number of similar ozone-depleting substances are now controlled by the 1990 Clean Air Act.

CFCs

In the 1930s, a new family of chemical refrigerants was developed by the DuPont Corporation, one of the oldest chemical manufacturers in the United States. They are simple chemicals containing chlorine, fluorine, and carbon called chlorofluorocarbons (CFCs). CFCs are nontoxic, nonflammable, stable substances that function significantly better than earlier refrigerants such as ammonia, a highly toxic gas. Several different CFC refrigerants were developed and marketed under the generic name, **Freon**. For example, Freon-12 (dichlorodifluoro-methane, CCl_2F_2) was used in automobile air conditioners and household refrigerators; it is also known as CFC-12. (The number assigned to CFCs is the key to their chemical formulas. Using the "rule of 90," 90 is added to the CFC number; the sum reflects numbers of carbon, hydrogen, and fluorine atoms. For example, in the case of Freon-12: 90 + 12 = 102, meaning there are 1 carbon atom, 0 hydrogen atoms, and 2 fluorine atoms. Missing atoms are assumed to be chlorine. Hence, the formula CCl_2F_2.)

The CFCs were also used as propellant gases in household spray-can products such as deodorants, paints, and hair products as well as in making plastic foam packaging and insulation materials. The electronics industry used CFCs for cleaning circuit boards and microchip units. The development of CFCs is an industrial success story because their inherent stability made home refrigerators safe and reliable.

Halons

Halons are chemicals typically used in fire extinguishers. They are related to the CFCs but contain at least one bromine atom. One of the common halons is bromochlorodifluoromethane, Halon-1211 ($CBrClF_2$).

(Continued)

Stratospheric chlorine level is a measure of the total level of ozone-depleting substances in the stratosphere. When this measure is used, bromine-containing substances are converted to equivalent levels of chlorine-containing substances. Totals are reported as parts per billion of chlorine by volume (ppbv). One ppbv is equivalent to adding one drop of a substance to a swimming pool 10 feet wide, 30 feet long, and 8 feet deep, filled with water.

Biological Impacts of UV-B Radiation

Many studies have been conducted to discover the effects of UV-B radiation on the earth's environment. Some of the results are given here:

- The impact of UV-B radiation on antarctic marine algae was studied by University of California biologists in 1993 at a time when south polar ozone depletion was at a maximum. Although algal growth rates were found to be 6 percent to 12 percent lower than normal, this is considered a significant but not catastrophic decline. It appears that increased south-polar UV-B radiation does not seriously impact antarctic marine plankton, fish, and mammals. (See Regional Perspectives at the end of the chapter for information about other studies in this area.)

- Studies on the effect of increased UV-B on 200 agricultural crops have shown that close to half of the species tested were adversely affected. That is, some species adapted to increased UV-B, while others suffered serious damage. One variety of soybean was not affected at all, but another variety displayed a 25 percent reduction in growth rate.

- UV-B radiation causes an increased incidence of eye cataracts in animals and humans. For example, there are more cataracts among people living at high elevations in Bolivia and Tibet where UV-B radiation is stronger than at lower elevations. Based on a projected 7 percent decline in

INSIGHT 13.1
(Concluded)

Halons are particularly valuable in fire-fighting and are used in commercial fire extinguishers in restaurants, factories, and aircraft. However, they are powerful ozone-depleting substances. A single bromine atom is 40 times more effective in destroying stratospheric ozone than a chlorine atom.

Methyl Chloride

Methyl chloride (CH_3Cl) is released to the earth's atmosphere by marine algae as a normal part of their life cycle. Some of this naturally occurring chemical rises to the stratosphere where it acts as an ozone-depleting substance. Its presence in the earth's atmosphere is part of the balance between ozone formation and ozone destruction in the stratosphere.

Methyl Bromide

Methyl bromide (CH_3Br) is produced naturally by marine algae and during forest fires when biomass containing small amounts of bromine burns. But of greater concern is its use worldwide as a broad-spectrum pesticide to control a variety of agricultural pests, including insects, nematodes (root worms), weeds, pathogens, and rodents. As much as 50 million kilograms ($\approx$ 54 thousand tons) of methyl bromide are used annually around the world, most of it to fumigate soils prior to planting crops such as strawberries and tomatoes. Soil fumigation generally involves injecting methyl bromide gas 12 to 24 inches below ground level to kill pest organisms. Sometimes treated soils are covered with plastic tarps to restrict pesticide escape to the air, but ultimately most of it enters the atmosphere.

In addition, methyl bromide is used as a post-harvest pest control agent for agricultural commodities such as grapes, raisins, cherries, and nuts during storage and treatment. The chemical is also used to sterilize a number of imported and exported food products during required quarantine periods. For example, the United States requires that all grapes imported from Chile be treated with methyl bromide before entering the U.S. marketplace. Japan requires that all U.S. cherries and apples be sprayed with methyl bromide. Other uses include fumigating buildings for termites, aircraft for rodents, and food processing facilities for insects.

Recent research has confirmed that methyl bromide is a powerful ozone-depleting chemical. Once released in the earth's lower atmosphere, it mixes with other gases, rises to higher elevations, and enters the stratosphere, where it triggers ozone-depleting chain reactions. It is estimated that agricultural uses of methyl bromide account for 15 percent of stratospheric ozone depletion.

---

ozone over Northern Hemisphere mid-latitudes, a 3 percent increase in occurrence of human eye cataracts is predicted.

- Because of its damaging effects on DNA, UV-B radiation can trigger reproductive malfunctions in exposed cells, especially skin tissue, as has been confirmed in mice. The most common outcome is non-melanoma skin cancers, mainly squamous and basal cell carcinomas that are usually treatable. It is predicted that a 1 percent decline in total ozone will lead to a 2 percent increase in the incidence of human skin cancers.

- Increased risk of melanoma skin cancer is also associated with UV radiation. Melanoma is a more serious type of skin cancer than non-melanoma cancers and is associated with increased mortality. Recent epidemiological studies link high levels of sunlight exposure during childhood with a greatly increased risk of developing melanoma skin cancer later in life.

- Laboratory studies involving mice have shown that exposure to UV-B radiation also diminishes their ability to combat skin cancers. The research suggests that an animal's immune system is suppressed by UV rays. Similarly, humans treated with immuno-suppressive, anti-rejection organ-transplant medication encounter increased risks of skin cancer with sun exposure.

The story of how the ozone hole was discovered—and reactions to the discovery—is traced in the following case study.

CASE STUDY: THE OZONE HOLE

Rowland and Molina's Research: The First Clue

In 1974, several years before the ozone hole was discovered, Sherwood Rowland and Mario Molina, research

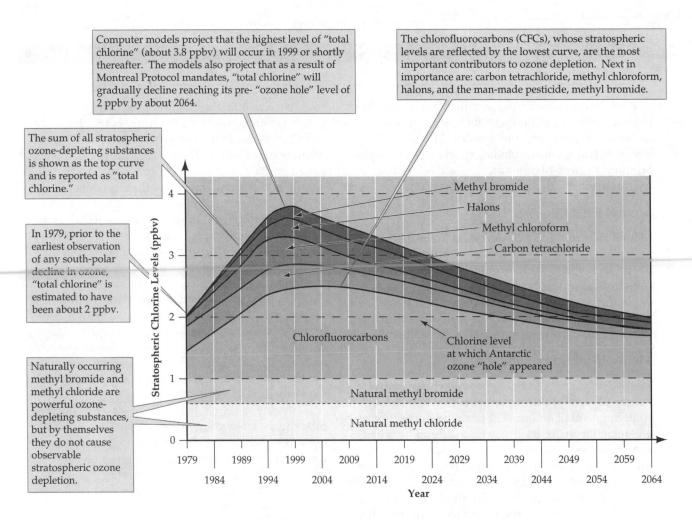

Computer models project that the highest level of "total chlorine" (about 3.8 ppbv) will occur in 1999 or shortly thereafter. The models also project that as a result of Montreal Protocol mandates, "total chlorine" will gradually decline reaching its pre- "ozone hole" level of 2 ppbv by about 2064.

The chlorofluorocarbons (CFCs), whose stratospheric levels are reflected by the lowest curve, are the most important contributors to ozone depletion. Next in importance are: carbon tetrachloride, methyl chloroform, halons, and the man-made pesticide, methyl bromide.

The sum of all stratospheric ozone-depleting substances is shown as the top curve and is reported as "total chlorine."

In 1979, prior to the earliest observation of any south-polar decline in ozone, "total chlorine" is estimated to have been about 2 ppbv.

Naturally occurring methyl bromide and methyl chloride are powerful ozone-depleting substances, but by themselves they do not cause observable stratospheric ozone depletion.

Figure 13.2 Estimates and projections of past and future levels of ozone-depleting substances in the earth's stratosphere. (*Source: Graphic adapted from original by F. A. Vogelsberg, Jr.*)

scientists at the University of California, reported the results of their studies of CFCs. They predicted that CFCs would rise in the troposphere and enter the stratosphere where they would attack and deplete ozone.

Over time, CFCs accidentally escape and disperse into the troposphere, and because they are stable chemicals, they persist there for decades. This persistence allows them to survive in the lower atmosphere long enough to rise to the stratosphere where intense UV-B radiation splits chlorine atoms from their molecules. For example, the refrigerant CFC-12, known as Freon-12, would react as follows

$$CCl_2F_2 + UV\text{-}B \rightarrow Cl + CClF_2$$

where CCl_2F_2 is the chemical formula for Freon-12.

UV-B is solar ultraviolet radiation in the stratosphere.

Cl is a highly energetic chlorine atom produced when Freon-12 splits.

$CClF_2$ is a molecular fragment produced by this reaction.

Rowland calculated that a single chlorine atom could initiate an ozone-destroying chain reaction that would result in the loss of at least 100,000 ozone molecules. The two reactions envisioned were:

$$Cl + O_3 \rightarrow ClO + O_2$$

$$ClO + O \rightarrow O_2 + Cl$$

In the above reactions, a chlorine atom (Cl) splits an ozone molecule, forming chlorine monoxide (ClO) and oxygen. Chlorine monoxide (ClO) then reacts with an oxygen atom to form an oxygen molecule and another chlorine atom. Notice what Rowland and Molina feared would happen: A chlorine atom would consume an ozone molecule and then be regenerated in a subsequent reaction to start the cycle over again—a chain reaction.

Rowland later confided, "Our immediate reaction to these findings was that there had to be some huge error." But they could not find an error in their work, and Rowland began to state publicly that CFC production, a $28-billion-a-year business at that time, should be stopped. The industry's response was, "This is pure nonsense." At that time, there was no scientific evidence of any actual ozone decline in the stratosphere, nor were there any signs of increasing UV-B radiation reaching the planet's surface. All Rowland and Molina had done was develop and publish a carefully researched scientific theory, but few scientists took notice.

Antarctic Researchers Detect Ozone Depletion

Since 1956, Joseph Farman and fellow scientists working with the British Antarctic Survey at Halley Bay, Antarctica, have been investigating the chemistry of the earth's atmosphere. Their work showed that ozone levels in the stratosphere above the South Pole typically averaged about 325 *Dobson units*. A relatively small decline occurred periodically between September and November each year, but ozone levels would then return to normal.

Dobson units are measured using the Dobson spectrophotometer, which determines atmospheric ozone levels from the ground up by measuring UV-B radiation at four wavelengths, two where ozone absorbs UV-B and two where it does not. One Dobson unit is defined as a 0.01 mm thickness of ozone at 0 °C and 1 atmosphere of pressure. For example, 100 Dobson units implies that the ozone level being measured would form a layer 1 mm thick if subjected to 0 °C and 1 atmosphere of pressure.

However, beginning in 1980, Farman's group observed greater than normal declines in stratospheric ozone over the South Pole, and the trend continued, with lower and lower ozone levels being found between September and November of each year. In October 1984, ozone levels above Halley Bay dropped to 200 Dobson units, a 38 percent decline. The decline continued in subsequent years, as shown in Figure 13.3

Reports of ozone depletion over Antarctica soon became widely known among scientists and world leaders. It seemed clear that a potentially serious problem was unfolding. Recurring declines in stratospheric ozone levels affecting geographic areas larger than Antarctica had the potential of allowing high levels of UV-B radiation to impact marine plankton, fisheries, wildlife, and perhaps even human populations.

As studies of the stratosphere above the South Pole continued, new insights were obtained. Chemical

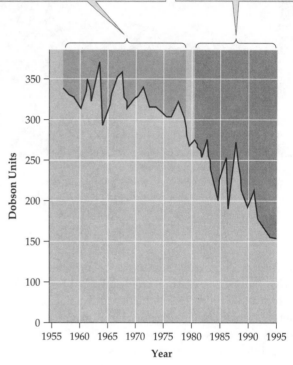

Up to 1979, British scientists working at Halley Bay, Antarctica observed periodic increases and decreases in stratospheric ozone levels from September to November each year. Ozone levels averaged close to 325 Dobson units and no trend toward lower ozone levels was seen.

In 1980, a significant south-polar ozone-level decline occurred; this proved to be the beginning of a trend toward lower and lower ozone levels. By 1995, ozone levels had fallen to about 150 Dobson units.

Figure 13.3 Stratospheric ozone levels above Halley Bay, Antarctica, reported in Dobson units averaged for September, October, and November of each year between 1957 and 1995. *(From R. D. Bojkov, "The Changing Ozone Layer," 1995, World Meteorological Organization.)*

analyses of stratospheric gases collected by high-flying aircraft demonstrated the presence of chlorine monoxide (ClO), a substance most scientists did not expect to find in the stratosphere. This discovery supported Rowland and Molina's theory, which had predicted the formation of chlorine monoxide as a part of human-caused ozone destruction.

On October 5, 1987, NASA reported unusually extensive ozone depletion above Antarctica (Figure 13.4). South polar stratospheric ozone had declined to 125 Dobson units, the lowest level recorded up to that time. In December, two months later, the ozone hole fragmented, and an ozone-depleted air mass drifted northward, passing over Melbourne, Australia, where it caused record-low stratospheric ozone levels.

Ground-based and earth-satellite monitoring of stratospheric ozone at meteorological stations around the world indicate that ozone depletion is continuing

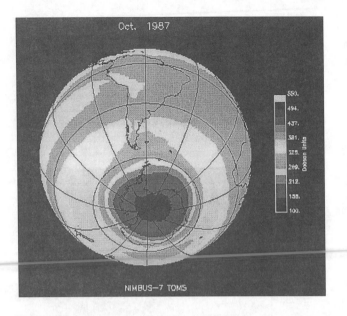

Figure 13.4 Depletion of south polar stratospheric ozone shown in terms of average Dobson unit values for the month of October 1987. *(Source: Russell Wooldridge, NASA Goddard Space Flight Center, Greenbelt, Maryland.)*

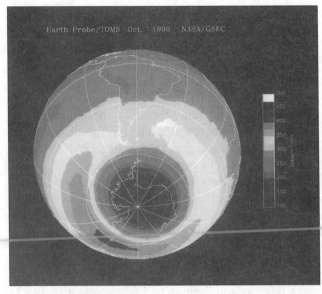

Figure 13.5 Depletion of south polar stratospheric ozone shown in terms of average Dobson unit values for the month of October 1998. *(Source: Russell Wooldridge, NASA Goddard Space Flight Center, Greenbelt, Maryland.)*

to occur, with the most significant declines taking place above Antarctica during the Southern Hemisphere's springtime. Figure 13.5 depicts south polar ozone depletion as of October 1998. By common agreement, scientists report a south polar "ozone hole" when ozone levels fall below 220 Dobson units.

The Problem Spreads

In 1989, ozone depletion was reported for the first time over the North Pole. This came as a surprise to many scientists who believed that the atmospheric conditions required for ozone depletion existed only at the South Pole. The discovery of losses in arctic ozone added to worries that ozone depletion could become a global problem affecting not only the polar regions but Northern and Southern Hemisphere mid-latitudes as well.

Stratospheric ozone levels above the Arctic are typically as high as 450 Dobson units, although somewhat lower levels are observed each Northern Hemisphere spring, starting in March. High arctic ozone levels are due to the fact that the very low temperatures required to form ice crystals needed to catalyze ozone destruction rarely occur in the stratosphere above the Arctic where winters are mild compared to those in Antarctica. Significant ozone declines were never observed there until 1989. But the winter of 1996-97 was particularly harsh in the Arctic, and on March 14, 1997, total ozone levels fell to about 350 Dobson

units, a decline of 22 percent. Satellite images confirmed the formation of an arctic "ozone hole."

In addition to ozone depletion above the earth's poles, lower ozone levels are now frequently observed throughout the year at Northern and Southern Hemisphere mid-latitudes. The Environmental Protection Agency (EPA) estimates that between 1986 and 1996, average UV-B exposure increased by about 5 percent at northern mid-latitudes and by about 3 percent at southern mid-latitudes as a result of declining ozone levels.

Actions to Protect Stratospheric Ozone

When the link between CFCs and ozone depletion was first made, the DuPont Corporation, developer of the CFCs, claimed that convincing scientific evidence did not exist. Corporation scientists did not accept Rowland and Molina's theory, nor were they impressed by the observations of antarctic ozone depletion reported by British scientists. However, in 1983, the National Academy of Sciences in the United States released a report on ozone that provided undeniable evidence of stratospheric ozone depletion. Based on that report, DuPont agreed to freeze CFC production levels. However, four years later, EPA data showed that DuPont's CFC production was *increasing* about 7 percent a year, a revelation that hurt the company's public image. In 1987, DuPont announced its commitment to research that would lead to ozone-friendly substitutes for the CFCs, with the goal of ending CFC production by 2000.

Development of HCFCs and Other Alternatives

By 1990, DuPont had developed a series of promising CFC substitutes. One of them, HCFC-22, was adopted by the food packaging industry to replace CFCs used as blowing agents in making plastic foam containers; it was also used in commercial air conditioners and heat pumps. These chemicals were called HCFCs because they contain hydrogen in place of one or more of the chlorine or fluorine atoms typically present in CFCs. HCFCs are more ozone friendly than CFCs because the hydrogen atoms cause a large fraction of their molecules to break down into harmless products when released to the atmosphere. Hydrogen atoms in these chemicals are subject to attack by highly reactive free radicals present in the earth's lower atmosphere, making the HCFCs less stable than the CFCs. Because they are less stable, a large fraction of HCFCs released to the atmosphere are destroyed in the lower atmosphere before rising to the stratosphere. However, it is estimated that as much as 10 percent of HCFC molecules survive long enough to enter the stratosphere and initiate ozone-destroying reactions.

The HCFCs are not a long-term answer to the ozone depletion problem, but they were used until totally safe CFC substitutes could be created and marketed. One of the new, totally ozone-safe chemicals developed by DuPont is HFC-134a, which laboratory studies and computer models have confirmed will not destroy stratospheric ozone. It is nonflammable, nontoxic, and appears to meet all of the requirements of a substitute for CFC refrigerants. By 1999, HFC-134a, along with required reengineered pumps and compressors, was being installed in new U.S. refrigerators and air conditioners, but a new concern was becoming evident: HFC-134a and other ozone-safe refrigerants are potent greenhouse gases. Their gradual release to the atmosphere, which is almost a certainty, will add to problems associated with global climate change (see Chapter 12).

Global Action on a Global Problem

The Vienna Convention   In 1981, before Joseph Farman's observations of south polar ozone depletion had become widely known, a number of scientists suspected there might be human-caused risks to the planet's ozone layer. A working group had been set up in Vienna, Austria, to outline appropriate measures to "protect human health and the environment against adverse effects resulting or likely to result from human activities which modify or are likely to modify the ozone layer."

The committee's goal was to develop an international plan (convention) to support research, encourage cooperation, facilitate information exchange, and work toward adopting measures to safeguard the planet's stratospheric ozone and thus preserve a vital shield against the potentially harmful impacts of solar UV-B radiation.

Thus, in 1988, the **Vienna Convention** was formulated and subsequently agreed-upon by 95 nations. One of its principal purposes was to establish an international organization to secure a general treaty that would effectively limit or control world nations' use of whatever chemicals, substances, or activities might pose a risk to the ozone layer.

The Vienna Convention was important because for the first time many nations had agreed to explore, and if necessary take control of, a global environmental problem before it had been clearly proven to exist. The work of the Vienna Convention led to an international meeting in Montreal, Canada, and the adoption of the Montreal Protocol.

The Montreal Protocol   World leaders met in Montreal in 1987 to develop plans and policy agreements aimed at protecting stratospheric ozone. By then, Rowland and Molina's theory and Farman's data on recurring south polar ozone depletion were well known. The treaty they drafted became known as the Montreal Protocol on Substances that Deplete the Ozone Layer. It called for a major reduction in the production of CFCs and halons, both of which were by then believed to be important ozone-depleting chemicals. The Montreal Protocol, initially signed by 24 nations and eventually ratified by 150, mandated that:

- CFC production in industrialized nations must be cut back to 50 percent of 1986 production levels by the year 1998. However, existing supplies could be recycled and reused.

- Halon production must be frozen at 1986 levels by 1992.

The **Montreal Protocol** went into effect on January 1, 1989. The fact that so many nations agreed to its terms made the Montreal Protocol the first truly important international environmental accord. However, those working with atmospheric computer models calculated that cutting CFC production by 50 percent (which the protocol mandated) would still leave unacceptably high CFC levels in the earth's atmosphere by the year 2000. In addition, other critically important ozone-depleting chlorine- and bromine-containing chemicals, such as carbon tetrachloride, methyl chloroform, and methyl bromide, were not addressed at all in Montreal. Furthermore, the protocol did not deal with CFC production in developing nations. Thus, deep concern over the fate of the ozone layer remained even after the Montreal Protocol was signed.

With new data reflecting worsening south polar ozone levels and reports of north polar ozone depletion, the nations that were party to the Montreal Protocol were convinced that more drastic actions had to be taken to protect the planet's ozone layer. Another international meeting was planned.

London Ozone Agreements   The framers of the Montreal Protocol and other world leaders met in London, England, in 1990 to consider the next steps needed to protect the ozone layer. The participants, reflecting differing viewpoints and political agendas, engaged in tough negotiations for 12 days. On June 29, the London Agreements to the Montreal Protocol were signed by a majority of attending delegates.

The **London Ozone Agreements** signaled a much tougher approach to CFC and halon control. Instead of merely cutting CFC production 50 percent by 1998 and freezing halon production at 1986 levels, both were to be totally phased out by the year 2000. In addition, the production of certain other ozone-depleting chemicals such as methyl chloroform was to be phased out by the year 2005. The HCFCs were to be phased out by 2040. United Nations Environmental Programme (UNEP) Executive Director Mostafa K. Tolba called the treaty a new chapter in the history of international relations. The *New York Times* called it "a landmark agreement."

Meanwhile, in the spring of 1990, a team of University of Chicago scientists reported a 45 percent rise in UV-B radiation above Ushuaia, a city near the southern tip of Argentina (at 55 °S latitude). It was the first time a major increase in solar UV-B radiation had been measured over a populated area. The following year, UV data collected at 45 °S latitude in New Zealand documented radiation levels twice as high as those known to exist at 48 °N latitude in Germany. This information indicated that sharp declines in stratospheric ozone above the South Pole were affecting populated land areas in the Southern Hemisphere.

While considerable progress in controlling ozone-depleting substances had been made at the London meetings, it was obvious to world leaders that even stricter CFC controls and phase-out timetables were needed to protect the planet from stratospheric ozone loss and rising UV-B levels.

Copenhagen Ozone Agreements   Representatives from more that 80 nations met in Copenhagen, Denmark, in November 1992 to address the worsening ozone problem. Their deliberations took account of data and reports contributed by scientists from all over the world. A sense of urgency prevailed. The agreements reached at Copenhagen mandated strict production phase-out schedules and timetables for most ozone-depleting substances:

- Production of CFCs in industrial nations had to be completely phased out by January 1, 1996.
- Production of halons, carbon tetrachloride, and methyl chloroform also had to be phased out by January 1, 1996. Developing nations were to be allowed a 10-year grace period during which they would phase out CFCs, halons, and other ozone depleting substances.
- Continuation of certain essential CFC uses, such as in asthma metered-dose medical inhalers, is allowed.
- A cap on HCFC production had to be established in 1996; HCFCs must then be phased out by 2030.
- A permanent Multilateral Ozone Defense Fund was established to help developing nations meet Protocol agreements by funding the acquisition of CFC-recycling equipment and developing CFC-free and CFC-friendly technologies.
- Production levels of methyl bromide, an important agricultural pesticide, were to be frozen at 1991 levels by 1995. However, no phase-out deadline was adopted.

The **Copenhagen Ozone Agreements** established important ozone-protecting initiatives, but serious declines in ozone continued to recur because of already existing levels of ozone-depleting chemicals in the stratosphere. For example, the ozone level above Antarctica fell to 88 Dobson units in October 1993, creating an ozone hole the size of North America. During the winter of that year, the World Meteorological Organization reported low levels of stratospheric ozone above Siberia—levels believed at the time to have been triggered by sulfate aerosols injected into the stratosphere during the Mount Pinatubo eruption of 1991. But two years later, during the winter of 1995, when these aerosols were no longer present, even lower Siberian ozone levels were observed. A UN Environmental Programme study, "Scientific Assessment of Ozone Depletion," concluded that ozone levels in Northern Hemisphere mid-latitudes had declined 4 percent per decade since 1979.

Sophisticated computer models of stratospheric ozone depletion show that by itself, the 1987 Montreal Protocol would have failed to reverse rising stratospheric chlorine levels. Ozone-depleting substances would have continued to accumulate, leading to lower and lower ozone levels. However, the models show that the Copenhagen agreements have advanced the date when ozone-depleting chemicals will no longer cause stratospheric ozone depletion.

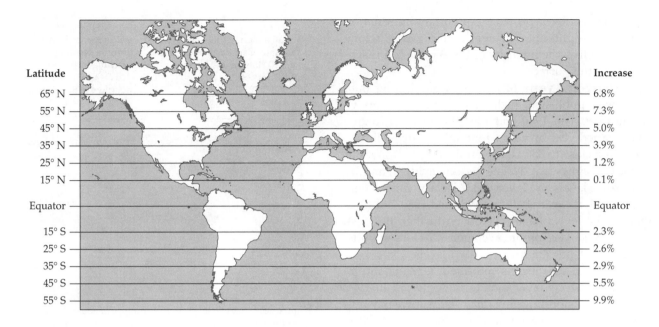

Figure 13.6   Global increases in ground-level UV-B radiation between 1986 and 1996 by latitude. *(Source: Data from the U.S. Environmental Protection Agency, 1996.)*

The Future of the Ozone Layer

Ozone depletion and UV-B radiation are now monitored worldwide on a regular basis. Figure 13.6 provides a global perspective on increases in UV-B radiation over a 10-year period, from 1986 to 1996, based on geographic latitude. Scientific attention continues to focus on stratospheric ozone depletion, especially near the planet's polar regions during their respective springtimes each year.

By analyzing the air in the upper troposphere, scientists can predict future trends because ozone-depleting substances must first pass through that region before entering the stratosphere. Research in 1996 showed that levels of ozone-depleting chemicals in the troposphere had peaked in 1994. Based on these data, computer models project that stratospheric chlorine- and bromine-containing substances will reach a maximum of about 3.9 ppbv in 1999. After that, they will gradually decline to 2.0 ppbv by about 2100, marking the end of stratospheric ozone depletion (see Figure 13.2).

International conferences among the many nations concerned about protecting the earth's ozone layer have led to the most successful treaty agreements of all time. They have established protocols and timetables based on reliable scientific and technical information, setting the stage for stratospheric ozone recovery. Worldwide CFC production, which peaked in 1988, declined by 60 percent in 1995 compared to 1986 production levels and is continuing to decline. As one example of the transition away from the CFCs, almost all new refrigerators, freezers, and automotive air conditioners are equipped with HFC-134a, a totally ozone-safe CFC substitute.

REGIONAL PERSPECTIVES

Canada: Monitoring UV-B Radiation

Canada was the first country to set up an ozone-monitoring network. Ground-based studies began in 1948, and by 1957 the network included Toronto, Edmonton, Goose Bay, Churchill, and Resolute Bay. In 1993, monitoring was expanded across Canada and the Canadian Arctic to include 12 stations.

In the 1980s, Canadian scientists developed the Brewer spectrophotometer, a sophisticated UV-measuring instrument capable of scanning a broad range of UV wavelengths. The Brewer instrument, now used in over 25 countries, is providing important information and new insights into changes in the earth's ozone layer as well as the intensity of UV-B radiation reaching the earth's surface.

Thirty-five years of Canadian monitoring data have been published, including ozone levels above Toronto between 1960 and 1994. Before 1981, ozone levels varied year by year within fairly narrow limits, averaging close to 350 Dobson units (Figure 13.7). But in 1981, a significant decline became evident, and by 1985 ozone levels had fallen to 337 Dobson units. There was a small recovery between 1985 and 1989, but since then ozone levels have fallen even further. By 1994, ozone levels over Toronto averaged 322 Dobson units, a decline of 8 percent from pre-1981 levels.

To keep its citizens informed, Canada implemented an ozone layer monitoring network in 1992 and reports its findings in *Ozone Watch*, a weekly bulletin. Figure 13.8, based on *Ozone Watch* for April 9, 1999, compares prevailing ozone levels with pre-1980 levels. The declines vary from −1.5 percent over Vancouver, British Columbia, to −16.5 percent over Toronto, Ontario. In addition, Canada's Ultraviolet Index Program forecasts the next day's UV-B levels on a scale of 1 to 10. A high number (greater than 8) warns of significant risk of sunburn from UV-B rays. Environment Canada estimates that in the 1990s summer stratospheric ozone levels above the country's mid-latitudes declined about 6 percent. It is estimated that as a result, exposure to UV-B radiation is about 12 percent higher during those months.

Antarctica: Effects of Increased UV-B Radiation on Phytoplankton

Many studies of the effects of increased UV-B radiation on marine plankton have been carried out in recent years. They are motivated by the knowledge that phytoplankton, a vital link in marine ecosystem food webs, can be harmed by exposure to UV rays. The potential for damage is greatest in antarctic waters because UV-B radiation reaching the earth's surface near Antarctica can vary dramatically. Increases as high as 1,500 percent (15-fold) above normal background levels occur seasonally as a result of stratospheric ozone depletion above the South Pole.

In 1992, studies designed to understand the response of marine algae, principally diatoms, to high levels of UV-B were carried out in the Bellingshausen Sea, an arm of the South Pacific Ocean off the coast of Antarctica. The work was conducted over a six-week period during the austral spring (September, October, and November). This is the time when ozone levels above the South Pole are at a minimum, thus allowing high levels of UV-B to reach the earth's surface. It was

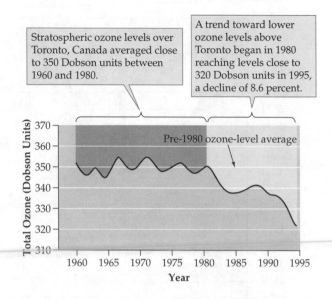

Figure 13.7 Ozone levels over Toronto, Canada, a 35-year record spanning 1960 to 1994.

found that phytoplankton exposed to UV-B radiation at that time exhibited an average decline of 12 percent in rates of photosynthesis.

A similar investigation aimed at measuring growth rates of phytoplankton species was conducted in November and December 1992 on board the British Antarctic Survey Vessel HMS *Discovery*. Elevated levels of UV-B reduced rates of photosynthesis by 17 percent. Most of the species studied displayed similar declines in primary production.

Other studies of aquatic and terrestrial organisms have shown that many plants and animals possess UV-B tolerance. Some are purely physical mechanisms. For example, nocturnal animals totally avoid solar radiation, while other plants and animals escape UV-B by living under rocks or at seawater or lake depths that filter harmful wavelengths of UV radiation. Surprisingly, some organisms actually manufacture sunscreen chemicals that shade out UV-B, thereby protecting the DNA in cell nuclei from the most harmful effects of UV-B. Certain other organisms possess highly specialized enzymes that facilitate the repair of cellular damage caused by UV rays.

While a seasonal decline of 12 percent to 17 percent in Antarctic primary production is significant, it is not catastrophic, and no major impacts on marine food webs are expected unless much more serious stratospheric ozone depletion events occur.

California: Using the Pesticide Methyl Bromide

As many as 4,000 farmers and food processors in California accounted for the use of 7.1 million kilograms

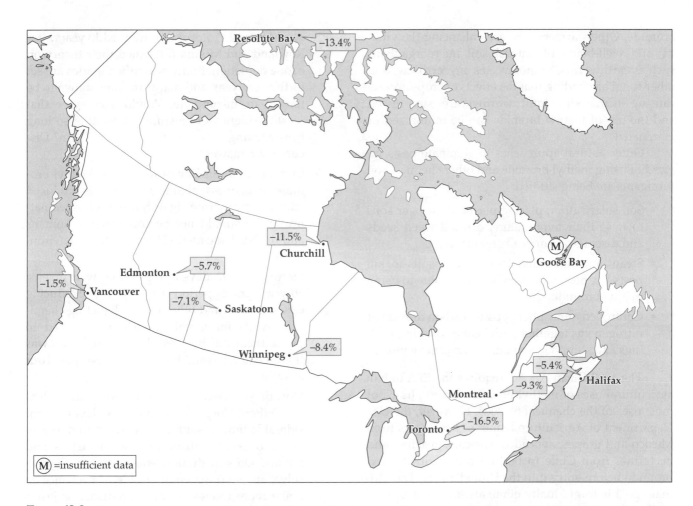

Figure 13.8 Canada's *Ozone Watch* map developed for April 9, 1999, showing declines in ozone compared to pre-1980 historical mean values.

(≈ 7,800 tons) of methyl bromide on 60 agricultural commodities in 1994, representing 14 percent of total worldwide use of the pesticide that year. The many applications of methyl bromide are described in Insight 13.1, but in California it is currently used principally to control pests affecting strawberries, ornamental plants, flowers, grapes, and almonds. The use of methyl bromide in California has increased sharply in recent years due to the phase-out of DDT, Aldrin, DDD, and other organochlorine pesticides in 1971 and 1972 (see Chapter 7) followed by the cancellation of chlordane in the late 1970s, DBCP in 1981, EDB in 1983, and Telone in the early 1990s. The dependence of California's agricultural producers and shippers on methyl bromide is a reflection of its continuing world-wide demand.

The large-scale use of methyl bromide is creating serious problems, mainly local toxic effects in people exposed to it and global effects resulting from man-made methyl bromide contributions to stratospheric ozone depletion:

- Methyl bromide is listed by the EPA as a Category I acute toxin having specific central nervous system and reproductive effects. Between 1982 and 1990 in California, at least 18 people died from methyl bromide exposure; in addition, at least 260 others suffered systemic illness, eye injuries, and skin damage.

- As discussed in Insight 13.1, it is estimated that global agricultural uses of methyl bromide presently account for 15 percent of stratospheric ozone depletion.

In 1995, the UN Methyl Bromide Technical Options Committee made up of experts from 23 countries reported that alternatives exist for at least 90 percent of agricultural methyl bromide applications. They cited the successful use of compost in Ohio nurseries in place of methyl bromide to suppress soil-borne plant disease. Large-scale strawberry farming in The Netherlands, both greenhouse and field operations based on soil-less potting mixes, are producing higher yields than in California. And, in California, crop rotation and integrated pest management are proving effective in controlling soil pests. For example, organic strawberry growers rotating to lower pH soils are reporting yields comparable to farmers relying on methyl bromide. Other farmers who are enhancing the vitality and well-being of natural soil microorganisms such as earthworms, actinomycetes, mycorrhizae, and other fungi are finding that the effects of crop-disease-causing nematodes (root worms) are suppressed and the use of methyl bromide can be minimized or eliminated.

Other natural approaches to eliminate pests instead of using methyl bromide and other chemical soil fumigants are being studied:

- Soil solarization, placing plastic tarps over soils to raise their temperature, can kill weed seeds and disease-causing soil organisms.
- Steam and hot water treatment of soils to kill pests has been shown to be comparable to using methyl bromide.
- Controlled atmospheres, using carbon dioxide or nitrogen gas together with heat or cold in buildings and shipping containers, can control pests.

The 1990 Clean Air Act requires the EPA to ban agricultural uses of methyl bromide and to phase out most uses of the chemical by 2005. However, the U.S. Department of Agriculture requires citrus fruits from Mexico and grapes, avocados, apricots, peaches, and nectarines from Chile to be fumigated with methyl bromide before entry into the United States. The ultimate goal is to gradually eliminate all major uses of methyl bromide, thereby minimizing its contribution to stratospheric ozone depletion.

KEY TERMS

Copenhagen Ozone
  Agreements
Dobson unit
Freon-12
London Ozone
  Agreements
Montreal Protocol
ozone
ozone-depleting
  substances

stratosphere
stratospheric chlorine
  level
troposphere
ultraviolet radiation
UV-A radiation
UV-B radiation
UV-C radiation
Vienna Convention
visible light

DISCUSSION QUESTIONS

1. Explain how the earth's ozone layer might have been formed in the planet's earlier history. You can assume that photosynthesis in the oceans was already underway.

2. The Montreal Protocol was adopted 13 years after Rowland and Molina published their theory on ozone depletion. Many scientific theories are set forth every year, and many are later shown to be flawed or incomplete. Would you agree that world leaders were justified in waiting so long before taking action on the ozone problem? Discuss your answer.

3. Comment on this line of reasoning: Methyl bromide, an ozone-depleting substance, is produced naturally by marine algae. Since it is formed naturally, we should not be concerned if humans add the pesticide methyl bromide to the environment.

4. Starting in 1992, developing nations were given a 10-year grace period to continue producing CFCs, halons, and other ozone-depleting chemicals, while industrial nations were forced to phase them out by 1996. In your opinion, was this an equitable and fair decision? Support your answer.

5. How do you view the efforts of industries such as the DuPont Corporation that have devoted considerable time, resources, and effort to developing ozone-safe substitutes for the CFCs and halons? Do you think it was a sincere effort to solve an environmental problem, a chance to make more money on new products, or just a good public relations move?

INDEPENDENT PROJECT

There is widespread agreement that ozone depletion in the stratosphere will result in increasing numbers of human skin cancers. Contact your state epidemiologist or some other source of health statistics and try to get data on total numbers of skin cancers reported in your state each year for several years—perhaps 10 or more. Plot these data on graph paper where the x axis will be particular years and the y-axis will be number of skin cancer cases reported. Study this information and describe how you would interpret the graph to interested citizens. Write a brief report.

INTERNET WEBSITES

Visit our web site at *http://www.mhhe.com/environmentalscience/* for specific resources and Internet links on the following topics:

Montreal Protocol

NASA ozone data

EPA ozone data

World Ozone and Ultraviolet Radiation Data Centre

Environment Canada: Atmospheric Environment Service

Total Ozone Mapping Program (TOMS)

Ozone levels over Halley Bay, Antarctica

EPA ozone depletion web site

SUGGESTED READINGS

Allen, Will, et al. 1995. *Out of the frying pan, avoiding the fire: Ending the use of methyl bromide.* Washington, D.C.: Ozone Action, Inc.

Bojkov, R. D. 1995. *The changing ozone layer.* Geneva, Switzerland: World Meteorological Organization.

Farman, J. C., B. G. Gardiner, and J. D. Shanklin. 1985. Large losses of total ozone in Antarctica reveal seasonal ClO$_x$/NO$_x$ interaction. British Antarctic Survey, National Environmental Research Council. *Nature* 315:207–10.

Kerr, James B. 1994. Decreasing ozone causes health concern: How Canada forecasts ultraviolet-B radiation. *Environmental Science and Technology* 28(12): 514A–18A.

Kirwin, Joe. 1993. Ozone update. *Our Planet, United Nations Environment Programme.* 5(1):56–58.

Montzka, Stephen A., et al. 1999. Recent changes in atmospheric levels of chlorofluorocarbons, methylchloroform, carbon tetrachloride, halons, and hydrochlorofluorocarbons. *Nature* 398:690.

Newman, Alan. 1993. What-ifs for a northern ozone hole. *Environmental Science and Technology* 27(8):1488–91.

———. 1994. CFC phase-out moving quickly. *Environmental Science and Technology* 28(1):35A–37A.

Rose, Julian. 1994. HCFCs may slow ozone layer recovery. *Environmental Science and Technology* 28(3):111A.

van der Leun, J. C., X. Tang, and M. Tevini. 1998. Environmental effects of ozone depletion: 1998 assessment. *Journal of Photochemistry and Photobiology B: Biology* 46.

Zurer, Pamela. 1995. Complexities of ozone loss continue to challenge scientists. *Chemical and Engineering News* (June 12):20–23.

———. 1995. Satellite data confirm CFC link to ozone hole. *Chemical and Engineering News* (January 2):9.

CHAPTER 14

Managing Hazardous Wastes

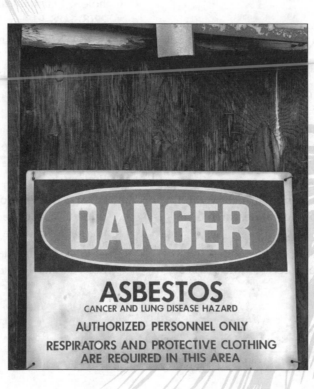

PhotoDisc, Inc./Vol. 44

We fear for our children, and we fear for their children . . .

the neighborhood lives in fear.

From Anne Anderson's testimony before the
U.S. Senate Committee on Public Works and
the Environment in 1980

Children were sick and dying in Woburn, Massachusetts, when Anne Anderson testified before Congress in 1980. Nineteen recent cases of childhood leukemia, along with other serious childhood illnesses, had been confirmed in the town. Most of the leukemia victims, one of whom was Anderson's son, eventually died. Scientific studies had established a compelling link between Woburn's public health crisis and the presence of hazardous chemicals in the town's drinking water. However, this realization had been several years in coming, and it would be several more years before the case was laid to rest.

In this chapter, we will trace the struggle of a group of private citizens to get attention and resolution for a serious environmental and health problem. We will also define hazardous wastes, study the issue of hazardous waste management, and examine the numerous human health problems associated with particular waste materials. Even before the Woburn incident, these harmful effects were known, and procedures were in place to manage hazardous wastes. However, as the case study shows, better policies were needed—and have now been established— to protect people and the environment from hazardous wastes. These new policies and procedures were prompted in large part by the tragedy at Woburn.

BACKGROUND

Hazardous Waste Management in the United States

Even before the Industrial Revolution, society had to deal with hazardous waste materials, and today the problem is much larger as industrialized societies use sophisticated materials and chemicals that often pose serious risks to humans, wildlife, and the environment. In recognition of these risks, the United States Congress enacted the Resource Conservation and Recovery Act (RCRA, pronounced RICK-RA) in 1976 to monitor hazardous wastes from point of generation to ultimate disposal, a "cradle-to-grave" management system. The goals of RCRA are:

1. To protect human health and the environment from the potential impacts of hazardous wastes.
2. To conserve energy and natural resources.
3. To reduce the quantities of hazardous wastes produced in the first place.

Identifying and Monitoring

The Environmental Protection Agency (EPA) reports that close to 13 billion tons of industrial, agricultural,

| TABLE 14.1 | Typical Hazardous Wastes Generated by Selected Industries |

Industry	Hazardous Waste
Chemical manufacturing	Strong acids and bases, highly reactive wastes, ignitable wastes, harmful by-products
Furniture and wood manufacturing	Wood preservatives, spent cleaning solvents, waste paint supplies, spent waxes and varnishes
Metal fabricating	Spent acids and bases, cyanide wastes, heavy metal sludges, degreasing agents, spent solvents and paints
Paper manufacturing	Ignitable wastes, corrosive materials, spent ink solvents
Printing	Photographic chemicals, heavy metals, spent solvents, acids and bases
Vehicle maintenance	Waste oils, waxes, and greases; spent solvents; lead and battery acids

SOURCE: U. S. Environmental Protection Agency: "RCRA: Reducing Risk from Waste." Solid Waste and Emergency Response: EPA530-K 97-004, September 1997.

commercial, and household wastes were produced annually in the United States in the 1990s. Most of these materials are not harmful, but about 2 percent are considered **hazardous wastes**—that is, according to RCRA's definition, substances having one or more of the following characteristics:

Ignitable   Materials that can fuel fires or are capable of spontaneous combustion. (Examples: gasoline, natural gas, oils, greases, and solvents.)

Corrosive   Acidic or basic chemicals that can attack and dissolve containers, such as metal barrels and storage tanks. (Examples: hydrochloric acid, sulfuric acid, lye, and spent batteries.)

Highly reactive   Unstable materials that can explode or react violently with air or water. (Examples: ammonium nitrate fertilizers, spent explosives, and peroxides.)

Toxic   Materials that can adversely affect humans or have serious impacts on the environment. (Examples: heavy metals such as mercury, pesticides such as parathion, and certain chemical wastes such as PCBs.)

Hazardous wastes may be solids, liquids, or gases. They may be by-products from chemical manufacturing processes, sludges and wastewaters from industries that use hazardous materials (Table 14.1), or solvents such as dry-cleaning fluids and degreasing agents. However, hazardous wastes do not generally include municipal garbage, domestic sewage, agricul-

tural residues, irrigation waters, demolition materials, or mine tailings; these are classified as **solid wastes**. Nuclear and radioactive materials are also excluded because they make up a special waste category regulated under the Atomic Energy Act.

Common Hazardous Substances and Their Effects on Humans   The Agency for Toxic Substances and Disease Registry (ATSDR), a unit of the U.S. Department of Health and Human Services, maintains a priority list of the most hazardous substances citizens are liable to encounter through contaminated air, water, soil, or food. Causes of contamination include:

- Improper storage of wastes at industrial sites.
- Illegal dumping at unauthorized sites.
- Deep-well injection at inappropriate sites.
- Inadequately constructed landfills.
- Substandard incinerators.
- Pesticides sprayed on agricultural lands.

Among the first recognized examples of hazardous materials are asbestos, known to cause respiratory damage and sometimes lung cancer; heavy metals such as lead, which can seriously harm the human central nervous system, especially in children; and industrial solvents, such as trichloroethylene, which are capable of infiltrating soils and contaminating groundwater if improperly disposed of.

The following top 10 substances hazardous to humans were identified by the EPA in 1997 (the most recent list):

1. *Arsenic*, a highly toxic substance, is fatal to humans at 60 parts per million (ppm) or higher.

It damages nerve, stomach, intestinal, and skin tissues. Long-term exposure causes dark skin discolorations, skin cancer, and tumors of the bladder, kidney, liver, and lungs.

2. *Lead* is toxic to the human central nervous system, especially in children. It damages the kidneys and the immune system. Exposure to lead causes premature births, diminished mental ability in infants and children, learning difficulties, and reduced rates of growth.

3. Chemical compounds of *mercury*, particularly methyl mercury, damage brain tissues, kidneys, and developing fetuses. Short-term exposure to high levels of mercury causes lung damage, skin rashes, and increased blood pressure and heart rate.

4. Breathing *vinyl chloride* vapors causes dizziness, unconsciousness, and possibly death. Long-term occupational exposure among those who make or use vinyl chloride has caused liver damage, injury to the nervous system, and immune reactions.

5. Humans who breathe high levels of *benzene* experience dizziness, rapid heart rate, headache, tremors, and death. Benzene, because it easily evaporates into the air, is an example of a **volatile organic compound (VOC).** Long-term exposure affects bone marrow, decreases numbers of red blood cells, triggers anemia, and may cause leukemia.

6. Human exposure to *polychlorinated biphenyls (PCBs)* leads to irritation of nose and lung tissues. Animals whose food contains small amounts of PCBs suffer liver and kidney damage, anemia, and reproductive system damage. Ingestion of PCBs has caused liver cancer in rats.

7. Breathing particulates containing *cadmium* severely damages the lungs and can cause death in humans. Long-term exposure leads to kidney failure and fragile bones. Animal testing shows that cadmium causes high blood pressure, iron-poor blood, liver tissue injury, and brain damage.

8. *Benzo[a]pyrene* is one of several complex hydrocarbons known as **polynuclear aromatic hydrocarbons (PAHs).** Benzo[a]pyrene causes cancer in laboratory animals and is believed to cause skin and lung cancer in humans.

9. *Benzo[b]fluoranthere* is another one of the PAHs. It is present in cigarette smoke and in foods cooked at high temperatures. Other sources include open burning, homes heated with wood and coal, coal tar, and asphalt. Benzo[b]fluoranthere is believed to be a human carcinogen.

10. Other PAHs can cause cancer in test animals. Studies in mice have shown that PAHs lower body weights, diminish reproductive ability, and cause birth defects. Therefore, the entire family of PAHs is considered hazardous.

Generation and Tracking

A generator of hazardous waste may be an individual, a small business, a large industrial plant, a university, or a research laboratory. Three categories of hazardous waste generators have been identified:

1. *Large-Quantity Generators*—Organizations and industries generating 1,000 kilograms ($\approx$ 2,200 pounds) or more of hazardous wastes in a calendar month. Examples include chemical manufacturers and pharmaceutical companies.

2. *Small-Quantity Generators*—Organizations and industries generating between 100 kilograms ($\approx$ 220 pounds) and 1,000 kilograms ($\approx$ 2,200 pounds) of hazardous wastes monthly. Examples include research laboratories, dry cleaners, and printing firms.

3. *Exempt Small-Quantity Generators*—Organizations and industries generating less than 100 kilograms ($<$ 220 pounds) of hazardous wastes monthly. Examples include photographic labs and dental offices. This class of generators is subject to minimal regulatory requirements.

All generators of hazardous wastes are required to keep records of the quantities and types of waste they produce. Under RCRA regulations, individual industries and organizations must determine whether the waste materials they produce should be considered hazardous, how to manage them, and where to dispose of them. Tracking and permitting procedures have been established to assure proper transport and disposal.

Close to 98 percent of all U.S. hazardous wastes are treated and disposed of at the site where they were generated; the rest are transported to commercial and public facilities. But the problem of safely handling, transporting, and disposing of hazardous materials has grown to immense proportions. Under RCRA, a tracking system records all hazardous waste shipments and requires that each one be accompanied by a one-page document identifying the waste generator, nature and amounts of waste materials being transported, shipping agent, intended disposal facility, and special handling that might be required. The tracking system is shown in Figure 14.1.

TABLE 14.2 Common Hazardous Waste Treatment Technologies

- *Incineration*   Controlled combustion of waste materials at elevated temperatures, burning organic substances and leaving behind mineral ash matter.
- *Biological treatment*   Using microorganisms to break down toxic organic compounds into less toxic decomposition products.
- *Activated carbon adsorption*   A physical process that adsorbs certain waste materials on the surface of specially prepared carbon.
- *Dechlorination*   A chemical process that removes chlorine from organic waste materials, making them less toxic.
- *Neutralization*   Treatment of highly acidic wastes with alkaline chemicals and/or treatment of highly basic wastes with acidic chemicals to render them less harmful.
- *Oxidation*   Reacting wastes with oxygen or other oxidizing agents to make them less toxic.
- *Solidification and stabilization*   Precipitation of particular wastes from a wastewater stream.
- *Solvent extraction*   A physical process whereby a solvent is employed to extract one or more solvent-soluble wastes from a larger volume of waste material to make it less toxic.

SOURCE: U. S. Environmental Protection Agency: "RCRA: Reducing Risk from Waste." Solid Waste and Emergency Response: EPA530-K 97-004, September 1997.

Treatment and Disposal

Each hazardous waste handling facility requires a permit issued by an individual state establishing operational parameters and allowable treatment methods. Treatment methods are designed to make hazardous waste materials less toxic and less harmful to the environment; the most common are described in Table 14.2. Two of the best-known techniques are incineration and activated carbon adsorption.

Incineration is the single most important treatment process because it can greatly reduce the volume and often the toxicity of different kinds of wastes. However, considerable controversy exists over its use. The problem is that at very high temperatures, chlorinated compounds called **dioxins** can form in incinerators and be discharged to the atmosphere. The dioxins consist of 75 related chemicals, one of which, 2,3,7,8-tetrachloro-p-dibenzodioxin, is extremely toxic and known to cause several different kinds of cancer in test animals. It is a suspected carcinogen in humans as well (see Regional Perspectives).

Activated carbon adsorption, another treatment method, is extremely versatile in its applications. Activated carbon granules, made from coal or charcoal using steam at elevated temperatures, possess very high surface areas that attract and adsorb a variety of chemical contaminants. For example, chloroform, and other hazardous substances formed when drinking water is chlorinated can be removed by filtering water through activated carbon. Pesticide contaminants in water can be eliminated in a similar way.

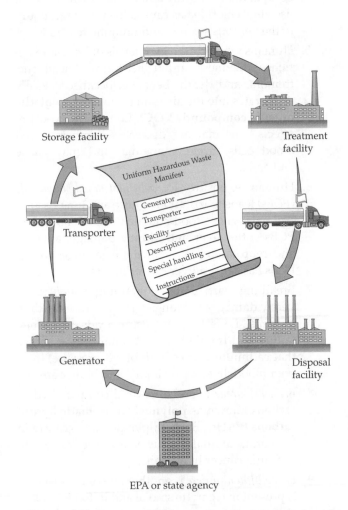

Figure 14.1 RCRA uniform manifest documents track each hazardous waste generator, transporter, storage facility, treatment facility, and disposal facility.

(Source: U.S. EPA: RCRA: Reducing Risk from Waste. Solid Waste and Emergency Response: EPA530-K 97-004, September 1997.)

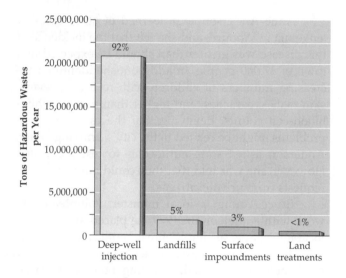

Figure 14.2 Principal means of hazardous waste disposal. *(Source: U.S. EPA: RCRA: Reducing Risk from Waste. Solid Waste and Emergency Response: EPA 530-K 97-004, 1997.)*

In 1985, it was discovered that a commonly occurring fungus known as **white rot fungus** could break down a number of hazardous environmental contaminants into simple, nontoxic products. The fungus belongs to a family of wood-rotting fungi that are prevalent across North America. An enzyme the fungi excrete decomposes lignin in wood along with other materials mixed with the wood. For example, white rot fungus can degrade the pesticide DDT, the herbicide 2,4,5-T, the wood preservative pentachlorophenol (PCP), creosote, and even fuel oils. A new company, Intech One-Eighty Corporation, has been set up in Logan, Utah, to commercialize this new approach to hazardous waste treatment.

The four most important hazardous waste disposal methods are:

1. Deep-well injection.
2. Disposal in landfills.
3. Surface impoundments.
4. Land treatment.

Deep-well injection, used by the oil industry, for example, to dispose of spent drilling muds (see Chapter 9), utilizes pressure to deposit primarily liquid hazardous wastes into steel-and-concrete-encased shafts. This technique poses potential harm to groundwater and to aquifers that serve or might serve as sources of drinking water.

Apart from injection wells, **landfills** continue to be the most common way to dispose of hazardous wastes. Wastes are placed into a dug-out land-surface depression engineered to collect **leachate,** liquid mate-

rials that may drain from waste materials. Methane, often produced by decomposing wastes, is also collected. If built and managed properly, landfills offer a low-cost and reasonably effective way to isolate hazardous waste materials from the environment. However, the risk remains that liners built into landfills to prevent leaching of hazardous materials into nearby surface and groundwater systems may fail.

Surface impoundments, also called pits, ponds, lagoons, and basins, are doubly lined natural or man-made holding tanks or land-surface cavities used to treat, temporarily store, and sometimes dispose of hazardous wastes. Strict EPA regulations govern the use of surface impoundments.

Land treatment is a disposal process in which hazardous wastes are applied onto or incorporated into surface soils. Natural soil biota are relied upon to break down or immobilize various hazardous components in the applied wastes.

Figure 14.2 reflects the relative importance of each of these techniques as of 1997.

Minimizing and Recycling

The National Academy of Sciences in Washington, D.C., has suggested that the best way to deal with the problem of hazardous wastes is to minimize their generation through waste reduction, recycling, and reuse. One firm that subscribes to this rationale, the Minnesota Mining and Manufacturing Company (3M), has had a waste reduction program in place since 1975 and estimates that it has reduced waste generation by 50 percent, saving the company $300 million each year.

A recent 3M innovation cuts hazardous waste at the company's Cordova, Illinois, plant where magnetic coatings for videotape are manufactured. The old process used ammonia and produced the by-product ammonium sulfate, a hazardous chemical discharged in the plant's wastewater. The new process uses sodium hydroxide; the by-product is sodium sulfate, which is similar to table salt, a nonhazardous chemical that has little environmental impact. Other examples of minimizing and recycling wastes include:

- Hughes Aircraft Company of Arlington, Virginia, was a heavy user of CFC-113 as a solvent to remove solder flux from printed electronic circuit board assemblies. But CFCs are hazardous to the global environment because they initiate ozone-destroying chain reactions in the earth's stratosphere (see Chapter 13). To overcome this problem, Hughes scientists developed a citric acid substitute for the CFC solvent. Citric acid is not only ozone safe, but cheap, plentiful, nontoxic, and nonpolluting. Hughes markets the solvent to electronics manufacturers throughout the world and in addition saves $4 million a year in its own operations.

- Hydrogen sulfide, present in crude oil and natural gas, is a hazardous substance mainly because of its toxicity. Furthermore, when burned as a component of these fuels, it contributes to acid rain. The petroleum industry removes hydrogen sulfide from crude oil and natural gas, converting it into sulfuric acid, an important industrial chemical used in making automotive batteries and in many other commercial applications.

- Methane, a by-product formed in landfills when organic wastes biodegrade, is collected and used as a fuel to generate electric power.

- Spent motor oils collected at automotive service centers are recycled into useful petrochemical products at refineries using reforming units and fractionating towers. This practice has virtually eliminated midnight backyard dumping of motor oils.

In 1985, the Canadian Chemical Producers' Association (CCPA) established a new program called Responsible Care aimed at safe and environmentally sound management of industrial chemicals over their life cycle. Responsible Care has now spread to over 40 countries, and the chemical manufacturers who are members are continually improving the safe management of the chemicals they use in ways that are responsive to the public interest and beneficial to the environment.

CASE STUDY: CHILDHOOD ILLNESS IN WOBURN, MASSACHUSETTS

Identifying and Resolving a Public Health Crisis

Anne Anderson and her family were living in Woburn, Massachusetts, 12 miles northwest of Boston, when her three-year-old son Jimmy was diagnosed with childhood *leukemia* in 1972. This was devastating news because there was no effective medical treatment for the disease at that time.

Leukemia is a debilitating, malignant disease of blood-forming tissues such as bone marrow that is characterized by a large increase in the numbers of leukocytes (white blood cells) in the bloodstream.

Anne Anderson soon learned of other cases of leukemia in Woburn, and she felt that the incidence of this disease was greater than should be expected in a town of 37,000 people. In addition, she identified increasing numbers of children with recurring fevers, skin rashes, earaches, and higher than normal white-blood-cell counts. Believing that the cause of these problems might be related to the city's drinking water, Anderson asked state authorities to test the water. However, she was told that they could not respond to requests from private citizens.

Bruce Young, a local minister, volunteered to work with Anderson. In 1979, he placed an ad in the local newspaper asking Woburn families for information about all known cases of leukemia. Within days, many people responded, reporting a total of 12 cases. When these were plotted on a map, half of them were *clustered* in east Woburn.

A **cluster** of cases is a statistically significant outbreak of an illness or disease in a defined area, not simply the occurrence of a number of cases.

After studying the map, Dr. John Truman, the Boston medical doctor who was treating Jimmy Anderson, alerted the Centers for Disease Control (CDC) in Atlanta, asking for a formal *epidemiological study* of Woburn's apparent leukemia cluster.

Epidemiology is a branch of public health involved in studying the origins and transmission of disease and the effects of human exposure to toxic substances. **Epidemiological studies** do not establish the cause and effect of a public health problem, but they often provide compelling evidence for the most likely cause.

Soon after the responses to Young's newspaper ad, 17 cases of childhood leukemia had been confirmed in Woburn, but public health officials still could not identify a cause, and citizens did not know what to do. The frightening medical implications of this illness, plus the fear, emotional trauma, psychological impact, and disruption within the affected families, were heart-rending. That same year, Young and Anderson organized a citizens' group called For A Cleaner Environment (FACE) to alert Woburn families to a possible growing public health crisis.

In 1980, the Massachusetts Department of Public Health (MDPH) released the results of a special study dealing with childhood leukemia in Woburn. The MDPH identified 18 incidents of the disease but reported that there was no actual cluster of cases. Their results further stated that the number of leukemia cases was not much higher than the normal back-

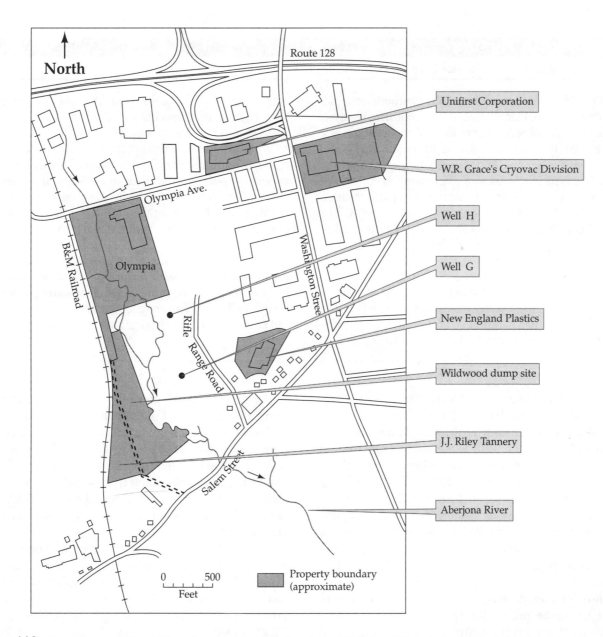

Figure 14.3 Map of eastern Woburn, Massachusetts, showing the locations of Wells G and H, the industrial sites, and the Aberjona River.
(Source: Environmental Protection Agency. "EPA Proposes Clean-Up Plan for Wells G and H Site" from Environmental Protection Agency Region I Office, February 1989, page 6.)

ground level of expected cases in a town the size of Woburn.

However, one year later, the CDC in Atlanta contradicted those results by issuing a report stating that a cluster of childhood leukemia did indeed exist in Woburn and that it was occurring at a much higher rate than normal. (Scientists at the Massachusetts Institute of Technology's Center for Environmental Health Sciences later demonstrated that between 1966 and 1986, the incidence of childhood leukemia in Woburn was four times higher than the national average.)

Anne Anderson had been right in believing that the incidence of the disease afflicting Woburn's children was excessive. Was she also right about its cause?

To find out, we must first understand the sources of the drinking water in the area around Woburn at that time and the connection between hazardous wastes and water-borne illnesses.

Woburn's Sources of Drinking Water

In the 1970s, Woburn was the home of several industries, including New England Plastics, UniFirst Corporation, W. R. Grace, and Beatrice Foods; the latter two are multinational corporations. UniFirst ran an industrial dry-cleaning business. W. R. Grace's Cryovac Division manufactured equipment for the food packaging industry. Beatrice Foods owned the John J. Riley

INSIGHT 14.1
Chemistry and Toxicology of TCE and PCE

TCE and PCE are nonflammable, synthetic-organic, oily liquids that smell a bit like chloroform. Because they can dissolve so many different substances, they are used as industrial solvents in paints, varnishes, degreasing agents, and dry-cleaning compounds. TCE is typically used as a degreasing agent or solvent to clean tools and sometimes as a paint thinner.

It is known that long-term exposure to TCE or PCE causes tumors and cancers to develop in laboratory animals such as mice and rats, but there is no definitive evidence that TCE or PCE can cause cancer in humans. Furthermore, neither chemical is known to cause leukemia in humans or test animals. Based on animal testing, the EPA classified TCE as a probable human carcinogen and set a drinking water limit of 5 parts per billion. While the 5-ppb fed-eral limit was not set until 1987, TCE was known to be a hazardous chemical in 1979 when Woburn's Wells G and H were shut down.

Routes of human exposure to TCE and PCE include contaminated drinking water, inhalation of vapors evaporated from their liquid state, and direct absorption through the skin (dermal exposure). Bathing and showering lead to both inhalation and dermal exposure. Persistent exposure to TCE, even at very low levels, can suppress the human immune system. This places susceptible individuals at risk for a number of health hazards, including cancer, because healthy immune systems destroy precancerous cells, while damaged immune systems may fail to control the multiplication of malignant cells.

Tannery, the last of a score of leather tanners that had operated in Woburn since the 1600s.

For decades, Woburn's drinking water had been pumped from several wells in the south-central part of town. But in the 1960s, population growth and industrial development outstripped the city's water supply, so additional sources had to be found. Thus, two new wells were drilled in east Woburn near the city's principal river, the Aberjona. Well G was drilled in 1964 on the river's east bank, and well H was drilled three years later about 500 feet north of Well G (Figure 14.3). The two new wells supplied about 30 percent of Woburn's water, drawing it from a bowl-shaped geologic formation under the river. As soon as the new wells went on line, Woburn residents complained about the water's bad smell and disagreeable taste, but no chemical tests of the water were carried out. It was later learned that hazardous chemicals had followed the natural flow of groundwater into this formation.

The Discovery of Water Pollution

In 1972 (the same year Jimmy Anderson was diagnosed with leukemia), the Aberjona River Commission reported that the river and its watershed were polluted. The commission based its report on common knowledge of indiscriminate dumping of industrial wastes in pits, lagoons, and trenches near the Riley Tannery and W. R. Grace's Cryovac plant. Workers had testified that trichloroethylene (TCE), a potentially harmful chemical solvent and degreasing agent (Insight 14.1), had been dumped at the W. R. Grace facility. (Until 1973, the W. R. Grace facility used about 200 gallons of TCE a year to remove oils and grease from metal parts that were to be painted and assembled to make food packaging equipment.) Discarded 55-gallon drums and piles of debris saturated with TCE were also found near the Riley Tannery, but there was no convincing evidence that the tannery had ever used TCE.

Despite this evidence, the commission's report was ignored. Several years later, in 1979, 184 discarded 55-gallon drums that had originally contained industrial solvents and other chemicals were discovered in a wooded vacant lot adjacent to the Aberjona River not far from Wells G and H (Figure 14.4). When the state's Department of Environmental Protection (its current name) began to study the area, TCE and other chemical contaminants were found in the water being pumped from both wells. The levels of TCE were:

Well G   267 parts per billion

Well H   183 parts per billion

The level of TCE found in Well G was more than 50 times the 5-ppb limit then being considered by the EPA (but not yet formally established), while the amount of TCE in Well H exceeded the informal limit by almost 40-fold. At this point in 1979, owing to the hazardous nature of TCE, the City of Woburn shut down Wells G and H, which had served Woburn for 15 years; they remain out of service today.

Figure 14.4 Discarded 55-gallon industrial-solvent drums discovered in a wooded vacant lot adjacent to the Aberjona River not far from Wells G and H.
(Source: EPA Site File, U.S. EPA, Region 1, New England.)

Water Pollution and Leukemia

In 1984, Professor Marvin Zelen at Harvard University's School of Public Health completed a three-year epidemiological study of leukemia in Woburn. Dr. Zelen, an expert in public health statistics, included all childhood illnesses that might possibly be linked to contaminated drinking water. He examined the health statistics of more than 5,000 children born between 1960 and 1982 in Woburn, along with the outcome of every pregnancy and birth. He reported that the combined evidence established a strong statistical link between hazardous chemical contamination in Wells G and H and the incidence of adverse health effects, including childhood leukemia, in Woburn.

Thus, it was shown that the incidence of childhood leukemia and other serious illness in Woburn correlated statistically with water pollution and water use associated with Wells G and H. Indeed, half of the city's leukemia cases occurred in east Woburn where most of the drinking water was pumped from Wells G and H. The highest rates of birth defects, respiratory disorders, and congenital eye, ear, kidney, and urinary tract defects occurred in the same area.

Looking back now, it is of special interest to note that between 1985 and 1997, only two new cases of childhood leukemia were diagnosed in Woburn, a rate almost the same as the normally expected background rate. This fact is consistent with the view held by many health scientists that TCE and perhaps other contaminants in Wells G and H were responsible for Woburn's cluster cases of leukemia.

Taking Legal Action

While the epidemiological study was going on, the Woburn lawsuit, Anne Anderson et al. versus W. R. Grace et al., was filed in Boston's U.S. District Court in May 1982. The principal defendants were W. R. Grace, Beatrice Foods, and the UniFirst Corporation.

The trial did not begin until 1986, largely because of its immense scope and the time needed by lawyers to develop evidence, depositions, scientific models, and theories. Pre-trial discovery work was unusually lengthy, requiring testing of soil and groundwater samples from the Grace, Beatrice, and UniFirst properties. In the meantime, UniFirst settled out-of-court for $1.05 million in 1985 after environmental testing revealed high levels of PCE in soil and groundwater at its site. Although it had been established that UniFirst had routinely used perchloroethylene (PCE), a chemical closely related to TCE, in its dry-cleaning operations, Unifirst settled without admitting any responsibility for the pollution on its property.

As the court case began, the fundamental questions posed to the jury by Judge Walter J. Skinner were: Is there clear and convincing evidence that TCE and certain related chemicals were disposed of on Beatrice and W.R. Grace land (each to be considered separately), and did these chemicals contribute to the contamination of Wells G and H? Eventually, the two suits were resolved as follows:

- In the case against Beatrice Foods and the John J. Riley Tannery, Judge Skinner determined that pre-1968 evidence of chemical dumping was not admissible because no knowledge existed until 1968 that contamination of the water table beneath the tannery would likely pollute Wells G and H. Thus, acting on the lack of convincing evidence at the time, the jury dismissed all charges against Beatrice Foods and the tannery, and the case was dropped. However, two years after the trial, the EPA reported that the Beatrice property "contains the most extensive area of contaminated soil [and] represents the area of highest contamination at the Wells G and H site."

- The case against W.R. Grace ended in 1990, eight years after it started, with an $8 million out-of-court settlement.

Legal documentation assembled by all parties involved in the court case amounted to 196 volumes. The airing of the Woburn case history on national television's "60 Minutes" and its write-up in the *Washington Post* and the *New York Times* made it a landmark legal case. But that was not the end of the story, because the Woburn incident was acting as a catalyst for new policies regarding hazardous wastes.

Superfund Legislation

The first notable hazardous waste disaster in North America occurred in 1970 at the Love Canal in upstate New York (see Regional Perspectives), but the Woburn incident served as an even more dramatic signal that existing laws and regulations were failing to control toxic chemicals and their potential impact on public health. In June 1980, the U.S. Senate began hearings to consider legislation to deal with hazardous wastes.

Senator Edward Kennedy invited Anne Anderson and Bruce Young to testify at the Senate hearings. They described what was happening in Woburn and why they believed illegal dumping of industrial wastes was contaminating the drinking water and compromising the town's public health. Anne Anderson's poignant remarks are cited in part at the beginning of this chapter. Here is a statement from Bruce Young:

> For seven years we were told that the burden of proof was on us as independent citizens to gather the statistics. . . . All our work was done independent of the Commonwealth of Massachusetts. They offered no support, and were in fact one of our adversaries in this battle to prove that we had a problem.

Following the hearings, Congress passed the **Superfund legislation,** also known as the Comprehensive Environmental Response, Compensation, and Liability Act (CERCLA) of 1980. A $1.6 billion trust fund was set aside based on new taxes to be levied on the chemical and petroleum industries. Superfund allows the EPA to respond to both immediate and long-term releases of hazardous substances that endanger public health or the environment. Before this, the EPA did not have this power or the funding to deal with major hazardous waste problems. Responses include rapid cleanup actions to mitigate crisis situations as well as extended remedial programs to handle protracted pollution problems.

Under Superfund, the EPA places severely contaminated hazardous waste sites on a National Priorities List; this procedure will be detailed later. Once a site has been listed, the following steps are taken:

1. Assess the degree and extent of contamination.
2. Determine environmental and health risks.
3. Select one or more cleanup remedies.
4. Develop specific technical details for cleanup actions.
5. Implement cleanup actions.

It is common for the EPA to compel responsible parties to perform necessary cleanup of hazardous waste sites, but this has often led to lengthy legal delays. If the EPA finds it necessary to start environmental monitoring and cleanup operations before legal issues are settled, the agency later seeks reimbursement through the courts to recover incurred costs.

SARA Title III

In 1986, Superfund legislation was strengthened through Congressional amendments and reauthorization. This action is referred to as the Superfund Amendments and Reauthorization Act (SARA Title III). Here are the principal changes and additions to Superfund brought about through SARA:

1. Superfund must recognize and utilize innovative treatment technologies and permanent remedies in cleaning up hazardous waste sites.
2. Superfund should consider and adopt environmental standards and regulations already in force through earlier waste cleanup actions undertaken by both federal and state agencies.
3. New enforcement powers and settlement tools are authorized.
4. Governmental authorities in affected states must be included in every phase of Superfund cleanup operations.
5. Superfund actions are to have increased focus on human health problems posed by hazardous waste sites.
6. Citizen participation in hazardous waste-site cleanup decisions and operations is emphasized.
7. The size of the Superfund trust fund was increased to $8.5 billion.

INSIGHT 14.2
EPA's Hazard Ranking System

EPA's **Hazard Ranking System** (HRS) is a numerically based screening procedure using preliminary site assessments and inspections to gauge the environmental and health risks posed by a particular hazardous waste site. HRS scores by themselves do not determine EPA cleanup priorities because the system cannot establish the extent of contamination or the kind of response needed. Follow-up studies and investigations are used to prioritize EPA actions. Nevertheless, HRS is useful in assigning numerical values to perceived risks such as:

- The likelihood that a site has released or will release hazardous substances into the environment.

- The characteristics of particular hazardous wastes, including toxicities and amounts released or likely to be released.

- The nature and proximity of the site to humans, fish, and wildlife.

In addition, HRS considers environmental pathways by which hazardous substances, if released, could be transported, including groundwater migration to drinking water aquifers, surface water migration to sensitive ecosystems such as fish and wildlife habitats, soil contamination near agricultural and residential areas, and atmospheric releases that might impact humans and wildlife.

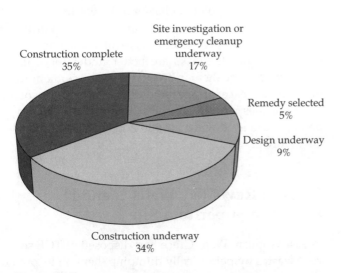

Figure 14.5 Relative stages of EPA actions with respect to 1998 Superfund sites.
(Source: http://www.epa.gov/superfund/whatissf/mgmtrpt.htm)

In addition, SARA required the EPA to revise its Hazard Ranking System (Insight 14.2) to assure that risks to human health and the environment are adequately considered before a new site is added to the National Priority List of hazardous waste sites.

National Priority List

It is estimated that as many as 25,000 abandoned or uncontrolled hazardous waste sites exist in North America. In the United States, sites determined to pose the greatest risks to public health and the environment are placed on the EPA's **National Priority List** (NPL). Placing a site on the NPL, a process called "listing,"

authorizes the EPA through Superfund to allocate federal funding for studies, planning, selecting restoration strategies, and ultimately constructing specific cleanup facilities to manage hazardous wastes at that site. Sites added to the NPL are often abandoned or uncontrolled chemical and industrial dumps. Sometimes there is no responsible party that the EPA can force cleanup actions upon, and individual states may not have the resources to cover the costs.

Hazardous waste sites that the EPA considers serious enough to list on the NPL are first published in the Federal Register along with the risks associated with them. The Federal Register is an official U.S. government periodical used to alert federal and state agencies and the public at large to proposed actions by federal agencies. After a period of public commentary, the EPA decides whether to add a new NPL site. The decision is made using these three mechanisms:

1. Proposed NPL sites are screened using the EPA's Hazard Ranking System. A numerical rating is assigned reflecting anticipated human health and environmental risks associated with the site.

2. Each U.S. state and territory is allowed to designate one top-priority site to be placed on the NPL.

3. A site is automatically listed if it meets these criteria:

   - The U.S. Public Health Service issues a health advisory recommending evacuation of people from that site, and

   - The EPA determines that the site poses a significant public health threat, and

• The EPA determines that it would be cost-effective to use its remedial authority (available only at NPL sites) rather than to use its emergency authority.

The number of NPL sites changes periodically as new ones are added and old ones are removed following remedial cleanups. Figure 14.5 reflects the stages of EPA actions with respect to 1998 sites. In May 1999, the EPA's NPL listed 1,211 hazardous sites.

Recent U.S. General Accounting Office (GAO) reports have focused on abandoned hazardous waste sites. One report analyzed 3,036 top-priority sites to consider whether it was necessary to clean them up under the Superfund program and found that fewer new sites will have to be placed on the NPL than previously estimated. The reason is that considerable cleanup progress has already been completed or is underway at many of these sites. By 2000, assuming adequate funding, it is expected that Superfund cleanup remedies will have been selected for close to 95 percent of existing nonfederal hazardous waste sites and 67 percent of existing federal sites.

Risk Assessment

Assessing the risks associated with hazardous waste sites has become more sophisticated and is now an essential part of environmental management. For example, a site that was once used to dispose of dry-cleaning solvents presents totally different risks from one where old electrical transformers containing PCBs were dumped. **Risk assessment** is now a major part of the Superfund program, and professionals working under Superfund are responsible for evaluating the potential risks to humans, fish, and wildlife from hazardous materials at abandoned or uncontrolled dump sites.

One of the first actions in risk assessment is to determine what the safe level would be for each chemical present at a particular site. Certain levels are considered tolerable because they are unlikely to cause adverse impacts to humans or the environment. The initial key questions at respective sites are:

1. What contaminants are present?
2. By what environmental routes might people and wildlife be exposed to these contaminants?
3. What specific dangers do these contaminants create, particularly to human health?
4. What concentrations of these contaminants are regarded as safe?

A "no action" analysis must be carried out at each site—in other words, what are the present and future risks if no cleanup actions are taken? Living near a Superfund site or a potential Superfund site does not by itself put people or the environment at risk. Risk must be assessed in terms of site-specific factors, including the nature of the hazards at the site, the number of people (particularly children) who might be affected, exposure of surface and groundwaters to possible contamination, and the likelihood of air transport to population centers.

Part of risk assessment is analyzing contamination in samples of soil, air, water, sediments, plants, fish, and animals collected at particular sites. Analyses confirm the presence or absence of specific chemicals and waste materials. Literature research reveals what is known and what further tests might be needed to understand the risks involved. Exposure to toxic materials is addressed in terms of levels, duration, and routes of exposure. Important questions include: Will children be exposed to hazardous wastes? Will drinking water be contaminated? Will fish and wildlife be at risk as a result of potential hazards at a waste site?

Environmental risk assessment is an important part of evaluating different types of environmental pollution. Once the risks are better understood, the goal is to manage them in acceptable ways. Sometimes this calls for excavating, removing, and treating contaminants, especially in soils. Totally different approaches are used when air and aquatic environments are involved.

Woburn Revisited: Wells G and H Become a Superfund Site

In east Woburn, W. R. Grace had disposed of TCE solvent wastes by periodically dumping them on its own property, a practice now believed to have contributed to the contamination of the Aberjona aquifer into which Wells G and H were drilled. In 1982, three years after hazardous industrial waste solvents were found in Wells G and H in Woburn, the two wells were placed on the NPL list for Superfund cleanup. The following year, the EPA ordered W. R. Grace, Beatrice Foods/J. J. Riley Tannery, and UniFirst Corporation to study the nature and extent of hazardous waste contamination at their respective facilities. UniFirst was instructed to remove TCE from a storage well on its property and to monitor the groundwater for possible contamination.

By 1988, when the EPA completed a survey of the properties near Wells G and H, it became clear that several Woburn industries had used similar dumping practices—practices that seriously polluted the soils within that site, the underlying groundwater, and sed-

Figure 14.6 EPA worker monitoring hazardous materials in an abandoned 55-gallon drum at a Superfund excavation site in Woburn, Massachusetts.
(Source: EPA Site File, U.S. EPA, Region 1, New England.)

iments in the Aberjona River. The presence of several hazardous materials was documented:

- TCE, trichloroethylene, and a number of solvents classified as VOCs in the Aberjona aquifer.
- VOCs, PAHs, PCBs, and pesticides in soils on properties near the two wells.
- PAHs and heavy metals, including chromium, zinc, mercury, and arsenic, in Aberjona River sediments.

Based on these findings, the EPA selected and implemented site cleanup remedies to eliminate or greatly reduce hazardous waste contamination at a 330-acre site that included the W. R. Grace property, the Wildwood dump site adjacent to the J. J. Riley Tannery, and the UniFirst property. Figure 14.6 shows an EPA worker monitoring hazardous materials in an abandoned 55-gallon drum at a Woburn Superfund excavation site.

By 1999, considerable progress had been made in meeting all Superfund cleanup goals. Close to 2,100 cubic yards of contaminated soils weighing approximately 200 tons were removed from the three sites near Wells G and H. Aquifer cleanup activities continue, with the goal of restoring water quality to meet EPA drinking water standards. It is estimated that Superfund cleanup of these sites will require several more decades and cost close to $70 million. Specific accomplishments so far include:

- **W. R. Grace**   All of the contaminated soils were excavated from the W. R. Grace property and transported to an off-site incinerator. By 1999, six years of Aberjona aquifer cleanup operations had been carried out. Water pumped from the aquifer is decontaminated using hydrogen peroxide and ultraviolet light and then returned to the aquifer by means of surface irrigation and groundwater recharge techniques. It is estimated that 24 kilograms ($\approx$ 53 pounds) of total VOCs have thus far been removed from 20 million gallons of groundwater.

- **Wildwood Property**   Cleanup of the J. J. Riley Tannery site (Beatrice Food property) and adjacent Wildwood property resulted in the removal of 60,800 kilograms ($\approx$ 67 tons) of hazardous sludge, 321,100 kilograms ($\approx$ 354 tons) of nonhazardous sludge, 231,300 kilograms ($\approx$ 255 tons) of debris, 45 discarded chemical drums, and 895,200 kilograms ($\approx$ 987 tons) of contaminated soils that were excavated for incineration. Soil and groundwater contamination still remains at Wildwood and is being handled by wells and chemical reactors to pump and treat groundwater, injection pumps to force air into the ground, and a special vacuum system to pull contaminants to the surface.

- **Unifirst Corporation**   Soils that were the most contaminated were dug up and removed, leaving only very low levels of contaminants behind. A system similar to that operating at the W. R. Grace facility is pumping and treating contaminated groundwater at the Unifirst site.

REGIONAL PERSPECTIVES

Minamata Bay: Mercury Contamination

A classic example of dumping hazardous waste and the subsequent health effects was first reported in 1956 in Minamata City in southwestern Japan along the coast of the Yatsushiro Sea. A second epidemic occurred the same year in the Agano River basin in central Japan's Niigata Prefecture. In both cases, a strange disease affected cats, dogs, birds, and people. Humans, especially children, experienced numbed limbs, lack of coordination, tremors, and slurred speech. Birth defects and physical deformities were common. Before it was over, there were 2,939 officially certified victims; in addition, 12,720 noncertified individuals were also afflicted. Many suffered seriously debilitating effects, and at least 79 people died.

Minamata disease, as it came to be called, was caused by methyl mercury contamination in fish, a common food in Japan. In Minamata City, the source of the methyl mercury was the nearby Chisso Chemical Company, which manufactured fertilizers, petrochemicals, and plastics. Chisso routinely dumped manufacturing wastes containing mercury into Minamata Bay. Methyl mercury, an organic compound formed by bacteria in marine and freshwaters, dissolved preferentially in lipid tissues of plankton, accumulated in the food chain, and concentrated in fish.

Prior to RCRA legislation in the United States and similar laws in other countries, it was common for industrial and manufacturing plants like Chisso to dispose of their waste materials in whatever ways were cheap and expedient. Often these wastes, whether hazardous or not, were vented to the air, buried at unsupervised dump sites, injected in deep wells drilled into porous subsurface geologic formations, or discharged to lakes, rivers, bays, or ocean waters.

Surprisingly, mercury contamination of freshwater and marine systems is still a significant problem in many parts of the world, illustrating a "business as usual" approach to disposing of hazardous wastes—that is, calculated neglect of hazardous waste regulations. For example, because of mercury contamination, a 1998–99 New York State Department of Health advisory cautions citizens to eat no more than one-half pound of fish a month from Lake Champlain and certain other lakes—and to consume no fish at all from Onondaga Lake.

Upstate New York: The Love Canal

In 1892, William Love, a nineteenth-century entrepreneur, began constructing a canal to connect the upper and lower sections of the Niagara River near the city of Niagara Falls in upstate New York. He envisioned a 7-mile-long navigable waterway that would bypass the historic falls, allow development of a hydroelectric facility, and become a site for building a model industrial city. But the onset of economic hard times ended Love's project, leaving behind a partially dug trench 3,000 feet long and 60 feet wide that became known as the Love Canal.

In 1942, Hooker Chemical Company, a subsidiary of Occidental Petroleum, acquired the Love Canal and used it as an industrial dump site. By 1953, Hooker had dumped an estimated 21,800 tons of hazardous materials into the canal. The site was then covered, and Hooker sold it to the City of Niagara Falls Board of Education for $1, disclaiming any responsibility for future damages due to the presence of buried chemicals.

In 1955, the 99th Street Elementary School was constructed on the Love Canal site, and before long a residential community, composed of hundreds of single-family homes, apartment buildings, city streets, and parks, had developed. But then residents of the area began to complain of peculiar odors and noxious matter seeping up onto their land and through their basement walls. Higher-than-normal rainfall in the 1970s raised the town's water table, accelerating the release of what would soon be identified as hundreds of hazardous chemicals, including 2,3,7,8-tetrachloro-p-dibenzodioxin, which had been buried in the Love Canal. Contaminants oozed into home basements, city streets, neighborhood parks, and even the 99th Street Elementary School.

In 1978, the New York State Department of Health documented increased rates of miscarriage, birth defects, crib deaths, and urinary tract disorders among 239 Love Canal families. The elementary school was closed, pregnant women and children under two years of age were evacuated, and residents were advised to avoid home basements and not to eat food grown in their gardens. The release of data by the Department of Health prompted the state of New York to purchase most of the homes near the Love Canal, although at first many nearby families were not included in the agreement. However, within days, the rising political visibility of the Love Canal disaster led President Carter to order an evacuation of the entire Love Canal community, and the federal government purchased abandoned residences at fair market value.

In 1994, Occidental Petroleum agreed to pay the State of New York $98 million to compensate the state's cleanup costs; the following year, Occidental agreed to pay an additional $129 million to the federal government for its cleanup costs. Currently, restoration of the Love Canal area is continuing, and there are even efforts to rebuild a residential community on the site.

Bangladesh: Arsenic in Well Water

Bangladesh, an Asian nation that once formed part of Pakistan, has suffered many crises. The most recent one is the discovery that more than half of its 125 million people are being exposed to drinking water contaminated with arsenic, a hazardous chemical.

Beginning in the late 1960s, millions of wells were drilled in Bangladesh under the auspices of the United Nations Children's Fund (UNICEF) to supply safe drinking water for the country's impoverished people. Until then, most people had obtained water from nearby rivers and lakes, which were grossly polluted. The wells were drilled because groundwater is generally a safer drinking water source than surface water. But in this case an estimated two million of the country's four million wells have been found to deliver water containing unacceptably high levels of naturally occurring arsenic. The drinking water limit for arsenic set by the World Health Organization (WHO) is 10 ppb, but at least two million of these wells have arsenic levels between 50 and 2,000 ppb.

Arsenic poisoning causes a number of health problems, depending on the chemical form of arsenic, the levels present, and the duration of exposure. Arsenic-related illnesses are referred to as *arsenicosis*. The early stage of chronic arsenicosis, marked by skin discoloration, is reversible using arsenic-free freshwater, vitamins, and certain drugs. But after 8 to 10 years of continued exposure, skin lesions, liver and kidney failure, and cancer result.

Arsenic in Bangladesh wells is not caused by illegal or improper hazardous waste management. In this case, it is occurring as a result of arsenic minerals leaching into groundwater aquifers through hydrogeologic mechanisms as yet unknown. Although arsenic has been known to be present in certain groundwaters for some time, the extent of the problem uncovered in Bangladesh is unprecedented. Now that the problem is gaining world attention, efforts are being made to discover techniques that can reduce the levels of arsenic in Bangladesh well water.

KEY TERMS

activated carbon adsorption

cluster

deep-well injection

dioxins

epidemiological studies

epidemiology

Hazard Ranking System (HRS)

hazardous wastes

incineration

landfills

land treatment

leachate

leukemia

National Priority List (NPL)

polynuclear aromatic hydrocarbons (PAHs)

risk assessment

solid wastes

Superfund legislation

surface impoundments

volatile organic compound (VOC)

white rot fungus

DISCUSSION QUESTIONS

1. Why is it so often the case that private citizens such as Anne Anderson and Bruce Young must assume the responsibility for focusing public attention on an environmental problem, documenting the extent and reality of the problem, and urging that action be taken to find a solution?

2. In industrialized societies, drinking water is tested frequently, sometimes daily, for possible bacterial contamination, but far less often, if at all, for hazardous chemicals such as TCE. Testing for these chemicals would add considerable cost to drinking water. Do you believe all drinking water, including bottled water, should be thoroughly tested before it is sold or provided to consumers? Discuss your answer.

3. In the United States, more than 90 percent of all hazardous wastes are disposed of by injecting them into deep wells, often drilled on the site of the industrial firms that generated the wastes. Comment on this practice, and if you don't approve of it, describe what you believe should be done.

4. Incinerators can produce and release dioxins into the air even if there are no dioxins in the waste materials delivered to the incinerator. Given this dilemma, what is your view on incineration of solid and hazardous wastes? What alternative to incineration would you advocate, and why?

5. TCE, a chemical known to cause tumors and cancers in test animals, was found in Woburn's drinking water as a result of contamination in Wells G and H. Although there is no animal or human evidence that this substance can cause leukemia, some public health scientists believe TCE suppresses the human immune system, thereby allowing susceptible individuals to contract leukemia and the other illnesses seen in Woburn. If you were on the jury that heard the

case brought against the Woburn industries, what would your opinion have been, and why would you have held that opinion?

INDEPENDENT PROJECT

Identify a hazardous waste site in your community, region, state, or province, and describe why it is classified as hazardous. What actions have been taken, are being taken, or are planned to deal with this site? Do significant numbers of people live nearby? To what towns or cities would prevailing winds possibly carry any hazardous materials if they were released? What toxic effects are associated with the particular waste materials at this site? Are you satisfied with the current state of action or inaction regarding this hazardous waste site? Write a short paper answering these questions.

INTERNET WEBSITES

Visit our website at *http://www.mhhe.com/environmentalscience/* for specific resources and Internet links on the following topics:

Woburn today

EPA Superfund

EPA Superfund risk assessment

EPA Integrated Risk Information System—List of substances

EPA National Priorities List

Toxics in the home

SUGGESTED READINGS

Brown, Phil. 1977. *No safe place: Toxic waste, leukemia, and community action.* Berkeley; Los Angeles: University of California Press.

Durant J. L., J. Chen, H. F. Hemond, and W. G. Thilly. 1996. *Elevated incidence of childhood leukemia in Woburn, Massachusetts: NIEHS Superfund basic research program searches for causes.* Cambridge, MA: Center for Environmental Health Sciences, Massachusetts Institute of Technology (April 29).

Harr, Jonathan. 1995. *A civil action.* New York: Random House.

Illman, Deborah L. 1993. Hazardous waste treatment using fungus enters marketplace. *Chemical and Engineering News* (July 12):26–29.

Lepkowski, Wil. 1998. Arsenic crisis in Bangladesh. *Chemical and Engineering News* (November 16):27–29.

Ott, Wayne R., and John W. Roberts. 1998. Everyday exposure to toxic pollutants. *Scientific American* (February): 86–91.

United States EPA. 1988. *Environmental progress and challenges: EPA's update.* U.S. Government Printing Office: 88.

———. 1997. *RCRA: Reducing Risk from Waste.* Solid Waste and Emergency Response: EPA530-K-97-004 (September).

———. 1998. *Industri-Plex Site, Woburn, Massachusetts.* Office of Emergency and Remedial Response: EPA 540/F-98/012, OSWER 9378.0-19FS, PB98-963214 (July).

———. 1999. *Guide for industrial waste management.* Solid waste and emergency response: EPA 530-R-99-001 (June).

Zurer, Pamela. 1999. Lasagna. Field-tested recipe for soil cleanup. *Chemical and Engineering News* (April 12):14.

Glossary

A

abatement A process of reducing or limiting the discharge of a pollutant to meet an environmental quality standard.

abiotic factors Non-living physical and chemical environmental factors such as temperature, humidity, precipitation, wind, light intensity, shading from light, soil chemistry, and atmospheric pressure. (p. 7)

acid rain Precipitation that is more acidic than normal, especially rain or snow having a pH less than 5.6. (p. 112)

activated carbon adsorption Treatment method that adsorbs certain waste materials on the surface of specially prepared carbon. (p. 284)

active solar heating Architectural home and commercial building strategies that use pumps, fans, and other devices to distribute heat captured through solar energy construction designs. (p. 229)

adiabatic cooling A cooling effect that occurs when an air mass rises and expands as a result of lower atmospheric pressure at higher elevations. (p. 52)

adsorb A physical process whereby a substance becomes attached to the surface of another substance such as activated carbon. (p. 111)

advanced wastewater treatment A process aimed at reducing nutrient levels in treated wastewaters, primarily phosphorus and sometimes nitrogen. Further reductions in BOD levels are also achieved. (p. 99)

aerobic respiration A cellular-level biochemical process whereby energy stored in living organisms is released through metabolism. Oxygen is consumed and carbon dioxide is formed. (p. 15)

aerosols Very small solid or liquid particulates, averaging about 10 μm in size, suspended in air. Aerosols occur in clouds, fogs, and mists. (p. 107)

age structure The relative proportions of different age groups in a population. (p. 133)

air pollutant *See* criteria air pollutant.

air quality standards *See* ambient air quality standards.

air toxics Hazardous air pollutants known to cause, or suspected of causing, cancer or other serious human health problems or ecosystem damage. (p. 120)

algae Photosynthetic unicellular, multicellular, or larger aquatic plants, including microalgae such as diatoms and macroalgae like kelp.

allelopathy Interactions between two plant species where one species releases a specific chemical substance that inhibits the growth of the other species. For example, certain grassland shrubs in southern California release aromatic chemicals that inhibit the growth of nearby grasses. (p. 13)

alpha particle A positively charged (2+) subatomic particle emitted by certain radioactive elements consisting of two protons and two neutrons.

ambient air quality standards Maximum allowable levels of specific air pollutants established by the U.S.E.P.A. under the Clean Air Act of 1970 and its subsequent amendments. (p. 117)

anadromous fish Fish that spend most of their life cycle in marine and estuarine waters but migrate to freshwaters to spawn and reproduce. Examples: Chesapeake Bay striped bass and Pacific and Atlantic salmon. (p. 84)

annual yield In agriculture, a country's or a region's production of a given crop per acre (or per hectare) in a given year. *Compare* total harvests. (p. 150)

anoxia A condition where no dissolved oxygen is present in an aquatic system. That is, DO = 0 mg/L. (p. 74)

anthropogenic Referring to human caused or influenced processes or events.

anti-greenhouse gas Referring to aerosol particulates such as sulfates produced from natural and anthropogenic emissions of sulfur dioxide; aerosols have the theoretical potential of cooling the earth's atmosphere. (p. 251)

anti-inflammatory agent A substance that lessens pain, irritation, and swelling in affected tissues.

aquifer A naturally occurring underground water resource associated with sand, gravel, or porous rock formations. Aquifers generally contain freshwater and are often excellent sources of drinking water and agricultural irrigation water. (p. 7)

arable Referring to agriculturally fertile land where crops can be grown. (p. 136)

aromatic   Referring to certain organic chemicals such as benzene, toluene, and naphthalene whose structures contain one or more benzene rings. The term was originally applied to fragrant, pleasant-smelling chemicals.

assimilative capacity   The natural ability of an ecosystem to decompose harmful wastes — self purification. A river, for example, can assimilate limited amounts of organic wastes (BOD) added to it. (p. 74)

atmospheric inversion   Generally an atmospheric condition where a layer of warm air lies above a cooler layer near the ground trapping stagnant air and preventing the dispersion of air pollutants. (p. 109)

atomic fission   The splitting of atoms such as uranium-235 into smaller atomic fragments. Fission may be accompanied by the release of large amounts of energy. (p. 207)

atomic number   The number of protons in a chemical element's atomic nucleus; designated by a subscript to the lower left of the element's chemical symbol. (p. 206)

atoms   Basic building blocks of all matter. Chemical elements and compounds are made up of atoms. (p. 206)

autotrophs   Organisms that produce organic nutrients from inorganic raw materials using an external energy source, such as light. Examples: primary producers such as algae, lichens, and grass. (p. 13)

B

baleen   A comb-like filter attached to the upper jaw bone of certain whale species (baleen whales) which enables them to filter plankton and other small organisms from sea water.

baseline   A data point or measurement used to estimate changes that occur relative to it.

base load   Average minimum daily demand for electric power, night and day, all year long. Base load is normally met through coal-fired and nuclear power plants. (p. 218)

benthic   Referring to bottom sediments such as mud, silt, and clay in aquatic environments.

benthic communities   Bottom-dwelling organisms such as worms living on or in bottom sediments.

best management agricultural practices   Federal, state, and county-mandated regulations and procedures aimed at minimizing non-point, agricultural nutrient, soil, and animal-waste runoff into surface waters. (p. 99)

beta particles   High-speed electrons emitted by certain radioactive elements.

bioaccumulation   The uptake and storage of environmental contaminants such as the pesticide DDT or the heavy metal mercury into an organism's tissues. (p. 159)

bioconcentration   *See* biomagnification.

biodegradable   Referring to waste matter and contaminants that can be broken down (decomposed) into simpler substances by microorganisms such as bacteria and fungi. Wood, paper, and manure are biodegradable, but heavy metals such as lead, cadmium, and zinc are not. (p. 74)

biodiversity   A measure of genetic, species, and ecological diversity in a given area. Species biodiversity is based on numbers of different species in an ecosystem (species richness) and population sizes of each species present (species equitability). The Shannon-Wiener biodiversity index is one of many indices used to quantify species biodiversity. (p. 11)

bioengineering   Techniques such as gene insertion used to modify DNA, the genetic blueprint of living organisms in order to introduce desired biological traits into future generations. (p. 162)

biogeochemical cycles   Environmental processes whereby chemical substances are perpetually circulated by living organisms, geologic weathering, and chemical reactions. Examples: cycling of nitrogen, carbon, sulfur, oxygen, phosphorus, and water. (p. 14)

biological oxygen demand (BOD)   A microbiological process that consumes dissolved oxygen in aquatic systems. It occurs when organic matter in municipal and industrial wastewaters is discharged to aquatic environments triggering excessive growth of bacterial populations. Respiration by increasing numbers of bacteria creates a rising demand for dissolved oxygen. (p. 74)

biomagnification   Increasing concentration of a contaminant in a food chain or food web leading to rising contaminant levels in the tissues of higher-level consumers. The concept is the same as bioconcentration. (p. 159)

biomass   Total weight of all living organisms or of a particular species in a given area. Biomass also refers to organic matter produced by plant and animal growth. Biomass is sometimes reported in terms of dry weight.

biome   The totality of similar terrestrial ecosystems that may span many of the planet's continents and islands. Biomes reflect distinctive assemblies of plant and animal life as well as distinguishing climatic, physical, and biological features. When represented on a map, earth's biomes appear as a planetary patchwork quilt of like ecosystems. (p. 8)

biomonitor   A plant or animal species that is relatively sensitive to pollution and whose populations are studied to assess the environmental status of an ecosystem. (p. 77)

bioremediation   Biological processes used to clean up environmental contamination. Example: fertilizer that was applied to Prince William Sound oiled coastal areas to expand bacterial populations capable of metabolizing oil. (p. 198)

biosphere   The relatively thin layer of soil, water, and atmosphere where life is found on the earth. It is made up of biotic factors (living organisms) and abiotic factors (materials and forces). Compared to the size of the earth itself, the biosphere is analogous to the skin of an apple, a very limited life-support system circling the planet. (p. 7)

biota   Living organisms. *Compare* biotic factors.

biotic factors   Living components of the earth's environment including monerans, protists, fungi, plants, and animals. (p. 7)

biotic potential   The maximum reproductive rate of an organism, given unlimited resources and ideal environmental conditions. *Compare* intrinsic capacity for increase. (p. 128)

birth rate (human)   More accurately, crude birth rate; the number of live human births in a given year per 1,000 members of a population. It is converted to a percentage by dividing by 10. (p. 132)

blue-green algae   *See* cyanobacteria.

BOD   *See* biological oxygen demand.

boiling water reactor (BWR)   A type of nuclear reactor where heat produced by atomic fission is used directly to boil water to make steam that drives electric-power generating turbines. *Compare* pressurized water reactor. (p. 210)

boom   A containment barrier used to corral an oil slick within its perimeter. (p. 198)

British Thermal Unit (BTU)   The amount of energy needed to raise the temperature of a pound of water by one degree Fahrenheit. (p. 200)

C

Calorie   The energy required to raise the temperature of 1,000 grams of water by one degree Celsius. 1 Calorie = 1,000 calories. Eating an apple supplies a person with about 60 Calories of energy. (p. 23)

campesinos (comp ah see' nos)   Indigenous, often landless, poor, peasant farmers who depend on nature for their survival.

canopy   Overarching umbrella-like tree branches and leaves that shade the forest floor. Canopies in tropical forests serve as habitat for algae, lichens, mosses, ferns, vines, bromeliads, insects, small mammals, and birds.

carbohydrate   A constituent of plant foods containing carbon, hydrogen and oxygen. Potato, corn and wheat are rich sources of carbohydrates.

carnivore   An animal that feeds on other animals. Carnivores generally occupy the third and higher trophic levels of a food web. (p. 13)

carrying capacity   Maximum species population size that can be supported in a given environment over a long time period; a population in equilibrium with its environment where biotic potential is in balance with environmental resistance. (p. 131)

catadromous fish   Fish that spend the greater part of their life in estuarine and freshwaters but migrate to ocean waters to spawn and reproduce. Chesapeake Bay eels are catadromous. (p. 84)

catalytic cracking   A process that splits (cracks) large hydrocarbon molecules in crude oil into smaller hydrocarbon molecules. (p. 190)

catalytic reforming   A process that links smaller hydrocarbon molecules to form more complex molecules such as those in gasoline. (p. 190)

certified organic foods   Food and food products approved as "organic" by a nationally recognized certification agency; approval generally requires that chemical inputs such as pesticides have not been used for at least three years and that farms are inspected annually. (p. 181)

CFCs   *See* chlorofluorocarbons.

chain reaction   A self-sustaining atomic reaction in which nuclear fission produces subatomic particles that initiate atomic fission in other nuclei. (p. 207)

chaparral   A temperate region biome. (p. 10)

chemical agriculture   Another name for green revolution agriculture; agriculture based heavily on the use of chemical inputs such as commercial fertilizers, pesticides, and food preservatives.

chemical dispersants   Chemical compounds and formulations designed to break up surface pools of spilled crude oil and redistribute the oil in the subsurface water column. (p. 197)

chemosynthesis   Biological primary production similar to photosynthesis but not involving light energy; autotrophic synthesis of organic compounds by certain bacteria using energy from inorganic compounds.

chlorinated hydrocarbon pesticides   Pesticides, such as DDT and chlordane, composed of chlorine, carbon, and hydrogen. (p. 158)

chlorofluorocarbons   Industrial chemicals containing chlorine, fluorine, and carbon used as cooling agents in refrigerators, air conditioners, and freezers and as blowing agents to make plastic foam materials; worldwide uses of CFCs are currently being phased out. (p. 122)

chlorophyll   A complex chemical pigment found mainly in green plants enabling them to capture solar energy and carry out photosynthesis.

chlorosis   Damage to chlorophyll in plant leaves and needles causing them to lose their green color and appear yellow. (p. 111)

cilia   Cellular hair-like projections capable of wave-like motions.

climax ecosystem   A self-perpetuating community of organisms that prevails in a defined habitat as long as environmental conditions under which it developed continue. The final stage in an ecological succession. (p. 50)

climax forest   A mature, stable, old-growth forest ecosystem; the final stage in a forest's ecological succession.

closed forests   Continuous, unbroken, dense vegetation that commonly blocks 90 percent or more of overhead sunlight from reaching the forest floor. (p. 42)

cluster   In epidemiology, a statistically significant outbreak of an illness or disease in a defined area, not simply the occurrence of a disease. (p. 286)

coal rank   A particular type of coal formed over geologic time intervals through coalification. Coal ranks include peat, lignite, sub-bituminous, bituminous, and anthracite coal. (p. 114)

coke   A type of purified, high-energy carbon obtained from bituminous coal and used in making steel. (p. 116)

cold desert   An arid, desertlike ecosystem found in cold climates. Example: arctic tundra.

command and control strategies   Conventional approaches to environmental problem solving where uniform standards are applied to individual industries, utilities, and production facilities to limit pollution. (p. 119)

commensalism   A symbiotic relationship between two species (symbionts) where one species may benefit while the other neither benefits nor is harmed. Example: orchids growing in tropical-tree canopies. (p. 12)

community   An association of interacting, self-sustaining, and self-regulating populations in a common environment. (p. 2)

competition   Interactions among organisms of the same or different species seeking food energy or shelter that may be limited or in short supply. (p. 12)

competitive exclusion principle   The concept that two species cannot occupy the same ecological niche at the same time.

condensation nuclei   Microscopic-sized atmospheric particles including dust, sea-salts, and sulfate aerosols, that can initiate rain or snowfall in moisture-laden air masses. (p. 66)

conifers   Needle-bearing trees that produce seeds in cones. Examples include pine, spruce, cedar, Douglas fir, yew, and hemlock.

conservation   Managing, protecting, and conserving natural resources while enjoying and using them in ways that guard against their exploitation, depletion, and pollution. (p. 3)

conservation tillage   Minimum tillage crop planting techniques that leave 30 percent or more of crop residues in and on the soil surface. *Compare* low-till agriculture and no-till agriculture. (p. 173)

consumers   Heterotrophic organisms that derive energy by feeding on other organisms or their products; primary consumers feed on primary producers while secondary consumers feed on primary consumers. (p. 13)

continental crust   Rock formations, typically about 35 km ($\approx$ 22 miles) thick underlying the earth's continents. (p. 188)

control rods   Neutron-absorbing devices positioned between fuel assemblies in nuclear reactors to regulate atomic fission reactions. (p. 211)

convection   Air or water currents caused by heat transfer from a warmer to a cooler body. Warmer air or water expands, becomes less dense, and rises; conversely, cooler air or water contracts, becomes more dense, and sinks. (p. 44)

Copenhagen Ozone Agreements   1992 ozone-depleting-substances agreements reached in Copenhagen mandating strict production phase-out schedules and timetables for most ozone-depleting substances. (p. 274)

copepods   Small, mostly herbivorous planktonic crustaceans, about 1 to 5 mm in length commonly found in freshwater, marine, and estuarine waters. The most numerous animal in the oceans. (p. 27)

coral polyps   Soft-bodied transparent sea animals that live symbiotically with marine algae. (p. 260)

coral reef bleaching   The loss of colorful symbiotic algae normally associated with coral reefs causing reefs to appear white as underlying limestone support structures become more easily seen. (p. 261)

cover crop   Plants, such as rye, alfalfa, or clover, that are planted immediately after harvest to hold and protect the soil. (p. 174)

criteria air pollutant   One of six air pollutants for which U.S. national ambient air quality standards were established by the U.S.E.P.A. under the Clean Air Act of 1970 and its subsequent amendments. (p. 117)

crossbreeding   Development of new plant varieties (cultivars) through cross-pollination that display desired traits such as preferred color, texture, and taste; resistance to drought, disease, and early frost; shorter growing season; increased yields; and higher nutritional values. (p. 148)

cross-pollination   Pollen transfer by wind, insects, or by hand from the anthers (pollen producing organs) of one flowering plant to the stigma (pollen receptor) of another plant of the same species. Fertilization leads to the development of new seeds. (p. 152)

crown roots   Roots in trees' high canopies. (p. 44)

crude birth rate   *See* birth rate.

crude death rate   *See* death rate.

crude oil   *See* petroleum.

crustaceans   Joint-footed, cold-blooded, invertebrate aquatic organisms (marine, freshwater, and estuarine) that form hard shells as they develop. Small crustaceans include copepods. Larger types include crabs, crayfish, shrimps, and lobsters. (p. 86)

cultivar   An improved crop variety consistently cultivated in a particular region. Cultivars display desirable genetic traits such as high yield, appealing taste and color, and disease resistance. (p. 148)

cultural eutrophication   Aging of surface waters like lakes resulting from human activities including agricultural and urban development. Cultural eutrophication leads to increased water turbidity, excessive growth of algae and weeds, decline in water quality, and loss of fish and wildlife. *Compare* natural eutrophication. (p. 87)

cyanobacteria (blue-green algae)   Prokaryotic, generally photosynthetic, bacteria classified as monerans. (p. 22)

D

death rate (human)   More accurately, crude death rate; the number of human deaths in a given year per 1,000 members of a population. It is converted into a percentage by dividing by 10. (p. 132)

debt-for-nature swaps   Forgiveness of international debt in exchange for nature protection in developing countries. Environmental groups and nongovernmental organizations often pay banks to write off uncollectable debts of developing coun-

tries at a steep discount in exchange for a promise by the debtor country to establish nature preserves. (p. 55)

deciduous   Referring to vegetation that loses its leaves to conserve moisture during cold seasons and times of drought. Deciduous plants include hardwood trees such as maple, oak, beech, teak, and cherry.

decomposers   Mainly heterotrophic bacteria and fungi that metabolize dead organic matter converting it to simple compounds that are recycled. (p. 44)

deep-well injection   Deposition under pressure of mainly liquid hazardous wastes, such as oil-industry drilling wastes, through deep steel and concrete-encased underground shafts. (p. 285)

deforestation   The action or process of clearing of forests. (p. 95)

demography   The statistical study of human populations including growth rates, age structures, geographic distribution, and their effects on social, economic, and environmental factors. (p. 132)

density   Refers to the number of organisms of the same species per unit area in a defined habitat. (p. 129)

density-dependent factors   Influences on individuals in a population that vary with population crowding (numbers of individuals per unit area). For example, competition for food increases with population density. (p. 129)

density-independent factors   Influences on individuals in a population that do not vary with population crowding (numbers of individuals per unit area). Example: periodic increases in UV-B radiation due to South Polar stratospheric ozone depletion adversely impacts antarctic ocean phytoplankton regardless of their population density. (p. 130)

detritus   Dead organic matter, such as decomposing leaves, bark, seeds, and animals, and the decomposers that live on it. (p. 13)

diatoms (die' a toms)   A common type of microscopic photosynthetic algae whose cell walls, which contain silica (SiO_2), are shaped like containers with lids.

dioxins   A group of related chlorinated organic compounds some of which are highly toxic. (p. 284)

dissolved oxygen (DO)   Oxygen gas physically dissolved in water whose concentration is expressed as milligrams (mg) of oxygen per liter of water. The range of dissolved oxygen is 0 to about 14 mg DO/L. (p. 74)

DNA   Deoxyribonucleic acid; the double-helix molecule in cell nuclei that contains genetically coded hereditary information that directs the development and functioning of living cells. (p. 162)

DO   *See* dissolved oxygen.

Dobson unit   A measure of solar ultraviolet-B radiation reaching the earth's surface; a measure of ozone levels in the stratosphere. (p. 271)

DO saturation   The maximum concentration of oxygen gas that typically dissolves in water depending on water temperature and salinity. (p. 74)

doubling time   The time required for a population to double in size. (p. 129)

dry forest   Refers generally to a tropical forest with an annual rainfall of 1,000 mm (≈ 40 inches) or less.

dryland farming   The practice of growing crops in dry regions receiving less than 50 cm (≈ 20 in) of rainfall a year; dryland farming normally produces crops without irrigation and depends mainly upon tillage methods that conserve soil moisture. (p. 176)

dustfall   Particulate fallout. (p. 115)

E

ecological niche   The functional role and position of a population within a community or ecosystem including its food resources, water, shelter, and how it interacts with other populations. (p. 12)

ecological succession   The sequential replacement of one vegetative community by another through a series of stages; succession ends when a climax community is established. *Compare* primary succession and secondary succession. (p. 49)

ecology   The study of interrelationships among living things and their environment. Ecologists study individual organisms, populations, and communities seeking to understand their natural occurrence, behavior, diversity, and relationships. (p. 2)

ecosystem   An ecological unit or community of living organisms interacting in a defined physical environment where all living things are linked by physical surroundings, chemical exchanges, biological interactions, and energy flow.

Ecosystems include biotic factors (living organisms) and abiotic factors (physical parameters). They are usually classified as terrestrial or aquatic and may be natural or artificial. A tropical forest is a natural ecosystem but a tree plantation is artificial. Other examples of artificial ecosystems include fish hatcheries, zoological gardens, agricultural systems, and urban environments. Ecosystems may also be described as undisturbed or disturbed. An undisturbed ecosystem is one largely unchanged or unaffected by human actions. Examples include old-growth forests, arctic tundra, protected grasslands, and remote lakes. Examples of disturbed ecosystems include urban lakes, rivers with hydroelectric dams, clear-cut forests, and cattle ranches. (p. 2)

ecosystem carrying capacity   *See* carrying capacity.

ecotourism   A kind of tourism in which environmental education has become a major theme. (p. 55)

edge effect   A change in species composition, physical conditions, or other ecological factors at the boundary between two ecosystems. Some organisms flourish at this boundary benefiting from increased edge area. Other organisms are harmed. (p. 44)

eelgrass meadows   Densely growing rooted submerged aquatic vegetation including, for example, eelgrass, water milfoil, and sago pondweed.

efficiency   The percent of available energy converted into useful work. Photosynthesis, for example, converts about 1 percent of available sunlight into food; the rest of the energy becomes so-called waste heat.

electrolysis   A process using electricity to split water molecules into hydrogen gas and oxygen gas. (p. 240)

electromagnetic spectrum   Energetic radiation including cosmic rays, x-rays, gamma rays, light rays, infra-red rays, ultra-violet rays, microwaves, and radio waves reported in terms of frequencies or wavelengths. (p. 88)

electron   A negatively charged (1−) subatomic particle with a mass approximately 1,840 times smaller than a proton. Electrons occupy orbitals outside an atom's nucleus and are responsible for each element's distinctive chemistry. (p. 206)

elements   Basic building blocks of all matter that cannot be broken down into simpler substances. There are 92 naturally occurring chemical elements including, for example, carbon, hydrogen, oxygen, nitrogen, sulfur, copper, and gold. A number of additional chemical elements formed through nuclear reactions are also known.

El Niño   A climatic change marked by shifting of a large warm-water pool from the western Pacific Ocean towards the east. Nutrient-rich upwelling currents along the coast of South America are blocked by this sea change and fisheries fail catastrophically. An El Niño event is normally accompanied by droughts in Australia and Southeast Asia together with heavy rain and snow in western North America. *Compare* La Niña. (p. 251)

environmental ethic   A way of thinking and making value judgements about the environment and the kind of world we want to live in and pass on to our children and grandchildren. (p. 6)

environmental externalities Harmful environmental, social, or economic impacts resulting from business, industrial, or utility activities not accounted for in the price of goods or services. Example: forest damage caused by acid rain. (p. 228)

environmental policy A legal framework based on environmental laws and regulations. Example: drinking water standards set by the U.S. Safe Drinking Water Act of 1974 and its amendments. (p. 6)

environmental resistance Forces and factors that limit a species' population size such as restricted food supply, lack of suitable habitat, toxic wastes, inadequate water, temperature extremes, disease, and predation. (p. 129)

environmental science Interdisciplinary scientific study of the environment and the nature and extent of human influences affecting environmental systems. Environmental scientists study natural and human influences in ecosystems; identify problems affecting living resources including fish, wildlife, and humans themselves; and work to find answers to environmental problems by applying knowledge and insights from many academic disciplines including chemistry, biology, toxicology, geology, economics, political science, and geography. (p. 2)

enzyme A complex chemical substance produced by living cells that accelerates specific biological reactions such as hydrolysis, oxidation, or reduction, but is unaltered itself in the process; a biological catalyst.

epidemiological studies Statistical investigations aimed at determining the probability of the origin, incidence, transmission, and prevalence of a particular disease or illness. (p. 286)

epidemiology A branch of public health involving the study of the origin, incidence, transmission, and prevalence of disease and the consequences of human exposure to toxic substances. (p. 72)

epiphytes Plant and animal life living on plants. Examples include algae and small crustaceans on aquatic vegetation as well as lichens on tree trunks and leaves. (p. 43)

erosion *See* soil erosion.

estuary A semi-enclosed body of water having a free connection with the open sea and within which seawater is measurably diluted by freshwater from land drainage. (p. 64)

ethics *See* environmental ethic.

ethylene A ripening agent, C_2H_4, produced by certain fruited plants; also, a petrochemical made from petroleum used to manufacture polyethylene plastics and other organic chemicals.

eukaryotes Organisms having distinct membrane-bound cell nuclei and cell organelles. *Compare* prokaryotes.

euphausiids Shrimp-like marine zooplankton up to 5 cm in length that are important in ocean food webs; called krill in antarctic waters.

eutrophication Aging of surface waters such as lakes that takes place naturally over thousands of years (natural eutrophication) or is accelerated by human activities (cultural eutrophication).

evaporation A physical process in which a liquid or solid is changed into a vapor. Example: liquid water evaporates into water vapor. Evaporation is facilitated by heat and air circulation. (p. 65)

evapo-transpiration The combined evaporation of water from leaf and needle surfaces including transpiration. (p. 66)

exponential population growth Repeated multiplication of a population's size by a constant factor over constant time intervals. Example: day 1, one bacterium; day 2, 10 bacteria; day 3, 100 bacteria; day 4, 1,000 bacteria, etc. (p. 128)

externality *See* environmental externalities.

F

fallow Plowed land that is allowed to lie idle during the growing season. (p. 177)

fauna Another name for animal life.

feedback A process that reestablishes a prior steady-state (equilibrium) condition (negative feedback) or facilitates changes leading to a new steady-state condition (positive feedback). (p. 89)

fermentation Anaerobic respiration during which biological matter is broken down into simple chemicals such as alcohol and methane by yeast or bacteria.

filter feeder An aquatic organism that filters its food from surrounding waters using specialized biological structures. Examples include clams, oysters, and mussels. (p. 88)

First Law of Thermodynamics The scientific principle that energy cannot be created or destroyed but can be transformed from one form to another. Electric energy, for example, can be transformed into light energy. (p. 26)

first trophic level The lowest feeding level in a food chain or food web where food is produced typically by photosynthetic green plants. *Compare* primary production. (p. 25)

fission *See* atomic fission.

flora Another name for plant life.

food chain   A sequence of organisms that derive energy and materials from one trophic level to another. An example would be an arctic wolf feeding on a caribou, which had eaten tundra grasses, mosses, and lichen. (p. 13)

food web   A network of interconnected food chains through which organisms obtain energy and materials they need to sustain life. (p. 13)

fossil fuel   Coal, oil, or natural gas; a hydrocarbon fuel formed over geologic time from living matter such as tree biomass or marine algae.

Freon   Any of a number of related CFC refrigerants. (p. 268)

freshwater   Water having few dissolved solids, minerals, or other substances in it. Water that is not too salty (saline) to drink or to use to irrigate crops. (p. 7)

fuel rods   Individual steel rods containing nuclear-fuel pellets such as uranium oxide that make up fuel assemblies in nuclear reactors. (p. 210)

fuelwood   Tree logs, branches, and woody debris collected and used to heat dwellings and cook meals. (p. 48)

fungi   Non-photosynthetic organisms with cell walls, filamentous bodies, and absorptive nutrition. Examples include molds, yeasts, mushrooms, and toadstools. Fungi, together with bacteria, are decomposers of organic matter.

fusion   *See* nuclear fusion.

G

gamma rays   Very high energy electromagnetic radiation emitted by certain radioactive elements.

gasohol   A mixture of about 90 percent ordinary gasoline and about 10 percent ethyl alcohol.

gene   A discrete biological unit occurring in cell DNA that transmits hereditary information from parents to offspring. (p. 148)

gene insertion   The basis of bioengineering. (p. 162)

genetically modified crops   Crops bioengineered from traditional crop varieties by altering DNA within cell nuclei. (p. 162)

global warming   Planetary warming superimposed on the greenhouse effect and believed to be due largely to human influences such as increases in atmospheric carbon dioxide levels through the combustion of fossil fuels. (p. 248)

greenhouse effect   Natural planetary warming due to heat-trapping by carbon dioxide, water vapor, and other naturally occurring greenhouse gases in the earth's lower atmosphere resulting in planetary temperatures that are beneficial to living things. (p. 7)

greenhouse gases   Natural- and human-derived gases resident in the earth's lower atmosphere that are responsible for trapping heat that would otherwise escape into outer space thereby warming the planet. (p. 122)

greenhouse warming   *See* global warming.

gross primary production   The rate at which solar energy is captured through photosynthesis in a given area. Part of this energy becomes plant biomass and part is used for plant metabolism, tissue repair, and reproduction. *Compare* net primary production. (p. 24)

growth rate   See population growth rate.

H

habitats   Particular environments where plants and animals live, find food, and reproduce. Example: the arctic coastal plain is reproductive habitat for Porcupine caribou. (p. 12)

hacienda   A large estate or plantation in a Spanish-speaking country.

half-life   The time required for one-half of a pesticide residue to decompose or for a radioactive substance to release one-half of its radioactivity. (p. 219)

halogen   Name given to any member of the family of elements that includes fluorine, chlorine, bromine, and iodine.

halons   Chemicals used in fire extinguishers and as fire-fighting materials.

Hazard Ranking System   System used to assign numerical ratings to hazardous waste sites proposed for listing on EPA's National Priority List reflecting human health and environmental risks associated with the site. (p. 291)

hazardous wastes   Waste materials produced by industry, business, government agencies, or private households that are ignitable, corrosive, highly reactive, and/or toxic.

HCFCs   Transitional chemical refrigerants and cooling agents that are less harmful than CFCs to stratospheric ozone.

heat island   *See* urban heat island.

heavy metals   A group of potentially toxic metals including lead, mercury, cadmium, copper, chromium, tin, and zinc. (p. 97)

heavy water   A chemical form of water in which the hydrogen atoms of each water molecule have been substituted by deuterium, an isotope of hydrogen; deuterium oxide. (p. 223)

heliostats   Sunlight-tracking mirrors. (p. 234)

herbicide   A chemical substance such as 2,4-D or Roundup used to control, manage, or kill unwanted vegetation (weeds).

herbivore   A second-trophic-level animal that feeds exclusively on first-trophic-level plants. (p. 13)

herbivory   A food-web relationship where animal life feeds on plant life.

heterotrophs   Organisms that are incapable of synthesizing their own food and must therefore feed on organic compounds produced by other organisms. *Compare* autotrophs. (p. 13)

high-level wastes   Waste materials that release significant amounts of potentially damaging radioactivity; spent fuel rods from nuclear power reactors are an example of high-level waste. *Compare* low-level wastes. (p. 220)

highly erodable land (HEL)   Soil having a high potential for erosion due to its slope, lack of vegetative cover, and deficiency of organic matter. (p. 155)

horizons   Horizonal layers displayed in mature soils, each layer having a visibly different composition. (p. 170)

humus   Dark, stable organic matter formed as bacteria and fungi decompose plant and animal matter. Humus gives soil its structure and serves as a major source of plant nutrients. (p. 170)

hybrid   A plant or animal variety resulting from a cross between two closely related species that do not normally cross. Hybrid crop varieties are created through controlled cross-pollination of inbred, genetically uniform, pure-breeding crop varieties. (p. 152)

hydric soils   Soils that are saturated, flooded, or inundated long enough during the growing season to develop anaerobic conditions near their surface. They are part of the criteria for identifying wetlands.

hydrocarbons   Organic compounds composed of the elements hydrogen and carbon. They are derived principally from fossilized plant matter that has been transformed over geologic time into gases like methane (natural gas), crude oils like petroleum, and waxy solids like paraffin. (p. 79)

hydroelectric power   Electricity generated by using the potential energy of water stored behind a dam to drive turbines in powerhouses built at the dam's base. (p. 230)

hydrologic cycle   Natural cycling of water within and among the earth's ecosystems through evaporation, atmospheric transport, precipitation, runoff, soil infiltration, and re-evaporation. Solar energy drives the hydrological cycle. (p. 7)

hydrophytic plants   Plants that grow in water or have a high tolerance for wet conditions. They are part of the criteria for identifying wetlands. (p. 86)

hypoxia   Low levels of dissolved oxygen (DO), generally 2 ppm or less, that are potentially harmful to aquatic life. (p. 93)

I

incineration   High-temperature combustion of waste materials that greatly reduces their volume and sometimes their toxicity; the single most important hazardous waste treatment. (p. 284)

incomplete combustion   Partial combustion of fossil fuels such as coal due to limited oxygen resulting in unburned particulate matter and carbon monoxide. (p. 110)

infrastructure   The complex of materials, equipment, personnel, and facilities needed to support or facilitate an activity or program. (p. 194)

inorganic   Referring to non-organic chemical compounds including salts such as sodium chloride, nutrients such as phosphates and nitrates, minerals such as sulfur, and metals such as iron. Many inorganic substances are required plant and animal nutrients. (p. 24)

integrated pest management (IPM)   Agricultural pest control based on strategies such as crop rotation and allowing soils to lie fallow that limit the need for chemical pesticides. (p. 161)

intertidal   Referring to marine habitats in coastal areas that are periodically exposed between high and low tides. (p. 195)

intrinsic capacity for increase (*r*)   The maximum rate at which a given species can increase in number assuming no limits to reproduction, growth, and survival. (p. 128)

Inuit (in' oo wit)   A common name for native people of the Arctic who live north of the tree line in Alaska and Canada. "Inuit" means "real people."

invertebrates   Animal life with no backbone. Aquatic invertebrates include insects, snails, clams, shrimp, and worms.

ionizing radiation   High-energy subatomic particles and rays that are capable of disrupting cellular-level DNA, damaging living tissues, and causing genetic defects. Examples: UV-B radiation, Xrays, and gamma rays.

isotope   One of possibly several different physical forms of a chemical element. Isotopes of an element have the same atomic number and exhibit essentially the same chemical behavior. However, they have different atomic mass numbers owing to differing numbers of neutrons in their nuclei. Carbon-12 is the most common isotope of the element carbon. (p. 206)

J

J curve An exponential growth curve resembling the letter J. (p. 129)

K

kelp (seaweed) Large, multicellular macroalgae often found in coastal marine environments.

K-strategists Organisms such as orca whales, monkeys, chimpanzees, and elephants that tend to maintain stable population sizes at or near environmental carrying capacity, *K*. They exhibit low *r* values (low fecundity), possess complex social patterns, are strongly competitive, and use resources efficiently. (p. 132)

L

lag phase Time period during which plant or animal population growth is very slow. A lag phase may precede exponential growth. (p. 129)

land conversion The systematic transformation of an undisturbed land area into a totally different environment such as a farm, cattle ranch, tree plantation, village, city, or industrial site. (p. 48)

landfill A dug-out-land-surface depression used for disposing hazardous wastes. (p. 285)

landraces Traditional or folk crop varieties selected from wild plants and cultivated locally. (p. 148)

land treatment Disposal process in which hazardous wastes are applied onto or incorporated into surface soils. (p. 285)

La Niña Colder than normal central and eastern Pacific Ocean waters due to eastward-blowing trade winds that intensify coastal upwelling near Peru and Equador causing monsoon rains in India; cool, wet winters in southeastern Africa; above-normal rainfall in eastern Australia; greater-than-normal snowfall west of the Great Lakes; and increased hurricane activity along the mid-Atlantic states. *Compare* El Niño. (p. 251)

L.A. smog *See* photochemical smog. *Compare* London smog. (p. 112)

LD$_{50}$ A chemical dose that is lethal to 50 percent of a test population. (p. 156)

leachate Liquid matter derived from municipal, industrial, or hazardous wastes that can potentially run off and pollute surface waters or seep through soils and contaminate groundwaters. (p. 285)

leeward Toward a sheltered area as, for example, toward an area not as exposed to prevailing wind and rain. (p. 52)

legume A member of the pea or bean family bearing pods that split along two seams. Leguminous plants include peas, alfalfa, clover, beans, and peanuts. (p. 46)

leukemia A debilitating, malignant disease of blood-forming tissues such as bone marrow characterized by a large increase in numbers of leukocytes (white blood cells) in the bloodstream. (p. 286)

lichen A symbiotic community of algae (or cyanobacteria) and fungi living together. There are many different kinds of lichen. (p. 22)

Liebig's Law of the Minimum The principle that one particular nutrient, the one in shortest supply, limits the growth of a plant or animal population.

lignite A type of brown coal often found in abundance but lacking high energy content; lignite combustion is a major source of air pollution. (p. 114)

limiting nutrient A particular nutrient or required substance that is least available to an organism relative to all of its nutrient requirements.

limiting nutrient principle The concept that a particular chemical nutrient is responsible for controlling the reproduction or growth of an organism or a population. A given nutrient may be the limiting factor at only certain times and places. (p. 89)

lodging The blowdown of certain crops in windstorms due to their long, thin, weak stalks. (p. 149)

logistic growth *See* sigma growth.

London Ozone Agreements 1990 ozone-depleting-substances agreements reached in London directing that CFCs and halon production would be phased out by 2000, methyl chloroform by 2005, and HCFCs by 2040. (p. 274)

London smog A type of air pollution in industrial areas consisting mainly of particulates, sulfur dioxide, and carbon monoxide. London smog is exacerbated by atmospheric inversions. *Compare* photochemical smog. (p. 112)

low-level wastes Waste materials contaminated with small amounts of radioactive residues; radioactive trash from hospitals and research laboratories are examples. *Compare* high-level wastes. (p. 220)

low-till agriculture Conservation tillage where little conventional plowing is carried out prior to crop planting.

M

maize   An early, primitive kind of corn sometimes called Indian corn; another name for corn.

market-based control strategies   Pollution control based on flexibility in the way individual industries and manufacturing firms are allowed to meet environmental standards. (p. 119)

mass number   The sum of the number of protons and neutrons in a chemical element's atomic nucleus; designated by a superscript to the upper left of the element's chemical symbol. (p. 206)

mesosphere   An atmospheric layer above the stratopause where temperatures fall with increasing elevation. (p. 108)

metabolite   A substance produced when a chemical breaks down or is metabolized; a necessary component in a metabolic process. (p. 160)

methanogenic bacteria   Bacteria that are responsible for the conversion of organic matter into methane gas. (p. 250)

Milankovitch cycles   Periodic variations in planetary cycles affecting the earth, such as the tilt of the earth's axis with respect to its orbit—variations that may affect the earth's climate. (p. 252)

moderator   A material used in a nuclear reactor to slow down neutrons released by fission reactions in order to maximize the probability of initiating subsequent fission reactions; water and graphite are moderators. (p. 208)

monerans (prokaryotes)   Organisms that lack distinct membrane-bound cell nuclei and cell organelles. Examples include bacteria and blue-green algae.

monoculture   An agricultural area where only a single crop or crop variety is grown. Such areas lack biodiversity and are vulnerable to disease and pest attack. (p. 153)

Montreal Protocol   1987 international agreement to work toward eliminating or curtailing stratospheric ozone depleting chemicals including CFCs, halons, and methyl chloroform. (p. 273)

mosses   Small, green, moisture-favoring plants that grow in clusters on rocks, trees, and soil.

mutagenic   Referring to substances or radiation that can alter an organism's DNA potentially causing inheritable changes (mutations). (p. 196)

mutualism   A symbiotic relationship between two species (symbionts) where each species may benefit. Example: termites cannot digest wood cellulose, but protozoa living in termite intestines break down and derive energy from cellulose which benefits both species. (p. 12)

N

National Priority List (NPL)   Official EPA listing of hazardous waste sites authorizing EPA through Superfund to allocate federal funds for studies, planning, restoration strategies, and construction of facilities to manage hazardous wastes at those sites. (p. 291)

natural eutrophication   Slow aging of surface waters, especially lakes, caused by the gradual runoff of soil and nutrients from undisturbed watersheds. No rapid or dramatic changes in water quality, fish populations, or aquatic wildlife normally occur. *Compare* cultural eutrophication. (p. 87)

natural gas   A gaseous fossil fuel composed mainly of methane but also containing ethane and propane. Natural gas is often found associated with petroleum, is relatively free of sulfur impurities and burns cleaner than coal or petroleum. (p. 189)

natural resources   Materials occurring in nature valued for their economic or environmental importance. Examples include air, topsoil, timber, groundwater, lakes and reservoirs, fish, and wildlife. (p. 3)

natural selection   A natural process borne of genetic diversity that enables organisms best adapted to an environment to survive while others die out. (p. 159)

negative feedback   *See* feedback.

negative lapse rate   The rate at which air temperature in the troposphere declines with increasing elevation: approximately 6.5 °C per kilometer (≈ 3.5 °F per 1,000 feet). (p. 108)

nekton   Larger animal life of the sea with locomotory abilities enabling them to swim against water currents. Examples: fish, squid, shrimp, sea turtles, and whales. (p. 31)

neritic   referring to near-coastal marine waters lying above a continental shelf. *Compare* pelagic. (p. 8)

net primary production   The rate at which solar energy is converted into plant biomass through photosynthesis in a given area. *Compare* gross primary production. (p. 24)

neutron   A neutral subatomic particle in atomic nuclei having a mass approximately 1,840 greater than an electron. (p. 206)

niche   *See* ecological niche.

nitrate (NO_3^-)   An essential plant nutrient composed of nitrogen and oxygen.

nitrogenase   A complex chemical enzyme that facilitates the conversion of atmospheric nitrogen gas into ammonia and nitrates.

nitrogen-fixing bacteria   Bacteria that produce the enzyme nitrogenase, which converts atmospheric nitrogen gas into plant nutrients such as ammonia and nitrates.

nonattainment area   A particular area or region that fails to meet one or more EPA criteria air pollutant standards. (p. 120)

non-point sources   Areas (as opposed to point sources) from which pollution runs off and enters sensitive environmental areas. Examples include farmland (fertilizer runoff); lawns and gardens (fertilizer and pesticide runoff); and urban centers (stormwater overflow and street debris). (p. 5)

nonrenewable   Referring to natural resources that are not replaced or replenished by natural processes at significant rates. (p. 228)

nontarget organism   A living organism that is not the intended target of a particular pesticide or pest control strategy but is nevertheless killed or impaired. *Compare* target organism. (p. 159)

no-till planting   Conservation-based agricultural practices where crop seeds are drilled directly into soils thereby supplanting conventional plowing. (p. 173)

nuclear enrichment   A physical process, such as gaseous diffusion, whereby one particular isotope of a chemical element is increased in concentration relative to the element's other isotopes. (p. 209)

nuclear fusion   A process whereby nuclear energy is released from very light atoms such as hydrogen when they are fused together under very high heat and pressure. The energy released by stars is nuclear-fusion energy. (p. 223)

nucleus   Dense central core of an atom. (p. 206)

O

oceanic crust   Rock layers beneath the ocean floor about 5–7 kilometers ($\approx$ 3–4 miles) thick. (p. 188)

old-growth forest   A forest with trees of all ages including some that are hundreds if not thousands of years old; an uncut virgin forest or a very old secondary growth forest.

omnivore   An animal that can eat at several different trophic levels. Bears and humans are omnivores. (p. 13)

One Percent Rule   The concept that on the average about one percent of sunlight energy is captured during photosynthesis and stored in plant biomass. (p. 25)

open forests   Forests with as little as 50 percent sunlight commonly shaded out. (p. 42)

organic   Referring to substances, both natural and human-made, living and non-living, that contain carbon and other elements such as hydrogen, oxygen, and nitrogen. Organic substances include alcohols, sugars, carbohydrates, amino acids, proteins, and lipids. Organic also refers to petroleum and petroleum products such as gasoline, diesel fuel, antifreeze, plastics, dyes, and pharmaceutical agents. In addition, organic identifies food and food products grown, processed, and shipped to market without the use of pesticides or other chemicals. (p. 24)

organic farming   Sustainable agricultural practices that do not include chemical inputs such as commercial fertilizers, herbicides, insecticides, fungicides, and food preservatives. (p. 180)

organo-phosphorus pesticide   A chemical pesticide containing the element phosphorus in its structure. Examples: parathion and malathion. (p. 158)

orographic precipitation   Referring to rain and snow that is generated when moist air is cooled as it is lifted to higher elevations by winds blowing over foothills and mountains. (p. 51)

ozone   A form of oxygen containing three oxygen atoms (O_3). Stratospheric ozone is highly beneficial because it blocks a high percentage of solar UV-B radiation from reaching the earth's surface while ozone in the lower atmosphere can be harmful due to its toxic and irritating effects. (p. 108)

ozone-depleting substances   Chemicals that persist in the troposphere, leak into the stratosphere, and react with ozone under the influence of intense solar radiation. (p. 268)

P

PAHs   *See* polynuclear aromatic hydrocarbons.

parasitism   A destructive symbiotic relationship between two species (symbionts) where the host organism may be harmed. Example: fleas on domestic dogs and cats.

particulates   Very small particles such as soot, smoke, and dust suspended in the air. Depending on their size and composition, particulates can be hazardous to the health and well-being of plants, animals, and humans.

passive solar heating   Architectural designs that maximize solar heating by using special building materials and solar-oriented window placement. Heat collected during the day is released within a dwelling at night. (p. 229)

pathogenic   Referring to organisms such as bacteria and viruses that can cause disease.

PCBs   *See* polychlorinated biphenyls.

peaking load Periodic daily and seasonal high demands for electric power reflecting, for example, power demands for evening cooking, summer air conditioning, and winter electric heating. (p. 218)

peat Spongy, water-saturated, organic material; peat is technically not "coal," but can be dried and used as a source of energy. (p. 114)

pedogenesis Natural formation of soils through weathering, decomposition of organic matter, and the actions of micro and macro soil organisms. (p. 169)

pelagic Referring to open-ocean or high seas waters extending beyond near-coastal continental shelf waters. *Compare* neritic. (p. 8)

permafrost Permanently frozen ground in polar regions occurring a few inches to a few feet below most tundra soils and extending in some places to depths as great as 600 meters ($\approx$ 2,000 feet). (p. 23)

permeable soils Soils having extensive air spaces and sufficient organic matter to absorb precipitation and hold water.

petrofuels Fuels such as gasoline, diesel oil, home heating oil, and kerosene (jet fuel) derived from petroleum.

petroleum Crude oil pumped from oil wells consisting of thousands of different organic chemicals; a natural resource key to meeting worldwide energy and chemical industry demands. (p. 188)

pH A measure of how acidic or basic a wet environment is. A pH of 7 is neutral. pH levels below 7 indicate acidic environments while pH levels above 7 indicate basic (alkaline) environments.

pheromone A chemical substance released by one organism that influences behavior or physiological processes in another organism. An example is a natural sex-attractant chemical released by an animal to signal its presence to other animals of the same species. (p. 162)

photic zone A water-column depth through which sufficient sunlight can penetrate to support photosynthesis. (p. 25)

photochemical smog A type of air pollution commonly occurring in urban areas such as Los Angeles where hydrocarbon fuels are widely used. Photochemical smog results when sunlight triggers chemical reactions between hydrocarbons and oxides of nitrogen creating toxic substances such as ozone. The problem is exacerbated by atmospheric inversions. *Compare* London smog. (p. 113)

photosynthesis A biochemical process involving plant pigments such as chlorophyll which capture solar energy and enable plants to convert carbon diox-

ide and water into organic matter such as glucose. (p. 13)

photovoltaic Energy-conversion where solar energy is converted directly into an electrical current. (p. 235)

phytoplankton Microscopic autotrophic plant organisms that drift with water and wind currents due to limited locomotory capabilities (drifters). Example: microalgae. (p. 30)

pioneer species The first organisms in a successional sequence of communities to colonize an area following a disturbance. In terrestrial succession, plants such as red alder may be a pioneer species. (p. 50)

plankton Very small plant and animal life that drift with water and wind currents due to limited locomotory capabilities (drifters); plankton include many different species of phytoplankton and zooplankton. (p. 30)

plant breeding The development of improved plant varieties as, for example, by cross-breeding landraces and/or cultivars that display desired genetic traits such as high yield, disease resistance, and short growing season.

point sources Identifiable, local sites where pollution is discharged to the environment. Examples: wastewater treatment plant effluents; coal-burning power-plant emissions; and industrial discharges. (p. 5)

polar vortex Stratospheric winds circling the South Pole and sustaining temperatures as low as −90 °C ($\approx$ −130 °F) thus favoring the formation of stratospheric ice crystals — catalysts that release chlorine or bromine atoms from ozone-depleting substances, which then deplete ozone. (p. 267)

policy *See* environmental policy.

pollution Detrimental changes in the physical, chemical, or biological characteristics of air, soil, water, or other environmental resources that can adversely affect fish, wildlife, humans, or other living organisms. *Compare* point sources and non-point sources. (p. 5)

polychlorinated biphenyls (PCBs) Synthetic organic chemicals made by chlorinating biphenyl, an organic compound consisting of two benzene rings linked together. PCBs are used in electrical transformers and capacitors.

polynuclear aromatic hydrocarbons (PAHs) A specific group of organic compounds, such as naphthalene, anthracene, and phenanthrene, with two or more benzene-like rings fused together. PAHs are found in soot, smoke, charcoal, creosote, and crude oil. (p. 110)

population   All of the individuals of a single species, whether monerans, protists, fungi, plants, or animals, living in a defined area or ecosystem. (p. 2)

population crash   A precipitous fall in the number of a particular species. (p. 93)

population growth rate (human populations)   Rate of natural increase plus numbers immigrating into a country (or region) minus numbers emigrating out of that country (or region) in a given year. (p. 132)

positive feedback   *See* feedback.

potable water   Freshwater that is largely free of harmful chemicals and disease-causing micro-organisms; water that is suitable for drinking. (p. 64)

prairie   Temperate grassland biome. (p. 10)

precipitation   Rain, snow, hail, dew, or other forms of water deposited on the earth's surface from the atmosphere. Also, chemical processes where water-insoluble solids form as a result of chemical reactions.

predation   A food-web relationship where an animal (predator) obtains energy by consuming other animals (prey) in whole or in part. Predation may also include herbivory. (p. 12)

preservation   Protecting unique natural environments from human developement. (p. 3)

pressurized water reactor (PWR)   A type of nuclear reactor where heat produced by atomic fission is used to superheat water under pressure; heat is transferred to a secondary water supply which boils, makes steam, and drives electric-power generating turbines. *Compare* boiling water reactor. (p. 210)

primary containment structure   Steel-reinforced concrete surrounding a reactor core. (p. 211)

primary energy sources   Fossil fuels, nuclear fuels, waterpower, geothermal energy, solar, and wind power. A source of energy that can be used directly without conversion to a secondary energy source such as electricity. (p. 114)

primary producer   An organism capable of building complex organic molecules such as living tissue from simple inorganic substances in the environment. Examples: aquatic and terrestrial plants such as algae, lichens, grasses, and agricultural crops. *Compare* autotroph. (p. 13)

primary production   Typically, the conversion of inorganic carbon in carbon dioxide into organic carbon by autotrophs. Example: photosynthesis where inorganic nutrients are transformed into plant biomass. (p. 13)

primary succession   Stages through which a mature ecosystem is gradually established in an area not previously occupied, such as a volcanic lava flow. *Compare* secondary succession. (p. 50)

primary wastewater treatment   The first stage of sewage and industrial waste treatment involving screening of grit and debris, grinding solid residues, allowing solids to settle out as primary sludge, and chlorinating treated wastes prior to their discharge to receiving waters. (p. 75)

pristine   Referring to an undisturbed natural environment noted for its wilderness values.

producer   *See* primary producer.

prokaryotes   Organisms such as bacteria and cyanobacteria that lack distinct membrane-bound cell nuclei and cell organelles. *Compare* eukaryotes.

protein   A macromolecule of carbon, hydrogen, oxygen, nitrogen and sometimes sulfur and phosphorus composed of chains of amino acids joined by peptide bonds and present in all living cells.

protists   Unicellular and colonial organisms having distinct, membrane-bound, cell nuclei and cell organelles (eukaryotes). Examples: most algae and protozoa.

proton   A positively changed (1+) subatomic particle in atomic nuclei having a mass approximately 1,840 times greater than an electron. The number of protons in an element's atomic nuclei defines its atomic number. (p. 206)

proven reserves   Quantities of a mineral resource such as crude oil which geologic and engineering data demonstrate with reasonable certainty to be recoverable from known sources under existing economic and operating conditions.

Q

quad   A quantity of energy equal to 10^{15} British Thermal Units (BTUs) and equivalent to the energy output of 42 typical coal-burning power plants or 42 typical nuclear reactors operating nonstop for one year. (p. 200)

R

r   *See* intrinsic capacity for increase.

radiative forcing   The relative ability of different gases in the earth's atmosphere to trap infra-red (heat) rays thereby adding to global warming. (p. 249)

radioactive   Referring to chemical isotopes whose atoms disintegrate naturally into lighter-weight elements and, at the same time, release energy in the form of subatomic particles and rays such as alpha particles, beta particles, and gamma rays.

radon A naturally occurring, chemically inert, potentially harmful radioactive gas formed by the radioactive decay of radium; radon can seep into buildings, particularly unventilated basement areas. (p. 121)

rate of natural increase (human) The annual rate of a population's increase (or decrease) without regard to immigration or emigration. It is equal to birth rate minus death rate and is generally considered as equal to r, a population's intrinsic capacity for increase. (p. 132)

reaeration A process whereby oxygen gas is dissolved in an aquatic system following a decline in DO level. Natural reaeration occurs in turbulent waters such as stream riffle areas and waterfalls. (p. 93)

refineries Oil-industry facilities where petroleum (crude oil) is separated into its various components using fractionating towers. Some components are chemically restructured into more useful products using petrochemical crackers and reformers.

remote sensing Techniques used to obtain environmental data and information from a distance employing, for example, aircraft and earth satellites.

renewable energy Energy generated without depleting the earth's natural resources. An example is hydroelectric power. (p. 228)

replacement level fertility The number of children needed to replace their parents without adding to population growth. Replacement level fertility is considered to be 2.1 children per family. (p. 139)

respirable Referring to very small particulate matter capable of being breathed into lung sacs. Respirable particles are generally 10 μm in diameter or smaller. (p. 111)

respiration *See* aerobic respiration.

Rhizobium **bacteria** A bacterial species that colonizes as nodules on the roots of certain legumes such as clover, alfalfa, peas, and beans. *Rhizobium* bacteria are able to fix nitrogen. *Compare* nitrogen-fixing bacteria. (p. 171)

risk assessment Evaluation of short-term and long-term risks associated with a particular activity or venture compared to its benefits in a cost-benefit analysis. (p. 292)

roe Fish eggs. Example: Caspian Sea sturgeon roe (caviar).

r-**strategists** Organisms such as lichens, mosses, bacteria, fungi, weeds, insects, and rodents with high r values indicating that they reproduce rapidly; r-strategists, which function as pioneer, colonizer organisms in disturbed environments, are con-

trolled mainly by density-independent factors. (p. 132)

runoff Precipitation that transports dissolved and suspended matter to receiving waters such as creeks, streams, rivers, and bays. Watershed runoff has the potential to exacerbate soil erosion and transport soils, nutrients, mineral matter, and pollutants to receiving waters. (p. 63)

rut The season of mating among wildlife such as caribou. (p. 29)

S

salinity A term describing how salty ocean and estuarine waters are. Salinity is the weight in grams of dissolved solids in one kilogram of a water sample. It is expressed as parts per thousand ($^o/_{oo}$). The average salinity of the world's oceans is about 35 $^o/_{oo}$ which is equal to 3.5 percent by weight. (p. 89)

salinization The accumulation of salt minerals in soils that occurs when cropland is irrigated with salty water. Salinization leads to soil infertility. (p. 176)

saltwater Referring to aquatic environments where dissolved salts such as sodium chloride are present at significant levels.

sand filter A facility for purifying water discovered by John Gibb. (p. 72)

SAV *See* submerged aquatic vegetation.

scientific method Making observations, gathering data, and formulating, testing, and altering hypotheses. (p. 2)

seagrass Submerged marine and estuarine rooted aquatic vegetation. *Compare* submerged aquatic vegetation.

seawater Marine water commonly found in oceans, seas, and high-salinity areas of estuaries.

secondary containment structure Concrete, dome-shaped structure that encloses a nuclear reactor's core, its concrete housing, and its steam generator. (p. 211)

secondary energy source Refers generally to electricity produced from a primary energy source such as a coal-burning power plant. (p. 114)

secondary succession Stages through which a mature ecosystem is gradually reestablished in an area where it previously existed but was cleared or transformed as, for example, by deforestation. *Compare* primary succession. (p. 50)

secondary wastewater treatment The second stage of sewage and industrial waste treatment involving mainly BOD-reduction using a trickling filter, activated sludge, or some other process. (p. 75)

Second Law of Thermodynamics   The scientific principle that the use of energy cannot be 100 percent efficient; part is always "lost" as waste heat which is discharged to the environment. An automobile engine, for example, is about 20 percent efficient and about 80 percent of the energy used ends up as waste heat. (p. 26)

second trophic level   A food-web feeding level where primary consumers obtain energy by consuming first-trophic-level primary producers. (p. 25)

sedimentary   Refers to silt, sand, and soil-like materials deposited at the bottom of an aquatic environment.

selective cutting   Harvesting only mature trees of certain species and size instead of clearcutting. (p. 182)

self-pollination   The transfer of pollen by pollinating insects or by hand from the anthers (pollen producing organs) of one flowering plant to the stigma (pollen receptor) of the same plant leading to fertilization and the development of new seeds. (p. 149)

shifting cultivation   *See* slash-and-burn agriculture.

sigma or logistic growth   An "s" curve showing rapid population growth followed by a leveling off of population size. (p. 131)

sinks   Natural processes that remove gases such as carbon dioxide from the atmosphere. (p. 250)

skimmers   Equipment used to transfer corralled oil into containment tanks using conveyor belts and vacuum pumps. (p. 198)

slash-and-burn agriculture   An agricultural practice where forest trees and shrubs are cut down and burned creating nutrient-rich ashes which are mixed with soils prior to crop cultivation. After three or more years crop yields generally decline, nutrient-impoverished soil areas are abandoned, and farmers must shift to another part of the forest. (p. 48)

soil   A complex mixture of weathered minerals derived from rocks, partially decomposed organic matter, and a wide diversity of living organisms.

soil compaction   The compression of porous soils often caused by farm implements leading to the collapse of soil pore spaces, reduced ability to hold air and water, and diminished productivity. (p. 178)

soil erosion   The wearing away, lifting, and transport of soil by natural processes such as precipitation and wind. (p. 155)

soil horizons   A vertical profile of horizontal layers indicating a soil's history, characteristics, and usefulness, for example, in agriculture. (p. 46)

soil husbandry   To continually sustain and even enhance the quality of agricultural soils in the hope of maximizing crop yields, minimizing pest control, and protecting against adversities such as drought. (p. 171)

solar constant   The fraction of total electromagnetic energy radiating from the sun that is intercepted by the earth. (p. 23)

solid wastes   Non-hazardous wastes including municipal garbage, domestic sewage, agricultural residues, irrigation waters, demolition materials, and mine tailings. (p. 282)

species biodiversity   *See* biodiversity.

stewardship   An ethical view that humans have a responsibility to care for and wisely manage the earth, its natural resources, and all living things. (p. 5)

stomata   Microscopic pores in tree leaves and needles that allow carbon dioxide intake and the release of oxygen and moisture.

stratosphere   An atmospheric layer lying above the earth's troposphere and extending about 12 km ($\approx$ 7.4 miles) to 52 km ($\approx$ 32 miles) above the earth's surface. Stratospheric temperatures rise with increasing elevation. (p. 108)

stratospheric chlorine level   A measure of total ozone-depleting substances in the stratosphere; bromine-containing substances are converted to equivalent levels of chlorine-containing substances. Totals are reported as parts per billion of chlorine by volume (ppbv). (p. 268)

stubble mulching   The practice of leaving crop residues and roots in the ground following harvesting and not plowing prior to the next season's planting. (p. 178)

submerged aquatic vegetation (SAV)   Underwater seagrass meadows comprised of many different grass species such as eelgrass, widgeon grass, wild celery, and pondweed.

subsistence agriculture   Harvesting naturally occurring wild plants allowing poor people almost everywhere to settle down in stable communities and eke out a living. (p. 148)

substrate   A place or location where plant and/or animal life can colonize.

subtidal   Referring to near-shore coastal marine habitats that generally lie below low-tide levels. (p. 195)

succession   A series of distinct changes in plant and animal communities over time in a particular environment; gradual replacement, through time, of one group of species in a community by other groups. (p. 50)

summer fallowing The practice of allowing agricultural land to go unplanted and sometimes unplowed during a growing season in order to capture and hold soil moisture. (p. 173)

Superfund legislation 1980 U.S. legislation known as the Comprehensive Environmental Response, Compensation, and Liability Act (CERCLA) establishing a large trust fund allowing the EPA to respond to immediate and long-term releases of hazardous substances that may endanger public health or the environment. (p. 290)

surface impoundments Doubly lined natural or manmade holding tanks or land-surface cavities used to treat, temporarily store, and sometimes dispose of hazardous wastes. (p. 285)

surface waters Aquatic environments such as streams, rivers, lakes, bays, estuaries, and oceans but not including groundwater aquifers.

sustainable agriculture Agricultural crop and livestock practices that maintain environmental balance in the face of intensive land use; agricultural practices that meet today's needs without compromising the ability of future generations to meet their needs. Sustainable implies that the inherent qualities, values, and capacities of natural systems are not being impaired or lost. (p. 173)

sustainable development Development that meets the needs of the present without compromising the ability of future generations to meet their own needs. (p. 5)

sustainable energy Energy generated without depleting natural resources and without causing developmental externalities. (p. 229)

sustained yield Limiting annual harvests of mature trees, regardless of market demand. (p. 182)

symbiosis An intimate biological relationship involving two species (symbionts) where both species may benefit (mutualism), the host species may be unaffected (commensalism), or the host species may be harmed (parasitism). (p. 22)

synthetic A human-made substance or material as opposed to a naturally occurring substance. DDT is a synthetic chemical while nicotine is a natural plant substance.

T

target organism A living organism that is the target of a particular pesticide or pest control strategy. *Compare* non-target organism. (p. 159)

tar sands Sedimentary materials rich in hydrocarbon deposits such as the Athabascan Tar Sands in Al-

berta. Tar sands are a potentially important future source of petroleum-like fuels and resources.

taxonomist A scientist skilled in the art of biological classification. (p. 11)

temperate Referring to mid-latitude moderate climate zones lying between the earth's equator and its poles.

Tennessee Valley Authority (TVA) An independent agency of the U.S. government established by the 1933 TVA Act and having broad regional powers in southeastern states including hydroelectric power, enhanced navigation on the Tennessee River, and economic development. (p. 212)

Ten Percent Rule The concept that on the average about ten percent of plant biomass is converted into animal matter at the second and higher trophic levels. The ten percent rule applies to every food-web energy-transfer step following primary production. (p. 25)

terminator seed technology Terminator seeds are genetically modified so that crops grown from them will produce seeds that cannot germinate. (p. 164)

terrestrial Referring to land-based, non-aquatic environments such as forests, deserts, grasslands, watersheds, woodlots, and parks.

tertiary wastewater treatment *See* advanced wastewater treatment.

thermal inversion *See* atmospheric inversion.

thermosphere An atmospheric layer lying above the earth's mesosphere characterized by temperatures as high as 1,200 °C ($\approx$ 2,200 °F). (p. 108)

topography A description of landform surface features, profiles, and contours including, for example, mountains, hills, valleys, rivers, and lakes.

top predator An animal that feeds at an ecosystem's top, or highest trophic level. Whales, polar bears, and eagles are top predators. (p. 13)

topsoil The uppermost, most fertile and valuable soil layer in a soil horizon typically rich in nutrients, organic matter, minerals, humus, moisture, and microorganisms. (p. 170)

total fertility rate The average number of children born to women in a given country or region during their childbearing years. (p. 132)

total harvests In agriculture, a country's or a region's total production of a given crop in a given year. *Compare* annual yields. (p. 150)

transgenic Referring to a genetically modified variety of plant or animal life created by the transfer of DNA from one species to another. (p. 163)

transpiration A biophysical process whereby water in vegetation is released as water vapor through pores (stomata) in leaves and needles. (p. 44)

tree line A transition zone between boreal conifer forest and arctic tundra. Few trees grow north of the tree line.

trophic level An ecosystem food-web feeding level; Examples: green plants (autotrophs) function as first trophic level primary producers; animals (heterotrophs) function as higher trophic level consumers. (p. 13)

tropical rainforests Wet forests lying between 23.5° N and 23.5° S latitude that receive at least 2,200 mm (≈ 87 inches) of annual rainfall.

troposphere The earth's lower atmosphere extending from ground level to an elevation of about 12 km (≈ 7.4 miles). Over 90 percent by weight of the earth's atmosphere is in the troposphere. (p. 107)

tundra A mostly flat, almost tree-less arctic ecosystem known also as coastal plain. Tundra is intensely cold most of the year but warms and becomes biologically productive during the Arctic's brief summers. (p. 23)

turbidity Cloudy, murky, or muddy water caused by soils eroded from disturbed watersheds or by algal blooms. Turbidity may limit light transmission to deeper waters. (p. 79)

TVA *See* Tennessee Valley Authority.

U

ultraviolet radiation Electromagnetic wavelengths between 200 and 400 nm. (p. 266)

understory A forest's lower-level trees, branches, and vegetation lying below its high canopy and sub-canopy.

urban heat island An urban area where re-radiation of solar heat energy from streets and concrete buildings warms the surrounding air causing higher than normal air temperatures. (p. 254)

UV-A radiation Ultraviolet radiation, usually from the sun, having wavelengths between 320 nm and 400 nm. UV-A wavelengths are longer, less energetic, and less harmful to life that UV-B radiation. Solar UV-A radiation is not absorbed by stratospheric ozone. (p. 266)

UV-B radiation Ultraviolet radiation, usually from the sun, having wavelengths between 280 nm and 320 nm. UV-B wavelengths are shorter, more energetic, and more harmful to life than UV-A radiation. Solar UV-B radiation is absorbed in large part by stratospheric ozone. (p. 266)

UV-C radiation Ultraviolet radiation having wavelengths between 200 nm and 280 nm. Solar UV-C radiation is totally absorbed in the stratosphere mainly by molecular oxygen. (p. 266)

V

vapor pressure The tendency of a substance to evaporate from its liquid or solid state forming a gas. Gasoline, for example, has a high vapor pressure and therefore evaporates easily into air.

Vienna Convention 1988 international agreement to secure a general treaty to limit chemicals, substances, and activities that might pose a risk to the stratospheric ozone layer. (p. 273)

visible light Electromagnetic radiation ranging from about 700 nm to about 400 nm (red to violet) (p. 266)

VOCs *See* volatile organic compounds.

volatile Referring to liquids (and sometimes solids) that easily evaporate causing their vapors to enter the air.

volatile organic compounds (VOCs) Certain solvents and other low-boiling chemicals that evaporate easily into air spaces and the open atmosphere. (p. 283)

Vostok ice core An ice core and its various segments drilled from the south-polar ice sheet in East Antarctica by Soviet scientists beginning in 1970. Analyses reveal that the deepest segments of this core are more than 160,000 years old. (p. 253)

W

watershed A catchment basin in which precipitation, mainly rain and snow, collects. Part of the precipitation evaporates back to the atmosphere, part is stored in watershed soils, trees, and vegetation, part infiltrates soils recharging groundwater aquifers, and part runs off into nearby surface waters such as brooks, streams, rivers, lakes, estuaries, and oceans. (p. 63)

water splitting A process that separates water into its two basic elements, hydrogen and oxygen. Example: electricity applied to water (electrolysis) splits water molecules forming hydrogen gas and oxygen gas.

wavelength A measure of electromagnetic energy describing characteristics such as the color of visible light; specifically, the distance between two waves of electromagnetic energy. *Compare* electromagnetic spectrum.

weathering   Biogeochemical processes influenced by wind, rain, light, freezing, and biological actions that slowly transform rock and mineral matter into soil components.

weir   A low-lying dam built to control water flow, manage flooding, and maintain freshwater reservoirs. (p. 65)

wetlands   Bogs, swamps, river deltas, and marshes whose soils and rooted vegetation are periodically inundated by standing water part of each year. (p. 85)

white rot fungus   A commonly occurring wood-rotting fungus that releases an enzyme capable of breaking down hazardous contaminants such as DDT into simple, nontoxic products. (p. 285)

wild plant varieties   Historically, early sources of grains, seeds, fruits, and tubers for most of the world's people.

wind turbine   A high-tech, aerodynamic, mechanically efficient, electronically sophisticated machine capable of capturing wind energy and producing electric power for human use. (p. 230)

windward   A direction from which the wind is blowing; areas that face prevailing winds. (p. 52)

X

x-rays   Very short wavelength electromagnetic radiation capable of penetrating and damaging materials including living tissues.

Z

zero emission vehicle   An automotive vehicle whose power source (such as a fuel cell or battery) releases no pollutants to the environment. (p. 241)

zero population growth   Referring to a country or region where annual increases in population (births and immigration) are offset by annual decreases in population (deaths and emigration).

zooplankton   Microscopic, free-floating, aquatic animal life with limited locomotory power causing them to drift with wind and water currents. Example: copepods. (p. 31)

Credits

Chapter 1

1.3: From I.D. White, D.N. Mottershead, and S.J. Harrison, *Environmental Systems, And Introductory Text,* 2nd edition 1992. Copyright © 1992 Chapman & Hall, Ltd. Reprinted by permission; **1.5:** © Alisa Kayser. Reprinted by permission

Chapter 3

3.1 and 3.4: Based on map provided through the courtesy of the Environment Agency, London, England. Reprinted by permission; **3.9:** From *Snow on Cholera,* The Commonwealth Fund, 1936. Reprinted by permission of The Commonwealth Fund, New York, NY; **3.11, 3.12,** and **3.14:** Adapted and reprinted by permission through the courtesy of the Environment Agency, London, England.

Chapter 6

6.3: Figure from *Environmental Science,* fourth edition by Amos Turk and Jonathan Turk. Copyright © 1988 by Saunders College Publishing, reproduced by permission of the publisher; **6.4:** From J. Stafford, "Heron Populations of England and Wales 1928-1970" in *Bird Study,* 18:218-221. Copyright © 1971 British Trust for Ornithology, England.

Chapter 8

8.5: Courtesy of the Prairie Farm Rehabilitation Administration (PFRA), Agriculture and Agri-Food Canada. Reprinted by permission.

Chapter 10

10.12: From *Chernobyl Ten Years on Radiological and Health Impact.* An Assessment by the NEA Committee on Radiation Protection and Public Health, November, 1995, OECD Nuclear Energy Agency. Copyright © 1995 OECD. Material available on OECD website: http://www.nea.fr/html/rp/chernobyl/chernoblyl.html.

Chapter 11

11.1: Reprinted by permission of Campbell Thomas & Co. Architects, Philadelphia, PA; **11.4b:** Courtesy of Kramer Junction Company, Boron, CA; **p. 314:** From Christopher Flavin, "Bull Market in Wind Energy" from *Worldwatch,* March/April, 1999, Vol. 12, No. 2. Copyright © 1999 Worldwatch Institute. Reprinted by permission.

Chapter 13

13.7: Reprinted with permission from James B. Kerr, "Decreasing Ozone Causes Health Concern: How Canada Forecasts Ultraviolet-B Radiation" in *Environmental Science and Technology,* Volume 28, Number 12, pp. 514A-518A, 1994. Copyright © 1994 American Chemical Society; **13.8:** Reprinted with permission from James B. Kerr, "Ozone Watch" Environment Canada, Friday, April 9, 1999.